In this book, one of community planning's leading thinkers pulls back the curtain on the intellectual and personal journey that has shaped four decades of scholarship. This collection will inspire anew those of us familiar with her work and be a touchstone text for future thinkers and practitioners of community planning.

Libby Porter, Professor, Centre for Urban Research, Royal Melbourne Institute of Technology University, Australia

In a book of imagination and wonder, Leonie Sandercock has interwoven politics and personal experience to surprise us all, to expand our senses of possibility, to give us an empowering vision of connection and responsibility, intimacy and critical politics too.

John Forester, Professor, Department of City and Regional Planning, Cornell University, USA

Sandercock provides an inside-out account of the ways of being, knowing, and acting that shaped her scholarship and practice, spanning the 1970s to the present. Her rich, reflective commentaries show how experience and academic insight co-evolve, so that the reader can deeply understand the fourteen seminal works included in the volume.

Richard Willson, Professor, Department of Urban and Regional Planning, California State Polytechnic University, Pomona, California, USA

Leonie brings to life forty years of debates in planning theory and practice before pointing to the next threshold: reimagining the soul of planning. Using her storytelling skills, this weaving of personal memoir and critical reflection on her own writings and film making is innovative, life affirming, and insightful, recognizing that we are not just talking heads.

Patricia A. Wilson, Professor, Graduate Program in Community and Regional Planning, School of Architecture, University of Texas, Austin, USA

MAPPING POSSIBILITY

Mapping Possibility traces the intertwined intellectual, professional, and emotional life of Leonie Sandercock. With an impressive career spanning nearly half a century as an educator, researcher, artist, and practitioner, Sandercock is one of the leading figures in community planning, dedicating her life to pursuing social, cultural, and environmental justice through her work.

In this book, Leonie Sandercock reflects on her past writings and films, which played an important role in redefining the field in more progressive directions, both in theory and practice. It includes previously published essays in conjunction with insightful commentaries prefacing each section, and four new essays, two discussing Sandercock's most recent work on a feature-film project with Indigenous partners. Innovative, visionary, and audacious, Leonie's community-based scholarship and practice in the fields of urban planning and community development have engaged some of the most intractable issues of our time – inequality, discrimination, and racism. Through award-winning books and films, she has influenced the planning field to become more culturally fluent, addressing diversity and difference through structural change.

This book draws a map of hope for emerging planners dedicated to equity, justice, and sustainability. It will inspire the next generation of community planners, as well as current practitioners and students in planning, cultural studies, urban studies, architecture, and community development.

Leonie Sandercock is a professor at the School of Community and Regional Planning at the University of British Columbia, Canada. Her main research interest is in working with First Nations through collaborative community planning, using the medium of film as a catalyst for dialogue on the possibilities of healing, reconciliation, and partnership. Other research interests include immigration, cultural diversity, and integration; the possibilities of a more therapeutic model of planning; the importance of stories and storytelling in planning theory and practice; and the role of multimedia in planning.

Routledge Equity, Justice and the Sustainable City

Series editors: Julian Agyeman and Stephen Zavestoski

This series positions equity and justice as central elements of the transition toward sustainable cities. The series introduces critical perspectives and new approaches to the practice and theory of urban planning and policy that ask how the world's cities can become 'greener' while becoming more fair, equitable and just.

The *Routledge Equity Justice and the Sustainable City* series addresses sustainable city trends in the global North and South and investigates them for their potential to ensure a transition to urban sustainability that is equitable and just for all. These trends include municipal climate action plans; resource scarcity as tipping points into a vortex of urban dysfunction; inclusive urbanization; "complete streets" as a tool for realizing more "livable cities"; the use of information and analytics toward the creation of "smart cities".

The series welcomes submissions for high-level cutting-edge research books that push thinking about sustainability, cities, justice and equity in new directions by challenging current conceptualizations and developing new ones. The series offers theoretical, methodological, and empirical advances that can be used by professionals and as supplementary reading in courses in urban geography, urban sociology, urban policy, environment and sustainability, development studies, planning, and a wide range of academic disciplines.

Living Detroit
Environmental Activism in an Age of Urban Crisis
Brandon M. Ward

The Green City and Social Injustice
21 Tales from North America and Europe
Edited by Isabelle Anguelovski and James J. T. Connolly

Sacred Civics
Building Seven Generation Cities
Edited by Jayne Engle, Julian Agyeman, and Tanya Chung-Tiam-Fook

Injustice in Urban Sustainability
Ten Core Drivers
Panagiota Kotsila, Isabelle Anguelovski, Melissa García-Lamarca and Filka Sekulova

Post-Industrial Urban Greenspace Ecology, Aesthetics and Justice
Jennifer Foster

Mapping Possibility
Finding Purpose and Hope in Community Planning
Leonie Sandercock

For more information about this series, please visit: www.routledge.com/Routledge-Equity-Justice-and-the-Sustainable-City-series/book-series/EJSC

MAPPING POSSIBILITY

Finding Purpose and Hope in Community Planning

Leonie Sandercock

Routledge
Taylor & Francis Group
LONDON AND NEW YORK

earthscan
from Routledge

Cover image: 'New Dawn', by kind permission of Musqueam artist Susan Point, who owns the copyright to the artwork. Image Copyright: courtesy of Coastal Peoples Fine Arts Gallery

First published 2023
by Routledge
4 Park Square, Milton Park, Abingdon, Oxon OX14 4RN

and by Routledge
605 Third Avenue, New York, NY 10158

Routledge is an imprint of the Taylor & Francis Group, an informa business

British Library Cataloguing-in-Publication Data
A catalogue record for this book is available from the British Library

ISBN: 978-1-032-35132-2 (hbk)
ISBN: 978-1-032-35129-2 (pbk)
ISBN: 978-1-003-32545-1 (ebk)

DOI: 10.4324/9781003325451

Typeset in Bembo
by Deanta Global Publishing Services, Chennai, India

Printed in the UK by Severn, Gloucester on responsibly sourced paper

CARBON NEUTRAL

MIX
Paper from responsible sources
FSC® C022174
FSC www.fsc.org

FIGURE 0.1 *In Memoriam, John Friedmann*. Portrait by Heather Ellyard. Photo, Leonie Sandercock

For
John (1926–2017)
in loving remembrance of times past

and

For
Ida May Hemphill Campbell (b. 2019) and Zebie Graham-Reeves (b. 2021)
who lift my sight into the future

FIGURE 0.2 Ida May Hemphill Campbell, b. 2019. Photo credit, Kimberley Kufaas

FIGURE 0.3 Zebie Graham-Reeves, b. 2021, Photo credit, Crystal Reeves

CONTENTS

FIGURES

PREFACE

Our 21st-century world is descending into aggression, fear, and separation. War, genocide, violence, racism, slavery, pandemic, poverty, natural disasters, are all now commonplace. Despair and withdrawal into self-preservation, prompted by feelings of powerlessness in the face of dysfunctional systems, are all-too-common responses to these multiple crises. In my 70-plus years on this exquisite small green-blue planet, the need for systemic change has never been more urgent than it is right now. We are living in a time of seeming chaos. But it is a chaos that is rich in potential for new possibilities: new ways of seeing, new ways of being, new relationships.

What can I/we do in the midst of these multiple crises? Ideally, we seek a path where our gifts and core values intersect with the needs of the world. As I prepare to retire from a lifelong involvement with teaching, writing about, and the practice of community and regional planning, I am looking back and taking stock as well as looking forward. What can we, as community planners, do to birth a healthier future? I believe that systemic change begins in small ways, with the choices of brave individuals whose actions demonstrate new possibilities. We never know how our small activities will affect others through the invisible fabric of our connectedness. The real engine of change, as Margaret Wheatley[1] has argued, begins with critical connections, not critical mass. Community planning is the work of building connections. It is the work of healing, repair, renewal.

I've been fortunate, since my mid-twenties, to work with people younger than I. At first, I had only a few years on my students, but now the age gap is encroaching on 50 years. It's been an incredible privilege, jointly generating the energy for personal and social change: from being mentored, to mentoring, and now coming full circle. My mentors saw something in me that I did not see in myself. They challenged me and cheered for me. Those elders who helped me find my path kept showing up into my thirties. Then I began a co-learning with my own students, and gradually became a mentor. Now, as my mentees flourish, some of them return the favor. Mentors and apprentices are partners in an ancient human dance. It is the dance of the spiraling generations, a mostly generative dance that reweaves the fabric of the human community as we touch and turn,[2] sometimes stumbling, sometimes gliding to new insights.

Acknowledging this dance of the spiraling generations, throughout this book I turn to the sages among youth, elders, and those in between: from Greta Thunberg to the late bell hooks,[3] to Valarie Kaur,[4] to Indigenous elders;[5] and to the wise ones, young and old, in the community planning field, who have devoted themselves to the work of community building, repair, and flourishing, to regenerative practices that heal by conjoining heart, mind, and spirit. This book follows my own journey of learning and unlearning, in pursuit of social justice, as it parallels the learning and unlearning of the planning profession.

The personal story that I offer here is the story of a journey to find a place for the heart and spirit as well as the mind in this work – what I now think of as reimagining the soul of planning. This is not a textbook about community planning. It is me turning the prism on my experiences in academia and planning. It is about my failures as much as my successes, my ignorance as well as my insight, my doubts and my convictions, my fears and my hopes. I stumbled into the field of planning over 50 years ago, when I was just 21. I have loved and left the field,[6] only to rejoin it, re-enthused about what this vocation can contribute to building community, connection, and caring, about the planet and each other. This book is for the spiraling generations: an invitation to my peers to reflect on where we are as a profession, and an offering for those seekers who are looking for a vocation that aligns with these core values.

I conceived this book as having three purposes. First, I wanted to pull together a collection of essays that best represent my thinking, have had significant impact, and form a set of coherent themes. But, second, I also wanted to reveal the personal and political contexts which gave birth to these new ways of thinking and acting, new possibilities for the planning field; to pay homage to all those who have contributed along the way; and to illustrate how the path is not a linear one. Third, I wanted to consider the currency and relevance of this reflexive gaze for today's world – for everyone dedicated to the shared political work of addressing inequality, systemic racism, White supremacy, and ecological catastrophy.

As I re-read most of my published essays and books, I was astonished by the authoritative academic voice I was able to adopt from quite an early age. That is how we are 'socialized' in academia. But that voice is so at odds with my lived experience through these 50 years that I want to draw back the curtain on it all, to talk about the context behind the texts: the relationships behind the emerging ideas and practices.

I've chosen the essays that best represent the evolution of my thinking, pedagogy, and practice since 1986 and organized them into four thematic sections. Part I captures my efforts to challenge 'the canon' by diversifying planning history, theory, and epistemology. Part II covers my utopian dreaming about the possibility of just and equitable co-existence in cities of difference: my 'cosmopolis project'. Part III, the storytelling project, explores various ways of expanding the 'language' of planning, focusing on three experiments with film as a form of community planning intervention. Part IV follows my inner and outer journeys, navigating Indigenous worlds, as I learned how to work with Indigenous communities. And the conclusion draws on all of the above to celebrate the collective practices through which citizens create their own meaningful life spaces in the face of larger social and historical forces. That is the basis of the hope I have for the future. Community planning can be both catalyst and support for these collective practices.

The commentaries that introduce each part/theme serve two purposes: first, to situate the ideas in a context of time and place and above all, of relationships; and second, to bring a critical eye to these writings. What was sharply in focus, back then? What was I failing to see?

Where do I think differently, now? What does a social justice project in planning look like today and beyond? Do I still see purpose and hope in community planning?

There are several new chapters about my most recent work with the Haida Nation on an Indigenous language film project. And the final chapter expresses where my thinking has arrived after 50 years, most notably, how what I've learned from Indigenous communities can help us think about the health of our collective life support system, Mother Earth.

The Introduction ('The Australian Years') and the five Commentaries are my attempt to trace a learning journey in its full embodiment: the life behind the work. I've divided my life into roughly two halves. The first half, lived in Australia until I was 36 (Introduction), traces an unlikely path from the wrong side of the tracks, an outer industrial suburb of a very conservative, colonial city, on a continent at the bottom of the world, through the experiences of classism during my undergraduate days, the sexism of my graduate years and beyond, to stepping at a young age into positions of apparent power that never quite felt powerful as a lived experience. I explore why that was so, what was troubling me? By my mid-thirties I was beginning to peel the wallpaper off the fairytale academic house, exposing a structure designed for the comfort and reward of White males. I had apparently 'made it' in that world, but there was something missing.

I left Australia in 1986 on a quest: for love, and for a more meaningful form of work, which is another form of love. I had fallen out of love with academia, was about to become a student again, in film school in Los Angeles, and to embark on a second marriage. The five Commentaries follow that emerging journey, in love and work, from the inside (lived experience) as well as the outside (career path/achievements) – that is, from the 'whole person' perspective. I've tried to braid the personal and professional journeys. I've tried to make visible the relationships and experiences along the way that have brought fresh insights, about myself and my chosen vocation, community planning. What stands out is how much of the learning and unlearning came as a result of very humbling experiences.

I have chosen not to include in this selection of my writings any of the work I did in Australia from 1974 to 1986, for three reasons. All of that work was specific to the political context of urban policy in Australia at the time, so it is necessarily dated. And, as I discovered on arriving at the University of California in Los Angeles in 1986, nobody was interested in events in this small faraway country (Commentary, Part 1). Above all, I now find it hard to identify with that young voice, with that masculinist[7] urban political economy language, with that authoritative posture. How could I have been so sure of myself in my writing, yet so unsure in my personhood? Well, this is how we were socialized to write at that time. The mission of my next 35 years was to unlearn some of that socializing, to allow heart and humility, empathy and kindness, joy and love, to find a home in my work. That did not happen overnight. The five Commentaries explain how and why it gradually did happen.

Living a life is a project of ongoing meaning-making. Writing is a way of sharing that meaning-making. The poet Mary Oliver asks us, 'what is it you plan to do/with your one wild and precious life?' (Oliver 1992, 94).[8] I am writing this for the generations coming behind me who may be as uncertain, insecure, and confused as I have been about finding my voice, finding purpose, meaning, and hope. Hope is not blind optimism or naïve faith in human nature. I do have an unshakeable confidence in the human spirit. But hope, as Vaclav Havel has taught us, is not the conviction that something will turn out well; it is the certainty that something is worth doing no matter how it turns out. That something, for me, is community planning.

LS
Spring 2022

Written as an uninvited guest who has had the privilege of living and working on the traditional, ancestral, and unceded lands of the xʷməθkwəy̓əm (Musqueam), Skwxwú7mesh (Squamish), and Səl̓ílwətaʔ/Selilwitulh (Tsleil-Waututh) peoples, who have occupied this land now known as Vancouver since time immemorial.

Notes

1 Margaret Wheatley writes about how changes in small places affect the global system, not through incrementalism but because every small system participates in an unbroken wholeness. Activities in one part of the whole create effects that appear in distant places. See *Leadership and the New Science* (Oakland, CA: Berrett-Koehler, 3rd ed, 2006).
2 Parker Palmer writes beautifully about this in *The Courage to Teach: Exploring the Inner Landscape of a Teacher's Life* (San Francisco: Jossey-Bass, 1997). One of my doctoral students, Maged Senbel, gifted me that book not long after I started teaching at UBC in the early 2000s. Maged has gone on to become a treasured colleague, and his interfaith work is a moral inspiration. Together, we later explored the importance of spirituality in planning in 'Spirituality, Urban Life, and the Urban Professions', an essay included in J. Beaumont & C. Baker (eds) *Postsecular Cities: Space, Theory and Practice* (London: Continuum, 2011).
3 See bell hooks, *All About Love* (New York: Harper Collins, 2000). She celebrates a spirit of neighbourliness: of fellowship, care, and respect for the dignity and wellbeing of the 'beloved community'; a way of thinking and behaving which honors principles of interbeing and interconnectedness. That, essentially, is how I would like to define community planning.
4 Valarie Kaur *See No Stranger: A Memoir and Manifesto of Revolutionary Love* (New York: Random House, 2021). Kaur is a visionary, a social movement lawyer, a 'warrior-sage' who draws on her Sikh ancestry's spiritual traditions. She challenges us to 'Imagine the world we will birth when we see no stranger' (Kaur 2021, 299).
5 See Chapter 14.
6 That story is told in the autobiographical Introduction: 'The Australian Years'.
7 I'm uncomfortable with this as an adjective because it essentializes 'masculine' as something undesirable, possibly to be eradicated, and separates masculine and feminine. That's not what I want to say, ultimately. In reimagining the soul of planning, I'd prefer to think in terms of the necessity and complementarity of yin and yang energies, masculine and feminine, as I suggest in the final Commentary. But I'm using it here because it's how I understood it at the time.
8 From 'The Summer Day', Mary Oliver, *New and Selected Poems*, vol. 1 (Boston: Beacon Press)

ACKNOWLEDGMENTS

The hardest thing about writing this book was deciding to do it, rather than just retiring quietly. So, special thanks to those who encouraged me to proceed with this when I was filled with self-doubt: John Forester, Tanya Winkler, Matti Siemiatycki, Magdalena Ugarte, Libby Porter, Aftab Erfan.

Immense gratitude to those friends and colleagues who 'volunteered' to read the new material and offer careful and constructive feedback. For detailed critical readings and conversations about the Haida film project (Chapters 10 and 12), thanks so much Scott Graham, Libby Porter, Jess Hallenbeck, and Jon Frantz. For courageously challenging me to think more deeply about the memoir-ish Introduction, I'm indebted to John Forester, Rick Willson, Libby Porter, Patricia Wilson, Michelle LeBaron, Jess Hallenbeck, and Lauren Brown. Elizabeth Walsh provided a thoughtful, appreciative response to the Commentary to Part One. Maged Senbel, Libby Porter, and Patsy Healey did the same for the Preface. And for last-minute conversations about sensitive passages, many thanks to Jessie Hemphill, Scott Graham, Dawn Smith, and Libby Porter.

My greatest debt is to Rick Willson, for wading through all of the new material and responding expeditiously and generously with detailed comments and suggestions and long email conversations. It was a delightful dialogue.

For cheering me on with enthusiastic responses to early drafts, heartfelt thanks to dear friends Giovanni Attili, Heather Ellyard, Alex Broons, Jamaine Campbell, and Paula Carr. And to those dear ones who've been my covid sisterhood support, eternal gratitude: Amy Kiara Ruth, Paula Carr, and Michelle LeBaron.

To Matthew Shobbrook and Grace Harrison at Routledge, thank you for being such helpful and supportive editors.

I am deeply grateful to Musqueam artist Susan Point for permission to include images of four of her paintings, and to the Coastal Peoples Fine Arts Gallery of Vancouver, BC, for permission to use their images of these paintings.

When I joined UBC in 2001, I had no idea that the campus occupied part of the traditional, ancestral, and unceded lands of the Musqueam Nation. I could never have predicted then that, a decade later, I would become involved in an educational partnership with Musqueam that has given me the opportunity to spend a lot of time in their community over the past decade, learning from their knowledge keepers and community members. For this most precious and profound experience, I am deeply grateful.

★

In choosing the following essays for inclusion, I have had to shorten many of them to keep the overall word length manageable. I have deliberately not 'updated' content or corrected any of the interpretations offered because I wanted the essays to stand as products of a particular moment in time, and to show an evolution in my thinking. In cases where I now think differently, I have noted and explained this in the Commentary to that particular section of the book.

Chapter 1 is an abridged version of 'Introduction: Framing Insurgent Historiographies for Planning', which appeared as the Introduction to my edited collection *Making the Invisible Visible. A Multicultural Planning History* (1998).

Permission: University of California Press

Chapter 2 is an abridged version of 'Who Knows? Exploring Planning's Knowledges', which appeared as Chapter 3 in *Cosmopolis 2: Mongrel Cities in the 21st Century* (2003).

Permission: Continuum Publishing, an imprint of Bloomsbury Publishing

Chapter 3 is an abridged version of 'Voices from the Borderlands: A Meditation on a Metaphor', which appeared in the *Journal of Planning Education & Research*, vol. 14, 2: 77–88, 1995: https://doi.org/10.1177/0739456X9501400201

Permission: Sage Publications

Chapter 4 is an abridged version of 'Towards Cosmopolis: A Postmodern Utopia', which appeared as Chapter 7 in *Towards Cosmopolis: Planning for Multicultural Cities* (1998)

Permission: Wiley

Chapter 5 was published as 'When Strangers Become Neighbours: Managing Cities of Difference' in *Planning Theory & Practice*, 1, 1: 13–20, 2000. DOI: 10.1080/14649350050135176.

Permission: Taylor & Francis

Chapter 7 was first published as 'Out of the Closet: The Importance Of Stories And Storytelling in Planning Practice' in *Planning Theory & Practice*, vol. 4, 1: 11–28, 2003. DOI: 10.1080/1464935032000057209.

Permission: Taylor & Francis.

Chapter 8, co-written with Giovanni Attili, is an abridged version of 'Digital Ethnography as Planning Praxis: An Experiment with Film as Social Research, Community Engagement, and Policy Dialogue', published in *Planning Theory & Practice*, vol. 11, 1: 232–45, 2010: DOI: 10.1080/14649350903538012

Permission: Taylor & Francis

Chapter 9, co-written with Giovanni Attili, was published as 'Changing the Lens: Film as Action Research and Therapeutic Planning Practice' in the *Journal of Planning Education & Research*, vol. 34, 1: 19–29, 2014. DOI: 10.1177/0739456X13516499

Permission: Sage Publications

Chapter 11 was published as 'Finding my Way: Emotions and Ethics in Community-Based Action Research with Indigenous Communities', which appeared as Chapter 1 in Christensen, Cox, & Szabo-Jones. Eds., *Activating the Heart*. (2018).

Permission: Wilfred Laurier University Press

Chapter 13 was published as 'Indigenous Justice: Dreaming Co-Existence' which appeared as the Epilogue to P. Gurstein & T. Hutton. Eds. *Planning on the Edge: Vancouver and the Challenges of Reconciliation, Social Justice, and Sustainable Development* (2019).

Permission: UBC Press

Chapter 14 is a revised version of 'Once Upon A Planet: Planning for the Transition from Ego-Driven to Eco-driven Economies', first published as Chapter 14 in A. Ferreira, K. Schonfeld & F. Savini. Eds. *Post-growth Planning: Cities Beyond the Market Economy* (2022).
Permission: Routledge, an imprint of Taylor & Francis

FIGURE I.1 'Farewell to Australia', artist Ann Forsyth. Photo, Leonie Sandercock

INTRODUCTION

The Australian Years: Finding and Losing Purpose in Planning

All good stories have a beginning, middle, and end. But you can start the tale anywhere and move back and forth. Let's start with this faux-movie poster image and the story it tells. I was 36. It was 1986. I had been a professor and head of a graduate urban studies program in Sydney since 1981, the youngest full professor in Australia, and one of only four women among the thousand or so full professors in Australian universities at this time. I was also an avid body surfer, hence 'Sydney's surfie sweetheart'. I was living in a two-room apartment on Bondi Beach, surfed every day, and expected to live here on the edge of the Pacific into my eighties, when I would die one day, blissfully, in the surf.

But my life was about to change dramatically. I had fallen in love, by correspondence, with someone who would become the second great love of my life and with whom I would spend the next 32 years. But he, John, lived in Los Angeles. And I had fallen out of love with academia, and was no longer interested in the career in politics that I had imagined for myself a few years earlier. So I quit my tenured full professorship, and was about to move to LA and become a student again, in film school. A new country. A possible new career. A second chance in love. Where my friends saw only risk, and told me so, all I could see was possibility.

This was a pivotal moment. There was life before LA, film school, and John. And then there was the adventure that followed. Definitely a 'moving adventure', as predicted in the farewell poster.[1] The 'surfie sweetheart' of the poster was about to reinvent herself, professionally. The story of that reinvention is told in the five Commentaries. But before I get to that story, I want to look back at how a very shy girl from the wrong side of the tracks in a small Australian city had come this far, and why she was now seeking radical change.

★

Pulling the Thread of Memory

There's the family we're born into, and the families we make along the way when we're ambivalent about the family we were born into. I come from a poor White family in an outer

DOI: 10.4324/9781003325451-1

industrial suburb of a conservative, colonial city (Adelaide), on a continent at the bottom of the world. My parents were children of the Depression years who never had the opportunity to go to high school, worked in domestic and factory jobs from the age of 12, were grateful to be offered a two-bedroom 900-square-foot house built by the public housing agency in 1949, and moved in a few months before I was born. There wasn't a book in the house I grew up in, or any music except for the radio (Bing Crosby, Pat Boone, Frank Sinatra). The only conversation in this house was gossip about neighbors and illnesses. There was constant bullying and belittling. The greatest absence was love, in any of its manifestations.

My father carried a heavy bag of prejudices about with him at all times. There were many categories of people whom I was forbidden to associate with: Catholics, immigrants, Aboriginal people, queer people. My mother voiced no opinions, and I was forbidden to have any, at least overtly. 'Don't you dare contradict me' was one of my father's most frequent reprimands. There was nobody in our neighborhood, in our known universe for that matter, who had ever gone beyond high school. I was in the fourth of five years of high school before 'university' entered my consciousness, and only then because my grades in the public exam system had won me a scholarship to this mysterious place.

But why would I want to go there? I just wanted to be a physical education teacher. Then I could be outdoors and active most of the day. In the absence of intellectual and cultural stimulation in my childhood, there were numerous land and water sports available through school and the local church. The two people I most admired in my teen years were Adelaide's own Dawn Fraser, the Olympic swimming champion, whose origins were hard scrabble; and my never-married aunt, who was not only the manageress of a bush pub but also owned a couple of bush race horses and a hair dressing salon. She smoked, drank, and swore. What a wild and glamorous life, I thought.

I was a bit of a wild kid in high school: too much energy, too easily bored, acting up in class because I couldn't in my highly disciplined home environment. I did well in exams without effort, and without any real interest, until ... two teachers, one of history, one of literature, saw something in me and alternately challenged and encouraged me to the point where I got genuinely interested in those subjects and started reading. Ron Gibbs, my history teacher, once picked up a chair, put it in front of him, lion-tamer style, and charged down the aisle at me when I was making mischief at the back of the classroom. That was his way of getting my attention. It worked. (And 'Leonie' comes from the Latin for 'lion'.) Mick Brewer, the English teacher, once took a small group of us to a poetry reading by Yevgeny Yevtushenko, and afterwards to a basement coffee lounge downtown. It was the coolest thing that had ever happened to me. These two wonderful teachers told me not to miss the opportunity to go to Uni (as everyone referred to it). With an Arts degree I could become a high school teacher and teach physical education if I wanted.

★

Undergraduate Days, 1967–1970

The most terrifying day of the first 17 years of my life was the day I first set foot on the campus of the University of Adelaide. What a contrast to my first day of high school, when I had delighted in telling anyone who'd listen that God was a big hoax. I'd come to that conclusion after being required to attend Sunday School every week since I was maybe seven. It might sound like I was raised in a strict fundamentalist family, but this was not the case. Strict, yes;

but neither of my parents ever attended church or professed any faith, as I discovered when I challenged them about having to attend Sunday School at the local Church of Christ. They just had a vague notion that this was the right way to raise a child.

But on that first day at Adelaide Uni, two questions came up repeatedly, in various settings: 'what does your father do?' and 'what school did you go to?' I quickly discovered I was one of a handful of kids who'd come from a public high school, as opposed to posh private schools with names like 'Methodist Ladies College', Presbyterian Girls College'... and that everyone else's father was some kind of 'professional'. That marked me as not belonging, all through my first year. I wasn't invited, included, spoken to. I went from rebellious, exuberant, outgoing, to solitary, inward, shy. I had no idea how to behave, what was expected. And there was nobody to guide me.

In high school I was popular, mischievous, a tad unruly, upbeat. I don't remember ever being gloomy, worried, anxious about anything, afraid of anyone, even when I was in trouble with teachers. Life seemed full of possibility, all of it related to sports. At Adelaide Uni I was an outsider, timid, too nervous to ever speak in tutorials, sitting alone in large lecture theatres, living in a state of perpetual self-doubt, terrified of being called on to speak.

I had the strangest sensation that year, and for many more to come, of both being invisible and yet having a spotlight on my head that said 'This young woman knows nothing'. Because I did know nothing. Nothing of *how to be* in this world I'd stumbled into, of academia, intellectuals, artists, snobs ... I first experienced this as a callow undergrad, then all through my PhD years, and then through my early 'brilliant career' in Australia. Doing so many things for the first time, like flying to conferences, dinners at posh places, lunches with 'famous' people. I feigned nonchalance, when really, I had no idea what was going on. I was creating a persona that was far braver than I actually felt ... always feeling like an impostor.

Something else was happening though, alongside all this discomfort. I was falling in love with reading: studying history, literature, politics and philosophy, studying obsessively, for the first time in my life. I got top distinctions in every subject in the exams at the end of first year – exams I was convinced I'd failed – and was offered the opportunity to do an extra 'Honors year' in all subjects. I chose history, because it was the history profs who showed a personal interest in me, even though I never spoke, based on my written work. Suddenly, fellow students wanted to know me. Now it was my turn to be aloof. (In Australia we call this behavior 'cutting off your nose to spite your face'.)

I've said that several history profs encouraged me. But there was one history professor who mesmerized me. Hugh Stretton was a brilliant lecturer who walked into a large lecture theatre and talked for an hour without notes, telling coherent stories about the Russian revolution, the French revolution, Bismarck, and so on. These lectures, in 1968, were the highlight of my week, for a whole year. I looked forward to them like one might look forward to the next episode of, say, *The Sopranos*, if one were addicted to that TV show in the era before binge-watching was possible. The following year, Prof Stretton published his first book, *The Political Sciences*, and everyone was shocked that all these years he'd been working on a book about the social sciences, not about history. In fact, he never wrote a history book as such. I found *The Political Sciences* difficult to read, not having any social science vocabulary, but I persisted, and it is probably the single most influential academic book I have ever read. It dared to argue that the social sciences were not scientific, not objective/neutral with respect to their subject matter, but always already political, always involving interpretation by an author. He demonstrated this by detailed textual analysis of some of the 'classics' in economics, sociology, political science, and history. Although I didn't know it then, this was daring because of the positivist hegemony in the social sciences. I think now of Stretton as Australia's C. Wright Mills, although at the time I didn't know Mills' *The Sociological Imagination* (1959) either.[2]

When, the following year, Professor Stretton became my Honors History thesis supervisor, I was both thrilled and terrified. How could I ever have a conversation with this towering intellect? For our first supervisory meeting I was asked to come to his house, walking distance from campus, in a neighborhood that was, literally and metaphorically, 'the other side of town', to me. I'd never been in a professor's home. How many ways were there that I could embarrass myself?

But what actually happened was that he greeted me at the front door and ushered me down a book-lined corridor to the kitchen, where his bubbly, friendly wife Patsy made us tea and made me feel comfortable, before leaving us to discuss the work of A.J.P. Taylor, the British historian about whom I was writing my thesis. Prof Stretton presented me with the first page of my first few pages of writing. He'd more or less obliterated it in red ink. My heart sank. I was about to be crushed. But he said the thoughts were sound. I just needed to learn how to write. So, he'd demonstrated how to do it better, with his red ink. This surgical editing was the first lesson I'd ever had on how to write. By this time, Stretton's second book had been published, *Ideas for Australian Cities*,[3] again not a history book. He'd written it in the last few months of his sabbatical, after finishing *The Political Sciences*, and I'd already read it and was in awe of how beautifully it was written, in Hemingway-esque plain language. I wanted to know how to do this.

I also absorbed my formative set of values from him, which could be described as Fabian socialism.

There were two other formative influences on my way of seeing the world at this time (the late 1960s). Australia's support of the USA in the Vietnam War, including the very unpopular introduction of conscription, was the target of major campus activism and street protests. I was swept up in all this, joining the protests and reading up on American imperialism. Closer to home, my father was fired from his minimum-wage job as a bartender at a downtown pub when he refused to have his work hours extended to 10pm. He'd been in this job for 30 years, and suddenly, in the face of this refusal, was given a week's notice. He hated his job, serving beer all day to an inebriated, rude clientele, but he was devastated. He was in his fifties and had no education. He didn't think he'd be able to find another job and, although my mum was by this time working in a factory, he could not live with the idea of being supported by his wife. He was powerless and crushed. I had been in major conflict with him ever since I could remember, a conflict heightened once I started Uni, because I had opinions he didn't appreciate. But for the six months that he was unemployed, I really felt for him. His sense of humiliation, although he never spoke about it, hung like a fog over our silent household. This blow to my father's precarious pride and masculinity brought home, literally, the critiques of capitalism I'd been reading in my political science courses.

Hugh Stretton was, as I've said, the single most important influence on my intellectual development, but he was more than that. He set an example in how he lived. I doubt he ever owned more than one suit. He drove a beat-up car. He became the deputy chairman of the South Australian Housing Trust (SAHT), the state's public housing agency, and devoted his energy to bettering the lives of working people. And he took a kindly interest in me that year when I was still so shy and unconfident that I couldn't speak in class. He came to my home (built by the SAHT) on the occasion of a fellow student's suicide, and sat with me in the backyard for many hours helping me to 'process' (as we'd now say) this traumatic event. And at the end of the year, once I'd received a First Class Honors degree, it was he who suggested that I do a PhD in Urban Research at the Australian National University in Canberra, as a platform for going into the urban policy field. I'd said I wanted to 'do something useful', but I didn't have a clue what that might be. But I'd read *Ideas for Australian Cities*, so his suggestion resonated with me. Among other things, the book argued that the fate of cities shouldn't be left entirely in the hands of architects, planners, and engineers, but rather needed some generalists from the humanities and social sciences who could think broadly and critically.

Four years before David Harvey's seminal *Social Justice and the City* (1973), Stretton had drawn attention to the fact that where you lived made a huge difference to your wellbeing and life chances. He was writing about locational inequality, but without the theoretical jargon. He simply pointed out how the working-class suburbs, mostly located on the periphery of the metropolitan areas, amidst the new post–World War Two automobile and white-goods factories (like the suburb I'd grown up in), were poorly serviced with respect to 'public goods' such as health, higher education, cultural facilities, and even parks and recreational opportunities. All of these were centrally located and involved long trips on inadequate public transport. He argued for investing more public funds in these neighborhoods as well as relocating public services out of the central cities and into the outer suburbs. He criticized the unbridled residential subdivision of the peripheries, often without basic sewage, let alone schools and shops, when families moved into these neighborhoods. And he emphasized that the people who suffered most from this unbridled suburban development process were women and children, who were isolated and dependent on public transportation to get anywhere, while that same transportation was designed to get men to their downtown jobs. His was a humanist vision, and the first radical critique of the pattern of and assumptions underpinning Australian urban development.

That was the vision that inspired me, the following year, having just turned 21, to continue my studies. In January 1971, desperately eager to get out from under my parents' roof, I took the train to Canberra, with a suitcase and a few boxes of books, and began my PhD years. On a farewell visit to that kitchen table before I left Adelaide, my mentor gave me two books: Lewis Mumford's *The City in History* and Ruth Benedict's *Patterns of Culture*, both of which are still on my shelves.

I have one very poignant memory of my 20-year-old self, walking from campus to the Stretton's house for one of our supervisory meetings. I was early, so I stood across the road from his house, sort of hiding behind a tree, and just staring at that house. I was filled with an unarticulated yearning: for this book-lined house, full of intelligent conversation around a kitchen table, and this loving couple. I suppose I was yearning for this to be my life one day.

As I grew into my emerging academic self in the next decade, I stayed in touch with Hugh Stretton. In fact, he 'rescued' me on more than one occasion from that recurring self-doubt, as I'll discuss later. Eventually our relationship shifted from mentor/mentee to one of friendship. I visited that kitchen table every time I returned to Adelaide, and it was one of the greatest joys of my life, decades later, to take John to that kitchen to meet and get to know Hugh and Patsy.

★

Doctoral Years, 1971–1975

And so began my PhD years in the Urban Research Unit (URU) in the Research School of Social Sciences (RSSS) at the Australian National University (ANU). The first two of these next three years were the worst years of my young life. If I had thought I was a lonely misfit at Adelaide Uni, that feeling was magnified at ANU. At Adelaide Uni there'd been profs who took an interest in me, saw some potential, encouraged me. URU folks were simply baffled about what I was doing there and what to do with me. I was a historian among engineers, economists, and a grumpy, retired professional planner, and it was the latter who became my supervisor. There was no structure to the PhD program, no courses to take. One was expected to be a more or less self-starting researcher, blessed with an annual meeting with one's supervisor. I was very young and immature, having lived under strict rules at home up until I left Adelaide. I had no idea how to boil an egg (not wanting to become a 'domestic slave' like my mum), let alone how to cook up a PhD topic. And no-one was about to give me an easy-to-follow recipe.

There was only one 'performance' requirement in the first year: a paper presentation outlining one's research topic, to be presented to an audience of RSSS scholars. My performance anxiety kicked in on cue on this occasion. I was supposed to open the seminar with a summary of my paper, distributed to attendees the previous week, so I anticipated that all present were armed with heavy artillery, aimed at me. After I was introduced by the chair, all eyes on me, I opened my mouth but nothing would come out. Heart pounding and hands shaking under the table, I simply could not proceed. After I'm not sure how many minutes of silence, I was rescued by a woman sitting close enough to see my panic attack. She started to make some comments, eventually formulating a question. This interlude gave me time to steady my breathing, and I was then able to respond to her.

I stumbled through the rest of the session somehow. I have no memory of what was said in that agonizing hour and a half, but my worst fears had been realized. I was unable to speak 'in public'. How could I survive in this environment, where even morning tea in the faculty tea room was a performance occasion? Afterwards, meeting the grumpy retired planner in his office, he tore shreds off me. How dare I criticize planners when I'd never been a planner or been near a planning agency. These radical ideas of mine were a product of being young and female, he told me. I should get a job in the public service for a while and see that out there in the 'real world' people were happy with their jobs and didn't need my critique to enlighten them.[4]

I was speechless. I left his office in tears, and what happened after that is a bit of a blur. I wandered around the campus for many hours, until late into the night, crying, and was eventually picked up by the campus police, who drove me to the hospital. That's what you did then with 'hysterical' women. For the next six months I did no further work on the PhD. I smoked a lot of pot with a gay American PhD student in anthropology, a fellow 'outsider', disillusioned with his field. I ate a lot and put on a lot of weight. I stopped exercising. I isolated myself. But I did start writing book reviews for various newspapers, a good and necessary outlet for my urge to write. I stopped going to those morning tea sessions in the RSSS. Then, mid-way through my second year, 1972, the head of the URU called me into his office. It had been noted that I was not making progress on the PhD, he said, but that I seemed to have time to write book reviews for national newspapers. Perhaps I should take a year's leave of absence, to figure out whether I really wanted a PhD, he suggested. If I didn't take a leave, they'd be forced to 'fire me', in so far as they wouldn't be recommending the renewal of my scholarship for a third year.

I left that office speechless, but not in tears this time. It was a fair question. Did I really want this PhD? I wasn't sure why I would want to endure any more of this (to me) cruel, toxic environment. This felt like a real crisis moment.

Four things came to my rescue.

First, I wrote to Hugh Stretton explaining my 'situation', confessing my self-doubt, and asking him why he had thought I was capable of doing a PhD. And why in this place? What did he see in me that I couldn't see in myself? His reply did two things for me. It was confidence boosting with respect to my intellectual abilities, bless him! And he reminded me that there was a federal election looming, and that the (moderately left wing) Labor Party was likely to become the government for the first time since 1949. The Party had a strong urban policy platform, so there'd be opportunities opening in the urban policy field. Hurry up and finish, he said, and get out there and 'be useful'. It was salutary to be reminded of my desired destination: a job with the soon-to-be federal Labor Government. That was indeed a motivator. I had never had any intention of hanging around the halls of academia.

For intellectual and emotional support, I had a fellow student, Peter Spearritt, who joined the URU the year after me, was also a historian, and came from a Sydney academic family. I enjoyed his irreverent sense of humor, and he helped me understand the academic world. He invited me to stay with his family in Sydney in their stunning harborside home, and introduced me to the city. And he suggested I reach out to R.W. Connell for intellectual guidance. Raewyn (R.W.) Connell was a young prof at Sydney Uni with a brilliant theoretical mind and strong left-wing beliefs.[5] She was extremely generous in 1972 in recommending readings in political and social theory that would help me to think about power, and thus how to 'frame' my emerging dissertation project: a critical history of Australian city planning. She offered to read my draft thesis, and subsequently recommended it to publishers. This was a crucial connection for me, and one that became a valued friendship in the years that followed.

My other crucial intellectual and moral support at this time began as a temporary substitution to my supervision committee. Robin (Bob) Gollan was a historian of the 'old left' persuasion. Bob and his wife Ann took me under their wing, inviting me to their home for meals and making me feel less isolated. Both Bob and Raewyn were supportive of the fact that I was aspiring to write a 'critical history' of Australian city planning, something about which my grumpy official supervisor/retired planner was skeptical. I was about to start field work in Melbourne and Sydney and had no connections in either city. Bob Gollan suggested I contact Ian Turner, a historian mate of his at Monash University in Melbourne. Ian, as I was to discover, was somewhat of a cult bohemian figure in Australian intellectual, artistic, and left-wing political firmaments. He too was extremely generous with his time when I arrived in Melbourne, immediately introducing me to a wide circle of folks who were thinking progressively and critically about the contemporary city. The most memorable were the investigative journalist Ben Hills, who wrote for Melbourne's *The Age* newspaper; and the endearing Ruthie and Maurie Crow, members of the Communist Party and co-authors of a three-volume *People's Plan for Melbourne* (1969). Ruthie and Maurie were humble, unpretentious, tireless in their devotion to 'the people', and totally devoted to each other. They also took me under their wing, introducing me to the disadvantaged inner-city neighborhoods of Melbourne, where battles were raging between working-class residents and state transportation and housing agencies that threatened those neighborhoods with displacement. They also showed me, by example, the power of community organizing and the possibility of a community-based planning. And Ben Hills taught me how to dig up dirt on politicians and planners involved in shady land deals. Life was starting to get quite exciting.

I shared a passion for Aussie Rules football that enabled me, in spite of my shyness, to connect with Ian Turner. Ian was known in Melbourne as (among many other things) the 'footy prof', because he wrote occasional columns for *The Age* about the game, and every year gave a humorous lecture about the game in the week before the Grand Final, for which he dressed up in footy gear and drank cans of beer while delivering his set piece. I had grown up watching footy and had won a car on a TV sports show at the age of 14 by correctly predicting game results throughout the season. Ian invited me to join his Saturday afternoon gaggle of friends who convened for beer and tucker at his home before walking to the Richmond Tigers game at the nearby footy field. That day was the most fun I'd had in the two years since I'd started the PhD program. Little did I know then, in 1972, that by 1974 he would become the first great love of my life.

My research trips to Sydney were exciting too. With Raewyn Connell's help I was introduced to a few people, who led me to others. I found the temerity to ask for a meeting with Jack Mundey, the president of the Builders Laborers Federation (BLF: the trade union representing construction workers). Jack, like the Crows, was a member of the Australian (i.e. non-aligned) Communist Party and a very progressive thinker. He'd been a star rugby player in his youth and twenties, and while he had very little formal education, he had become a self-educated 'greenie'. Under his charismatic leadership of the BLF, that trade union had just become the front line of urban environmental battles in Sydney by placing 'Green Bans' on construction projects they deemed environmentally and/or socially unsound. It was heady stuff. Jack and I bonded over sports, politics, and beers. I aligned myself with the BLF and offered to support them in any way I could, writing pamphlets and such like. I also joined the Communist Party. My politics had shifted from the Fabian socialism I'd absorbed from Hugh Stretton to an Australian version of Euro-communism. I sold the Party newspaper on street corners, and felt like the (peaceful) revolution was just around the corner.

The revolution, for some of us, was DURD, the new federal Labor Government's Department of Urban and Regional Development. It attracted an eclectic bunch of young zealots who saw an opportunity to address urban inequalities through DURD's ambitious range of policies: public sector land banking to keep land and thus housing prices affordable; decentralization of urban growth to new towns; relocation of government offices to outer suburbs; rehabilitation of inner-city working-class neighborhoods, for the working class. All of this had been inspired by the clear vision communicated in Stretton's *Ideas for Australian Cities*. There didn't seem to me to be any contradiction between being a member of the Communist Party and also being a near-fanatical supporter of DURD.

I quickly shifted from zero to 100 miles an hour on the PhD research, having persuaded the head of the URU that I was willing to knuckle down and finish it in the expected three years. And finish I did, exactly three years after arriving at ANU. In addition to the fieldwork in three cities, interviewing folks about recent history, I'd spent six months in the basement of the ANU library, poring through 100 years of parliamentary debates and newspapers to analyze the path of ideas about city planning and the political forcefields in which those ideas were hashed out, sinking or rising according to whom was in power, where, and with what leverage. As I started to write it all up, I discovered that I could assimilate a vast amount of material, make sense of it, and write faster than I cared to admit to anyone. The examiners loved it. My oral exam, about which I was predictably terrified, turned out to be more of a collegial chat, and all the examiners recommended revising it for publication. In March 1974

I graduated and was offered a two-year Post-doctoral Fellowship in the URU, a year of which could be used for travel abroad. I'd gone from being the misfit of the unit to their new and all-of-a-sudden respected colleague.

I had begun a relationship with Ian Turner by this time. He had a sabbatical due in 1976 and wanted to go to London. Serendipitously, I had met David Donnison, director of the Centre for Environmental Studies (CES) in London, who'd been a visiting prof at the Urban Research Unit in 1974. David had read my dissertation, offered advice on how to turn it into a book, and invited me to come to the CES for a year. I was able to use my travelling post-doc to coincide with Ian's sabbatical. David also connected me with a British publisher, Heinemann, who did a British edition of the forthcoming book *Cities for Sale* (1975). That book turned out to have legs long enough to enjoy a readership for several decades. It was a book that wrestled with an age-old question: is urban social reform possible in a property-owning capitalist liberal democracy? The first paragraph will give a hint of its orientation.

> This book is about failure. City planning in Australia this century has failed to improve the welfare of our city dwellers and unless we understand why, our urban planners will continue to devise counter-productive land-use plans and our urban policy makers will perpetuate the series of income transfers from the poor to the rich which has accompanied the urban planning process in the past. To the extent that the planning movement has been concerned with redistributive social justice, it has been frustrated since its emergence by the established power structures and ideology of Australia's urban, industrial, bureaucratic society – by the old property-based power holders, by the more recent influence of technocrats and by the dominant doctrine of modern western societies that continuing economic growth, based on industrial expansion and the spread of technology, is not only inevitable but also highly desirable. Our urban problems are the spatial expression of the inequalities of a capitalist economy. But that is not the whole story.
>
> *(Sandercock 1975a, 1)*

I had accepted the post-doc rather than going straight to work for DURD, partly because I was flattered that clever people thought my dissertation was worth publishing, and I wanted to devote time to turning it into the book. But really, it was love. I had fallen in love and wanted to be able to synchronize my life with Ian's by going to London. I spent six months revising the dissertation for publication, then took on a consulting job for the South Australian government, a modernizing, progressive Labor Government with a similar urban agenda to that of the federal government of the time. That agenda included embracing the policy of building new towns to take the pressure off existing metropolitan areas and to experiment with different versions of city building than the ubiquitous suburban single-family home. Legislation had been passed providing funds for the building of a new town about 50 miles from Adelaide. I was tasked with advising the government on how to incorporate public participation in this project. Public participation had become one of the new mantras of the early 1970s in the wake of critiques of the damage done by top-down, expert-driven 'rational' planning. I had hammered home its importance for the democratizing of planning in my dissertation/book. Now I got to spend six months reading everything available on experiments with public consultation in the UK and USA, from Sherry Arnstein's critique to the UK's Buchanan Report.

I produced a 150-page report which was enough to turn me into the expert on the topic for the next decade in Australia (Sandercock 1975a).

I was 25 and I'd never been out of Australia. Hugh Stretton now stepped in to introduce me to folks he knew in the USA. He wrote to Mel Webber at Berkeley and Peter Marris at Harvard. I flew out of Sydney in November 1975, just before the shock dismissal of the federal Labor Government, after only three years in office, by the Governor General, acting on behalf of the Queen – a truly bizarre event in Australian political history. Capitalism was striking back, putting an end to talk of restricting foreign ownership of Australia's mineral resources, among other numerous perceived offenses against the natural (capitalist) order of things that Labor had ever-so-gently challenged. I was relieved that I hadn't started a job with the DURD Ministry, only to find it being disbanded by the incoming conservative government. Instead, I found myself spending much of the next decade, one dominated by conservative forces in Australian politics and society (mirroring Thatcherism in the UK and Reaganism in the USA), trying to understand what had gone wrong with that beautiful vision. Was it capitalism striking back? And if so, what did that imply for future prospects of peaceful urban social reform? Or, were there fatal flaws in DURD's theory of locational inequality? Or, was it a case of too much righteous ideological enthusiasm, coupled with lack of experience working inside government bureaucracies, that led to DURD idealists/zealots being scuttled by seasoned senior bureaucrats in other departments who saw it as an upstart threat to their own empires? Could the new Marxist urbanism provide an explanation?[6]

★

Flying out of Australia and into the unknown, again, was another of those terrifying yet exhilarating experiences. I spent a few heady days wandering the streets of San Francisco and was hosted at Berkeley by Prof Mel Webber, who picked me up from my hotel in a car that was at least 60 feet long, or so it seemed to me. This was 1975, before rising gas prices started a re-think about fuel efficiency. Mel took me to his home for dinner and made an effort to show interest in my new book, but I could tell it wasn't 'his thing'. On the opposite coast, my effort to present my book to a seminar of luminaries at the Joint Centre for Urban Studies at Harvard/MIT was a disaster – not quite as bad as that first presentation at ANU as a neophyte PhD student, but still inarticulate. I can still feel how hot my face was the whole time, how nervous and unconfident I was. It was torture for me to have to do this performance stuff. I had never had any coaching about how to 'present'. All I had was a big ball of fear in my throat that was almost a physical obstruction, although it was obviously psychological,[7] connected presumably to being silenced by my father as a child and all through my undergrad years.

My interaction with Peter Marris, though, over dinner with he and his wife, Dolores Hayden, was memorable. He was such a gentle, kind soul: a profound thinker who could communicate his ideas without the need for jargon or intellectual intimidation. I admired their life together, the way they supported each other's work, the way Peter treated Dolores with such respect. This was the third relationship I'd witnessed (after the Strettons and the Crows) that seemed to offer a model of genuine partnership and equality in love and work. I didn't quite articulate this to myself at the time, but I was unconsciously searching for such examples of how to live in a loving relationship in a good way.

★

London as a Post-doc: December 1975–December 1976

If 1975 had felt like the best year of my young life – as I emerged from the PhD with a book, a consulting job, a two-year post-doc, and a first taste of having some 'identity' instead of being the invisible girl who knew nothing – then my year at the Centre for Environmental Studies (CES) was like going back to my first year at Adelaide Uni. I knew nobody except David Donnison, and he had just left CES. I was so shy that I avoided people as much as possible and just tried to read and absorb. Everyone was speaking a new language that I didn't comprehend: the language of Marxist urbanism. The new guru was the Spaniard in exile in France, Manuel Castells, who had just published *The Urban Question*. I tried to read it, and couldn't understand it. He came in person to CES and was treated with reverence by the researchers there. The CES had its own stars, such as Doreen Massey and Michael Harloe. And there was a visiting American prof from UCLA, John Friedmann, whose work I hadn't yet come across. John was the only person who reached out to me, inviting me to have a conversation about his book project, *The Good Society*, which no-one else at CES was interested in because it wasn't on the Marxist bandwagon. This is still my favorite of all of John's books: a daring exercise in critical utopian thinking, incorporating his own and others' poetry and aphorisms. But I was too shy to respond to this invitation and started to avoid John for fear of being 'exposed' as having nothing to say. My impostor syndrome re-appeared with a vengeance during this year.

Fortunately for my equilibrium, Ian arrived in London in February and we resumed our relationship, hanging out at pubs, galleries, second-hand bookstores, and jazz venues such as Ronnie Scott's, where we once heard Dizzy Gillespie play. Ian knew London, and had various friends and networks there, including the lefty historians Eric Hobsbawm and E.P. Thompson and the Australian feminist and expat Germaine Greer, whose work I had devoured in my early twenties. Ian and Eric shared a passion for and incredible knowledge of jazz, and it was a treat to listen to the two of them argue, over lagers, about whether Louis Armstrong's Hot Five was superior to his Hot Seven band, or whether Billie Holiday could hold a candle to Bessie Smith. My initiation into left-wing intellectual and political circles, which had begun with R.W. Connell, the Crows, and Jack Mundey back in Australia, was proceeding apace. I was not yet clear where I stood, between the Castellsian neo-Marxist urbanists and their critique of urban planning; Jack Mundey's direct action tactics in the built environment, using union power informed by Communist ideology; and Ian's brand of pragmatic Australian Labor Party politics. I wanted to embrace them all.

In the summer of that year in London, Ian and I went to Paris for a week, on a shoestring budget, and then to Rome, exploring those magnificent cities, museum after museum, church after church. It was exhausting in the best way. Ian had never been to Italy, so we were experiencing this wonderful culture together. Italy quickly became my favorite place.[8] The year that had started in a rocky way had transformed into something magical. Ian asked me to come live with him in Melbourne. We returned to Italy at the end of the year to explore the treasures of Florence and Venice, before returning to Australia together.

★

First Academic Position: 1977–1981

Now I needed to find a job in Melbourne. I had never intended to become an academic, in part because of my performance anxiety, in part because I didn't see a lot of great role models, and certainly no women. But the Labor Party was no longer in power federally, nor were there any progressive state Labor governments, so my initial plan of working in urban policy had to be put on the back burner. These were dark times for those of us with reformist social justice agendas. It was the beginning of a rollback of the policies and institutions of the post-war welfare state. Academia looked like a reasonable short-term plan. And when I thought about the teachers who had helped me to get this far in my life – a very great distance, economically and culturally, from the neighborhood where I grew up – I wanted to be a good teacher.

I applied successfully for a lecturer (assistant professor) position in Urban Studies at Footscray Institute of Technology (FIT). FIT suited me perfectly. Far from being an elite research university, it was a humble institute of technology, located in the working-class western suburbs of Melbourne. The Urban Studies program was small, and most of its students were the children of post-war immigrants from Greece, Italy, Croatia: the first generation in their families to attend post-secondary education, just like me. I knew I could relate to their insecurities. But could I overcome my own?

I was so afraid of having to stand up and lecture in front of a class of 50 or so 20-year-olds that I wrote every lecture, longhand, and sat and read it to the very bored students. That was about 30 pages of longhand lecture notes per lecture, which meant staying up very late several nights a week. Ian, a brilliant and entertaining lecturer, could not understand why I needed to do this. He encouraged me to make notes that I could talk from. I was eventually so sleep-deprived that I had to give that a shot. I dreamed of one day giving a lecture with just one index card of headings, and then maybe no notes at all. I eventually achieved that dream, but it took a decade, and the anxiety and nausea before each lecture never went away.

During the next couple of years, I started work on a second book, *The Land Racket: The Real Costs of Property Speculation* (1979). This book was both an exposé of corruption among politicians and planners in Melbourne and a diatribe against the very institution of private property, which, I argued, thwarted any efforts at redistributive urban planning. My intent was to bring down the state government. The person who spoke at the book launching event was John Cain, the leader of the Labor Party in Victoria. A few years later, he led the Party to victory, but his government did nothing about land speculation. Taking on property/finance capital was seen by the Labor Party as political suicide, surely proving my argument about the power of private property as an institution.

I had hoped, with these first two books, to influence the direction of urban policy in Australia, but I hadn't yet articulated for myself a theory of change that would explain why writing books might or might not change the world. At this moment in history, academic work seemed like the only thing I could do, having given up on the Communist Party, under Ian's influence. Ian had been a member of the Communist Party back when it was aligned with the Soviet Union. When the Soviets invaded Hungary in 1956, he had dared to condemn this, and for that was expelled from the Party. He was now a very active member of the Australian Labor Party (ALP), and I joined the local branch, in solidarity. He had rejected the ideological purity and self-righteousness of that earlier form of political praxis for a more pragmatic and messy involvement with the ALP, a Party which had an electoral base in the trade union movement and the working class, and occasionally stood a chance of making a difference in people's lives in the here and now. I was wrestling with this pragmatic politics versus the new left intellectual discourse about the role of the state (and thus of the planners who worked for the state) in a capitalist society. Were these planners always, inevitably, the handmaidens of capital and the processes of capital accumulation? A part of me was afraid to answer that question, for fear of losing all sense of purpose.

Ian had written a history of the Australian trade union movement, *In Union Is Strength*. He was also passionate about popular culture: jazz, footy, cartoons, graffiti. He wrote a wonderful book about graffiti, elevating it to an art form of the disenfranchised and disenchanted. I'm sure that book influenced my thinking, two decades later, when I started to write about multiple ways of seeing and knowing the world. Ian loved teaching. He loved good conversation. Our dinner parties were memorable both for the quality of conversation and the quantity of wine consumed. His/our circle of friends was eclectic: painters, poets, playwrights, politicians, cartoonists. I was starting to come out of my shell, find my voice. It was an intoxicating life, in the best sense.

And then it ended. On Boxing Day of 1978, while playing a game of beach cricket with our mates on our annual holiday pilgrimage to Erith Island, Ian had a heart attack, and dropped dead in front of us all. Several mates tried to revive him. We were on an island in the middle of Bass Strait, an hour's ride by helicopter, eight hours by fishing boat, from the mainland. I had just turned 29.

★

I dealt with this loss in several ways. I wrote to Hugh Stretton and to Peter Marris, both of whom responded with beautiful condolence letters. Hugh had known Ian; Peter had not. But I remembered that Peter had written a book about loss, and I read it. That book, *Loss and Change* (1975), stands alongside Stretton's *The Political Sciences* as one of the two most important academic book I've ever read. At the time, it was personally consoling to me to read about the stages of grieving and to understand it as a process, even as I was living it. The enduring value of the book, though, was the way in which Peter had drawn parallels between individual grieving and the collective grieving of communities in the wake of their displacement by planning decisions. The notion that communities could grieve returned to me with explanatory power decades later, when I listened to practitioners talking in a baffled and irritated way about the highly emotional forms of resistance that they were facing from communities threatened with change. That was when I began to develop the concept of a therapeutic planning practice. And once I started working with Indigenous communities and learning about the intergenerational trauma resulting from the genocidal policies of the Canadian state, I was able to recognize the many layers of grief that were and are still alive in those communities. *Loss and Change* attuned me to the need for an expanded language for planning, a language of the emotions, which became one of my main projects in the second half of my life (Parts 1–4).

I decided to resume teaching six weeks after this loss, when the academic year started. That seemed like a better option than staying home alone, crying all day *and* all night. My two closest women friends, Judy Holding and Jenny Love, helped me through the blur of the first year. These two women showed me the value of deep friendship. I cannot now imagine life without it. bell hooks captured this (and so much else) so well: 'friendship is the place in which a great majority of us have our first glimpse of redemptive love and caring community' (hooks 2000, 134).

A year into the grieving process, I made a gutsy decision. Ian had been under contract to the publisher, Granada, to write a book about Aussie Rules football and had accepted several advances on royalties. The editor at Granada now asked if I would be able to finish the book. I said yes, and for the next 18 months that's what I focused on.

And life continued: as it does.

Two years after Ian's death, I had just finished the footy book[9] when John McLaren (my boss at FIT) suggested that I apply for a job at Macquarie University in Sydney, as professor and head of the Graduate Urban Studies program. It seemed highly unlikely to me that I could go from lecturer to full professor in a single bound, leap-frogging the usual rungs on the ladder, but that's what happened. I packed up my house, my books, and my cat and moved to Sydney in 1981. Another new beginning. In sadness, and in anticipation.

Second Academic Position: 1981–1986

I began my new role as head of Graduate Urban Studies with the intention of spending the first year just listening to my colleagues and to the students and not doing anything to rock the boat. All my colleagues were male, and all were older than I was, so that seemed wise. But within a year I had views about what could be improved, especially with regard to teaching and to the curriculum. I soon discovered that nobody wanted to be told by this young woman that 'things needed to change'. I was naïve enough that I had not anticipated, and failed to understand, this resistance to change. Maybe I was arrogant, in my quiet way, even as (most of) my colleagues, in their more vocal way, were stubborn. I believed I had Reason and Truth on my side. If only I had read John Friedmann on the importance of dialogue, in *Retracking America* (1973), or John Forester's still-to-be-written revelatory chapter on 'Listening' in *Planning in the Face of Power* (1989). It took (too) many more years for me to understand the importance of these communicative skills, in every domain of living and loving.

When I gave my professorial inaugural lecture, I could find no better person to quote in support of my pedagogical vision than Hugh Stretton:

> Students come to university at an age when they are most vulnerable to suggestions that they are immature, that it's time to let go of the apron strings and toughen up. It was at that moment that their positivist teachers and text books in planning, engineering, sociology, economics etc., told them that values had no place in science. They must leave all their soft, childish, womanish values outside the classroom. Worst of all, for its effects on their societies (and even perhaps for their own general joy in life) they got the insistent, authoritative message that technical expertise was in, moral thought and social purpose was out.[10]

For the next five years at Macquarie University, I searched for an appropriate curriculum balance between 'technical expertise and moral thought and social purpose' (Sandercock 1983e). I spent more and more time popularizing arguments about the need for continued radical intervention in our cities on behalf of the 'have-nots'. I was becoming a public intellectual, invited to speak on radio and TV and write in newspapers and magazines about urban social and environmental issues. And I was working on a new book with a Melbourne colleague, Mike Berry. *Urban Political Economy: The Australian Case* (1983) brought together each of our writings about the state of urban policy in the decade that had begun with such high hopes under DURD, hopes that came crashing down after the defeat of the Whitlam Labor Government and the demolition of federal urban policy. My essays were *realpolitik* (rather than theoretical) analyses of the fatal flaws of federal urban policy as conceived by and implemented through DURD. Mike's essays on various aspects of 'the housing question' were seeking a theoretical framework for a Marxist analysis of housing.

The question I had posed in the concluding chapter of *Cities for Sale* – 'was urban/social reform possible in a property-owning capitalist democracy?' – was the question that continued to exercise me during these years. I was still trying to relate to the critical writings of those Marxist urbanists I'd encountered in London, yet not seeing any practical way forward in their critiques. The humanist approach of Hugh Stretton had been overshadowed in academic circles by the current discourse dominated by these neo-Marxist urbanists, for whom urban planning in a capitalist society was both 'necessary and impossible' (Fogelsong 1986). The pragmatic politics of urban social reform, DURD-style, kept bumping up against this discourse. Through Ian's influence and example, as well as Stretton's, I continued to advocate for progressive, redistributive, and participatory urban planning policies, working within the state. But I wanted to have it both ways. I wanted to belong to the 'urban political economy' club whose language I had devoted so much intellectual energy to 'mastering'. It was certainly a part of me. But there was something missing that I hadn't yet identified.

My urge to bring about changes in the Graduate Urban Studies program got put on the back burner in 1983, when the Labor Party was re-elected as the federal government under the leadership of the former trade union chief, Bob Hawke. Hawke was not interested in resurrecting DURD, so I didn't think I'd be getting involved with the new government. But within months, I was recruited to three different federal government initiatives. First, I was invited to join Australia's Commission to UNESCO. Then the Minister for Science asked me to join his new brainchild, a think tank with the ambitious title of Commission for the Future. Third, I was asked by the Minister of Education to chair a new committee that would be allocating funds to post-doctoral researchers working on themes of interest to the new government. I accepted all of these, eager as ever to contribute in whatever way I could. This exposure to some of the inner workings of the federal bureaucracy, along with my friendship with several elected Labor politicians now enjoying a stint in government (one a senator, the other a member of the prime minister's cabinet), was instructive in a critical way. When I'd accepted the chair at Macquarie University, I'd told myself I'd do it for five years. Then I would go into politics, to be where I'd thought I always wanted to be: at the coalface of social change, not armchair theorizing it. But the more I learned about the daily lives of politicians and federal bureaucrats, the more I realized this was not for me. The anecdotes of sexism and sexual harassment were hair-raising. The closeness to power had a certain seductiveness, but I didn't want, literally, to be 'seduced'.

The work in Canberra, though, did seem like a good complement to my work at Macquarie, which felt stuck in some way. What was that sense of 'stuckness' about? I was a full professor already, and so many doors seemed to be opening, in so many directions. I was living a life that would have been unimaginable to me even five years earlier, let alone ten. Leaving aside the absence of a loving partner, wasn't I now living the dream I'd subconsciously dreamed when I stood opposite Hugh Stretton's house in Adelaide as a 20-year-old, overcome with an unarticulated yearning? I could not put it into words, in spite of having read all the major feminist tracts of the previous decade. But the Italian novelist, Elena Ferrante, writing fictionally about a woman in academic life in Italy during this same decade, has given me the words for what I was living:

> And no one knew better than I did what it meant to make your own head masculine so that it would be accepted by the culture of men. I had done it. I was doing it. … I had striven to give myself male capacities. I thought I had to know everything, be concerned with everything. I wanted to make a good impression on men, be at their level. … Such persistence in memorizing fashionable jargon, wasted effort. I had been conditioned by my education, which had shaped my mind, my voice. To what secret pact with myself had I consented, just to excel? And now, after the hard work of learning, what must I unlearn?
>
> *(Ferrante 2014, 281–2)*[11]

What I am saying is that although I was now in an apparent position of (minor) power, as a head of a graduate program, all of my colleagues were White, male, and older, and not interested in my views if those views threatened to rock the boat, the taken-for-granted ways of doing things. Ever since my undergraduate days, I'd had to impress and win the approval of male colleagues and superiors. I had absorbed from these men the rules of the game, which they had defined or had had to learn for themselves. (Yes, they were trapped too, in limited ways of seeing and knowing). There were no other role models that I had encountered. To be sure, some of these men, my teachers, were my mentors, and had given me a different set of values. But we were all living in the same room of that 'fairy tale' academic house, and no-one desired to start stripping off the wallpaper – myself included. I was, nevertheless, starting to feel the straightjacket, the claustrophobia.

But something was about to shift for me, first personally and then professionally.

In 1984, emerging from the final stage of grieving, I decided to sell my 1920s three-bedroom house in the inner-city neighborhood of Balmain and move to a brand-new two-room apartment, all glass and light, overlooking Bondi Beach, where I could bodysurf every day. My happy place, the ocean.[12] I had been wrestling with how to write about the research I was doing with unemployed steel workers and coal miners in Wollongong. This was where I was 'stuck'. My reading in political economy had led me to Bluestone and Harrison's seminal work *The Deindustrialization of America* (1981). Buried somewhere in a footnote was a reference to a Working Paper by John Friedmann at UCLA that sounded relevant to my path of inquiry. But the paper hadn't been published, except as a departmental Working Paper, so I decided to write to John, remind him that we'd met in London, tell him what I was working on, and request a copy of his paper. I got a very quick response, by snail mail, enclosing the paper and, to my surprise, an invitation to become 'pen pals'. I took up this invitation a few weeks later, on the day I moved into my Bondi Beach apartment. That day was symbolically significant for me, in that I was moving out of a dark, interior-oriented house that resembled Ian's and my house in Melbourne, into this light-filled, ocean-oriented place, and I felt like I was moving out of the darkness of grieving and into the new light, embracing life again in the fullest sense. So, when I sat down on my deck overlooking the Pacific Ocean to write to John, instead of writing about my research, I wrote at some length about what I'd been through, emotionally, in the past six years since Ian died. And John replied in kind, writing about his difficult divorce and his bachelor life since then. For the next six months we wrote each other in this very personal vein, sharing our love of poetry and music. Hand-written letters crisscrossed the Pacific, initially fortnightly, then weekly, then daily.

After eight months of this correspondence, and several months of phone calls, on an impulse I bought a ticket on Qantas for a three-day visit to LA and phoned John to tell him

I was arriving the next Friday morning. There was quite a long silence on the other end of the line …

When he met me at LA airport, on another wild impulse I cartwheeled in the carpark. And the rest, as they say, is history.

After that three-day visit, John invited me to spend my upcoming six-month sabbatical at UCLA and live with him for those months. At the end of that six-month 'test-drive', we decided to join our lives. I returned to Sydney to give the appropriate six-month notice at Macquarie, and began to wrap up my life there. The decision to leave Australia was also a decision to potentially leave academia, in search of something more satisfying, less of a straightjacket than I'd begun to feel academic writing, academic life, was. And that's what led me to become a student again, applying to the Master of Fine Arts in scriptwriting at UCLA's fabled Film School.

In July 1986, my new life began. Los Angeles. Film School. John.

Notes

1 Designed by young Sydney architecture graduate Ann Forsyth, whom I didn't know at the time. She was working with my close friend, Wendy Sarkissian. A few years later, though, Ann arrived at UCLA as a student in the masters program at the Graduate School of Architecture and Urban Planning, where I was teaching. We subsequently co-authored a paper on Gender and Planning Theory (Sandercock and Forsyth 1992) before she went on to do her PhD at Cornell. Ann is now a distinguished professor at Harvard.

2 Mills and Stretton shared something that I admired from my earliest days as a Uni student. Mills was not just a sociologist but also a humanist and a social critic, who focused his energetic intellect on the vital issues of his time.

3 This became a best seller, reaching well beyond academia, and was revised in second and third editions. Its dust jacket mantra was 'women and children first'.

4 My critique drew entirely on 'foreign' authors like Richard Sennett (*The Uses of Disorder: Personal Identity and City Life*, 1970), Robert Goodman (*After the Planners*, 1971), Herb Gans (*The Levittowners,* 1967; *People and Plans*, 1968), Jane Jacobs (*The Death and Life of Great American Cities*, 1961), Paul Davidoff ('Advocacy and Pluralism in Planning', 1965), and Paul and Percival Goodman (*Communitas: Means of Livelihood and Ways of Life*, 1947), none of whom my supervisor would have read. Still, he had a point. I knew nothing of the real world of planning agencies in Australia.

5 Her first book, *Ruling Class, Ruling Culture: Studies of Conflict, Power and Hegemony in Australian Life* (1977), had yet to be written, but became pivotal in an emerging neo-Marxist (new left) critique of Australian society and governance. A later work, *Gender and Power* (1987), became seminal on an international stage in the evolving discourse about gender relations, patriarchy, and capitalism.

6 Once I returned to Australia in 1977, this was what preoccupied me for a decade. See Sandercock 1979, 1983a, 1983b, 1983c, 1983d; Sandercock and Berry 1983, 1983a; Sandercock and Orchard 1989.

7 For the record, and for the benefit of those who suffer the same terror, this fear is still with me today, although somewhat less intense and not quite so obvious; but I still have moments when I get up in front of an audience and go blank, unable to remember what it was I wanted to say.

8 So it was delightful when the work I started doing in the 1990s was well-received in Italy. Many invitations followed to visit professionally, and my 1998 book, *Towards Cosmopolis*, was translated into Italian.

9 Ian had written a third of it, and had a few notes for another chapter. I wrote the rest. The book is a social history and political economy of the game. The hardback edition of 10,000 copies sold out quickly.

10 This was more or less exactly what had happened to me in my first year in the PhD program in 1971. The quote is from H. Stretton, 'The Corruption of the Intellectuals', Oscar Mendelsohn Lecture, Monash

University, 1980, unpublished manuscript, pp. 19–20, cited in L. Sandercock, 'Thinking in Pictures' in L. Orchard & R. Dare (eds) *Markets, Morals, and Public Policy* (Sydney: Federation Press, 1989).

11 I quote from Ann Goldstein's translation of Elena Ferrante's novel *Those Who Leave and Those Who Stay* (New York: Europa Editions, 2014). The novel's main character, Elena, comes from a dysfunctional working-class family on the periphery of Naples. Excelling in school, she wins a chance to attend university and goes on to become a writer who marries into an upper-class academic family from northern Italy.

12 I had started swimming and surfing regularly and discovered that this was an excellent way to deal with sorrow and strife, a habit that has served me well ever since.

References

Fogelsong, R. 1986. *Planning the Capitalist City: the Colonial Era to the 1920s* (New Haven: Princeton University Press).

Sandercock, L. 1975a. *Cities for Sale: Property, Politics and Urban Planning* (Melbourne: Melbourne University Press).

Sandercock, L. 1975b. *Public Participation in Planning* (Adelaide: South Australian Government) Printer).

Sandercock, l. 1979. Urban policy. In: Patience, A. and Head, B. (Eds) *From Whitlam to Fraser: Reform and Reaction in Australian Politics* (Melbourne: Oxford University Press).

Sandercock, L. and Berry, M. 1983a. *Urban Political Economy: The Australian Case* (Sydney: Allen and Unwin).

Sandercock, L. and Berry, M. 1983b. Cities without jobs? Prospects for the 1980s. In Sandercock, L. (Ed), *Law and Medicine Berry. Urban Political Economy: The Australian Case* (Sydney: Allen and Unwin).

Sandercock, L. 1983a. Urban policy: from Whitlam to Fraser. In Sandercock, L. (Ed) *Law and Medicine Berry. Urban Political Economy: The Australian Case* (Sydney: Allen and Unwin).

Sandercock, L. 1983b. A socialist city in a capitalist society? in Sandercock, L. (Ed) *Law and Medicine Berry. Urban Political Economy: The Australian Case* (Sydney: Allen and Unwin).

Sandercock, L. 1983c. Who gets what out of public participation? In Sandercock, L. (Ed) *Law and Medicine Berry. Urban Political Economy: The Australian Case* (Sydney: Allen and Unwin).

Sandercock, L. 1983d. Reforming land policy. In Sandercock, L. (Ed) *Law and Medicine Berry. Urban Political Economy: The Australian Case* (Sydney: Allen and Unwin).

Sandercock, L. 1983e. Educating planners: From physical determinism to economic crisis. In Sandercock, L. (Ed) *Law and Medicine Berry. Urban Political Economy: The Australian Case* (Sydney: Allen and Unwin).

Sandercock, L. and Forsyth, A. 1992. A gender agenda. New directions for planning theory, *Journal of the American Planning Association* 58(1), 49–59.

Sandercock, L. and Orchard, L. 1989. Urban policy. In: Head, B. and Patience, A. (Eds) *From Fraser to Hawke* (Melbourne: Longman) Cheshire).

PART I

Diversifying Planning's History, Theory, and Epistemology

FIGURE 1.1 *Tree of Life*, by kind permission of Musqueam artist Susan Point, who owns the copyright to the artwork. Image Copyright: Coastal Peoples Fine Art Gallery, Vancouver, BC, Canada.

Commentary: The Los Angeles Years: 1986–1996

There are moments in life when one feels so suddenly *alive* that it's as though *real life* were just beginning. The first time I felt that was in my mid-twenties when I'd fallen in love with the first of the two great loves in my life, Ian Turner. I had moved in with him in Melbourne's inner city, had just started my first academic job, and, through Ian, was suddenly immersed in a world of politics, jazz, football, and cultural life (Introduction). The second such experience was my move to Los Angeles in my mid-thirties to join the second great love of my life, John Friedmann; to become a student again, in the renowned Film School at the University of California, Los Angeles (UCLA); and to teach part-time in the Graduate School of Architecture and Urban Planning (GSAUP), where John had been teaching since 1969 Figure 1.1).

It is hard to imagine a more stimulating environment than what I experienced there for the next decade. During that time, I led parallel lives, in both of which I was more than fully invested. In film school, I was tapping into a creativity I never knew I possessed. I was, again, a fish out of water: I knew nothing about the creative or business side of scriptwriting. My nine fellow students were all in their early twenties and wanted to write for TV shows I'd never heard of. But the learning by doing was intense, demanding, fun, and rewarding, on so many levels. It was work, but it was playful. In teaching planning history to grad students in GSAUP, I was being exposed, through them, to new historiographies that challenged my own earlier work, especially my first book, *Cities for Sale* (1975). And at home, with John, it was 24/7 engagement with ideas, classical music, movies, visual art. There was hiking in the San Gabriel Mountains and late-night jazz clubs. All of it was daunting for the still shy girl from Adelaide, and at the same time, exhilarating. I can't help but wonder, looking back to half a lifetime ago, where all the energy came from for these parallel lives. I think it must have been love.

This first Commentary has the difficult task of tracking this decade of a double life. I wrote ten screenplays during this time and sold one of them, which was made into a TV movie of the week. I also published a second, American edition of *Cities for Sale* (Sandercock 1990); wrote a few academic journal articles; and, by the end of the decade (when we moved to Australia), had two book manuscripts in preparation. Because this Commentary is introducing the academic writing of this decade, I'll begin by discussing how that new direction in my thinking evolved. But I'll return to the screenwriting story, and the love story, to close this Commentary. In charting the evolution of my academic self over this decade, 'the LA years', I want to focus as much on the context as on the text/s: that is, the environment, the relationships, and the conversations out of which emerged, at the end of the 1990s, those two academic books that found an international audience (Sandercock 1998a, b). Perhaps the following is a bit too tidy, thanks to the benefit of hindsight and reflection, but I see four catalysts for the new directions my work took once I left Australia in search of something still to be defined.

The first was a result of a talk I gave, shortly after joining GSAUP, about my research into the social and psychological impacts of the economic restructuring of the steel and coal industries in the town of Wollongong, south of Sydney. While the theme of global economic restructuring was a hot topic in US academic circles at the time (Bluestone and Harrison's 1982 book *The Deindustrialization of America* had been seminal), nobody seemed very interested in my Australian case study. Or perhaps they just weren't interested in the methodological dilemma I had not been able to solve: how to combine a macro-political economy perspective with detailed social/psychological profiles of unemployed coal miners and steelworkers. This was, after all, one of the reasons I was now enrolled in film school: to figure out how to tell that kind of story in a different, more involving way. Whatever was going on with my GSAUP audience that day, my self-reflection after that talk was a simple one: 'I have to re-invent myself if I want to continue to be academically active here'. I wasn't certain I wanted to be academically involved anymore, but I did know that I needed to create new, more liberating possibilities for my self-expression, for my spirit. That's why I'd uprooted myself and crossed the Pacific. When I accepted an invitation to teach planning history in GSAUP,[1] I had no inkling that discussions in those seminars with some wonderful students would begin a rejuvenating process of re-thinking my approach to the writing of history, specifically of planning history.

The second catalyst was an invitation from the publisher Transaction Press (Rutgers University) to publish a second edition of *Cities for Sale* (1975) for an American audience, with a new Introduction making it relevant to that audience. The subsequent 33-page Introduction to the Transaction edition, re-titled *Property, Politics and Urban Planning* (Sandercock 1990), marked the re-emergence of myself as a critical historiographer, influenced by new currents of thought in feminist history as well as the broader field of historiography, where the works of Hayden White and Dominic La Capra were especially important. In that Introduction, I not only cheekily critiqued the entire output of American planning historians but also took apart my own first edition, and outlined how I would now write a very different book. My self-critique was informed by two sources. The first was Joan Kelly's radical re-thinking of history from a feminist perspective, arguing that gender (along with class and race) must be a category of analysis in any writing of history (Kelly 1984). The second was the pathbreaking Australian historian Henry Reynolds, who was rewriting Australian history from the perspective of the dispossession of its First Peoples (Reynolds 1981, 1987, 1989). So much new reading and re-thinking seemed necessary, and yet I was simultaneously a full-time student in film school. This would not have been possible were it not for being pushed by the very bright grad students noted above, especially Iain Borden, Taina Rikala de Noriega, Robin Bloch, and Larry Barth. Teaching in that environment created for me the kind of stimulus that had never been available when I was a lonely, isolated, unconfident PhD student who was not taken seriously by her supervisors.

The third catalyst during this LA decade was a continuation of the second, but with new faces among the students. Through the early 1990s, my planning history seminars continued to be a hotbed of (sometimes hotheaded) discussions about the ways that some of the new social and political theorizing could inform the writing of planning history. Thus, in 1993, when Luigi Mazza, professor and editor of the Italian journal *Planning Theory*, invited me to guest edit a special issue of the journal, examining the relationship between history and theory, I leapt at the opportunity. Not surprisingly, given the context I've just outlined, four of the six resulting essays were written by doctoral students in my course (Clyde Woods, Moira Kenney, Gail Dubrow, Barbara Hooper), and another by the brilliant cultural anthropologist and new friend, James Holston. My own introductory essay (Sandercock 1995a) built on what I'd begun to explore in 1990 in the Introduction to the American (second) edition of *Cities for Sale*.

Such was the interest aroused by that special issue of *Planning Theory* in 1995 that I decided to expand the circle of contributors and put a book together. I convened a paper session for the 1995 Association of Collegiate Schools of Planning (ACSP) conference with the four PhD student contributors, and it was in that session that Ted Jojola introduced himself and asked whether I had thought of including an Indigenous perspective. That was one of those moments (described further in the Commentary to Part IV) of acute embarrassment *but also* the beginning of a beautiful friendship. I had not considered this, but that was about to change after an initial conversation with Ted. I had also gotten to know Bob Beauregard through these ACSP sessions, and held him in high regard for his contributions to planning theory. Bob was to become a friend and colleague, reading the manuscripts of several of my future books and providing incisive and insightful feedback. To this day I regard Bob, along with John Forester and Patsy Healey (see Commentary, Part III), as role models of the best that a scholar/academic can be: generous to a fault in devoting time to reading the work of others, and always responding from a place of integrity rather than ideology or competitiveness. Bob, and James Holston, both already established scholars, became contributors to the edited collection in progress, as did June Manning Thomas, then a prof at Michigan State University, whom I was getting to know and love, again thanks to the ACSP network. But the rest of the contributors to *Making the Invisible Visible* were all doctoral students or recent graduates of GSAUP, and all of them made my own intellectual journey an absolute delight and such a collaborative effort: Iain Borden, Susie Wirka, Olivier Kramsch, and Dora Epstein, alongside Clyde Woods, Moira Kenney, Gail Dubrow, and Barbara Hooper. I dedicated that book 'To My UCLA students, 1986–1996', and my memories of times shared with them are among the most precious of my long academic life. My introductory chapter to that book (*Making the Invisible Visible*, 1998) is Chapter 1 in this collection, because the thought process behind it was the beginning of my new academic/intellectual self (my reinvention), and was a foretaste of my emerging intellectual project: to diversify planning history and theory.

It was James Holston's concept of 'insurgent citizenship' in that initial *Planning Theory* collection that helped crystallize my own thinking about insurgent historiographies for planning, and, more broadly, a concept of insurgent (or radical) planning (a concept since taken up and developed by the next two generations of planning scholar/activists: Faranak Miraftab, Vicki Beard, Raksha Vasudevan, and Magdalena Novoa). And it was also at this time that I found my thinking congruent with that of bell hooks, then a new rock star in the academic firmament, whom I quoted at the opening of my Introduction to that collection:

> Subversive historiography connects oppositional practices from the past and forms of resistance in the present, thus creating spaces of possibility where the future can be imagined differently – imagined in such a way that we can witness ourselves dreaming, moving forward and beyond the limits of confines of fixed locations.
>
> *(hooks 1994)*

What I still love about this quote is that it connects critical thinking about the past with dreaming, or utopian thinking, about the present and future, 'creating spaces of possibility'. That is exactly what I was doing in the field of planning during this decade and beyond. The dreaming of a different societal future, on my part, appears here as Chapter 4 in Part II.

That time at UCLA was such a mutual learning environment, the likes of which I'd never experienced in Australia. In part this can be explained by differences in the structure of PhD programs in Australia (modelled on the UK) and the US/Canada. In Australia, the doctoral student is a lone researcher, not required to take any coursework, somehow expected to be a self-starter, and minimally supervised (I had one meeting a year with my PhD supervisors). Finishing a doctorate generally takes several years longer in the US/Canadian systems because of all the coursework requirements, but the extra time is surely worth it in terms of potential camaraderie, and the opportunity to find one's intellectual kindred spirits. I think too, that some of this joyful mutual learning environment was my own creation. I did not put myself above the grad students. I didn't see any great distance between them and me. I too was learning year by year, and re-thinking everything I'd been taught or thought I knew. These were the best of times. But they were also troubled times…

The fourth and most profound catalyst of my new directions was a much less comfortable one. The 'LA decade' was not all wine and roses, either in terms of what was happening on the 'other' side of the city[2] or my lived experience of that context. Remember Rodney King's near-death experience at the brutal hands of the LA Police Department in March 1991? Remember the not-guilty verdict in April 1992, from an all-White jury, for the LA police officers on trial for that assault on King? Remember LA burning in the immediate aftermath of that jury verdict?

CUT TO: my UCLA classroom, Fall 1992.

A few months after the LA uprising in the wake of the not guilty verdict, I was teaching my graduate Planning Theory class. (Around 1990, John decided he'd had enough of teaching said course. I had a very different take on the then-dominant White male discourse of planning theory, so I was excited to be offered the opportunity to take over that class.) That fateful Fall, there were about 30 students in the class, at least half of whom were students of color: African American, Asian American, Chicano/a, and Central American. For several years I'd been encouraging and facilitating discussions about gender, race, ethnicity, and sexual preference – of 'difference' – in this class, despite the lack of apparently relevant readings. (I drew on autobiographical, literary writings, as well as students' own life stories.) I knew that such discussions, this semester, would be tense and potentially divisive, but it hardly seemed appropriate to keep the city's troubles out of the classroom. Still, I wasn't prepared for the response I got when I showed a video, pirated from public television, titled 'Race Matters'.

This 90-minute video was a discussion with a studio audience, moderated by the White talk-show host Phil Donahue, but featuring lengthy presentations by three African Americans of very different political persuasions: one a Harvard intellectual, another a businessman, the third a Chicago journalist who'd grown up in the notorious public housing projects there. Each of their presentations gave a different analysis of the 'problems' of Black America, and proposed different solutions: from the rainbow coalition, to economic self-help and self-sufficiency, to education. There was also an invited panel of half a dozen people of color, again from a variety of backgrounds and political persuasions.

The program was animated and very controversial. The three African American presentations were thoughtful and constructive. The panel members all made constructive interventions. Donohue was the only person who came out of it badly, in my view, in that he repeatedly reacted in a very personal and defensive way whenever a person of color mentioned 'White racism', and kept insisting that he had many Black friends. He embarrassed himself. I thought he was inept and a fool, and wondered why the public TV network had chosen him as moderator. But it hadn't occurred to me that he was offensive.

My intention in showing the video was to get discussion going around the issue 'Race Matters', and to think about it in relation to Los Angeles and its planning challenges. Each of the three African American presenters had been given the resources to make a 10-minute video of activities within their communities which seemed to offer hope for the future. I thought that was a constructive place to start. I thought I was 'doing the right thing', bringing voices of people of color directly into the classroom via the program. We all watched it, then took a coffee break before reassembling for discussion. I then called for thoughts on what we'd just seen.

There was an awkward silence at first, and then an African American woman spoke, very emotionally. She said that her fellow Black students were appalled and offended by the program and deeply disappointed in me for showing it; that it was a racist program that perpetuated the idea that Blacks were 'the problem' when in fact it was White racism that was the problem; and that the behavior of the moderator was proof of the bad taste of the program. She said that the Black students had decided, during the coffee break, that they were not prepared to take part in any discussion based on this video, at which point she teared up, and was unable to continue speaking.

I was stunned: totally unprepared for such a reaction. I was literally speechless. A deafening silence descended over the class. I looked around the room: hoping for rescue, hoping for a miracle? It didn't happen. I was naked and burning. Minutes later, recovering my voice, I asked if others felt the same way. No-one wanted to speak. I sat there, slowly absorbing my own blindness. Looking back, I was not only naïve in my expectation of the kind of discussion that might ensue from seeing such a program; I was also *blind* – that is, unable to see that there would be different ways of 'reading' that program. While I, in my White privilege, could casually dismiss Donohue's behavior as dumb, African Americans were deeply offended. While I thought the program had merit in giving voice to people of color, my students saw, rather, that it was being run by a White man. And while I experienced the three presentations as constructive contributions to an understanding of what could be done in their communities, my students saw this as a total misrepresentation of the real issue: White racism.

After what felt like an eternity, but may have been 10 minutes of us all sitting in silence (no-one left the room), I offered to organize a discussion for the following week on the subject of White racism. At that point, other students began to find their voices. Students of color other than African Americans complained that the program gave the impression that the only issue was Black versus White, whereas the LA uprising was a much more complex, multi-ethnic and multi-racial event. Tentatively, a couple of White students asked whether there was, or ever could be, any role for them working in Black communities or communities of color. Gradually some discussion emerged. It was awkward and painful (in different ways) for everybody. But it was a start. Out of initial anger and hostility, we collectively, tentatively, opened a space and made an effort. I learned that I had a lot to learn about 'difference'. And that was when I *really* began the project of diversifying planning theory: out of this experience of humiliation.[3]

The enterprise of planning theory at this time (1992) was an almost exclusively male and definitely White domain. (Patsy Healey and Judith Innes were yet to emerge as significant contributors.) As a feminist, I had recently argued, in a paper jointly written with then masters student Ann Forsyth, that planning could not be theorized as though its practices were neutral with respect to gender (Sandercock and Forsyth 1992). Lesbian/gay/queer theorists were beginning to make a similar argument, from their positionality. A parallel argument clearly needed to be made with respect to race and ethnicity. What this signified to me was the need to shift from simply insisting on the difference that theory makes – that is, the importance of the practice of theory, and of how theory informs all practice, visibly or less visibly[4] – to an understanding of the importance of *theorizing difference*. In other words, I started to argue for 'difference' as a category of analysis. But what was accepted as planning theory literature had thus far shown no interest in the debates around 'difference' in social theory, feminist theory, cultural studies. There was nothing within our field to build on. I had to make a bit of a bold leap, going outside the planning literature to find inspiration for this new enterprise. And that is how Chapters 2 and 3 came into being.

I had been reading authors of color outside the planning field for some time, again partially guided by some of my students of color. Now I found a way to include these voices, which at the time I called (as they were calling themselves) 'the voices from the borderlands' (Sandercock 1995b). That essay, published in the *Journal of Planning Education & Research*, caused baffled consternation among many colleagues. Typical comments went like this: 'It's beautifully written, but I just don't see what it has to do with planning'. That response suggested I had to develop a parallel argument about ways of knowing: how there are many different ways of knowing, partially at least determined by culture, race, class, and gender. That became my 'epistemology of multiplicity' (Chapter 2). All of this eventually came together in *Towards Cosmopolis: Planning for Multicultural Cities* (1998). By the time it was published, I had returned to Australia to start the next phase of this intellectual and academic journey. But that book was the culmination of my LA years and owes everything to the folks who pushed me during that tumultuous and transformative decade.

Looking back from 2022, that 1995 'Voices from the Borderlands' essay (Chapter 3) was not only a direct response to my classroom crisis of 1992 but also was setting my agenda for the next 25 years. Even though I was drawing on authors well outside the planning field – bell hooks, Cornel West, June Jordan, Gloria Anzaldua, Guillermo Gomez-Pena – I connected with their emphasis on *praxis*, as well as the themes of home and belonging, meaning and action, ways of being, knowing, acting, and loving. I'm sure too that my own origins, as a working-class girl who was made to feel like an outsider as an undergraduate, coming from the wrong high school in the wrong part of town (Introduction), helped me identify with the folks who saw themselves as outsiders, inhabiting the margins.

This LA decade was a period of theoretical restructuring and epistemological questioning. The critical historiographical work led me into dreaming about alternative futures for both the planning profession and the cities we live in. I had begun to see the city through the eyes of my students of color, and to dream of a different kind of city, my *cosmopolis* of the mind, a city of democratic inclusion, where difference is celebrated. That dreaming (as well as the theoretical restructuring and epistemological exploration) culminated in the book *Towards Cosmopolis* (1998). And that theme, how planning could contribute to the realization of cosmopolis, was to preoccupy me for the next decade (1996–2006). That is the focus of Part II.

This combination of dreaming and praxis drew me to people who had a praxis in the city, as opposed to a (merely) theoretical praxis. In the 90s, GSAUP initiated something it called the Community Scholars program, inviting community organizers (whom I thought of as activist planners) to spend a semester in residence, sharing what they were doing in their (marginalized) communities. The two folks whose work most inspired me were LA's Gilda Haas and Boston-based, Barbados-born Mel King. In the empirical part of *Towards Cosmopolis*, I discuss their work as exemplars of *insurgent planning*. Mel King had written a beautiful book about this work, *Chains of Change* (1981), which I quote *to this day* in my classes. His assertion that 'there's no such thing as a poor community', in the sense that every community has human resources and every human has gifts, was a neat way of saying what John McKnight subsequently theorized as the asset-based approach to community development, and which I was to see exemplified in the Collingwood Neighborhood House in Vancouver (Chapter 8) as well as embodied in Indigenous community planning. I was fortunate to get to visit Mel

King's home in the south end of Boston and appreciate how he lived his philosophy in his everyday life. His back door was literally always open and there was always a pot of soup on the stove and a loaf of bread on the kitchen table. I was in awe of everything about that man.

Closer to home, my GSAUP colleague Jackie Leavitt was also doing awesome community-based work in the predominantly Black public housing projects in LA, after completing her PhD on abandoned housing in the Bronx and writing a terrific book about that, *From Abandonment to Hope* (Leavitt and Saegert 1990). Dolores Hayden was one of the 'stars' at GSAUP at this time, with a fan following among feminist students. Her revisionist historical research was pathbreaking,[5] but it was her husband's books that had the most profound impact on me. I had met Peter Marris and Dolores at Harvard/MIT in 1975 thanks to an introduction from my undergraduate mentor Hugh Stretton (Introduction). Peter had published *Loss and Change* in 1975, and it became and remains one of my favorite books in our field, and one that eventually helped me to formulate the concept of *therapeutic planning* (Chapter 9). In the 1980s he published *Meaning and Action* (1987), a book that transformed how I thought about power and about theories of change. In academic life at that time, I don't think I'd come across another male as gentle and caring as Peter was. I considered him a dear friend, partly because I'd been able to discuss personal loss with him back in 1979 (by snail mail). Now we found ourselves discussing the possibilities and merits of fictionalizing planning, as opposed to doing more empirical or theoretical work. I had been working on a screenplay about an actual 19th-century utopian community in Kansas (Octagon City), and Peter was working on a novel. I was impressed that he discarded an entire first draft, started all over again, and eventually published the riveting *Dreams of General Jerusalem* (1988), a novel about the politics and personalities of White outsiders attempting to do planning in an unnamed east African country. My copy has been loaned to a whole generation of PhD students in Vancouver, where I delight in introducing his entire *oeuvre* in my advanced planning theory seminar.

In this Commentary I've privileged my relationships with grad students as the most generative ones in my life in this LA decade. But Peter's influence came long before and continued beyond that decade. He was the kind of thinker I aspired to be. And obviously there was something in the air, so to speak, in GSAUP at that time. Not only were there resident 'stars' like Ed Soja and Michael Storper, Marty Wachs, Hayden, Marris, and Friedmann, but there was also a constant stream of visitors. I was never able to connect with Soja's work, which was too abstruse for me. But Marty and Helen Wachs were very dear friends of ours, and although I didn't connect much with any work in transportation planning, I revered Marty Wachs for his ethic of service to the academic community – actually, more than that: his ethic of service to the human community. And we were both huge sports fans.

Of the stream of folks who passed through GSAUP, two became friends and made a lasting impression as scholars: Bent Flyvbjerg and Janet Abu-Lughod. Bent and I shared a windowless office for a year in 1988 when he was transcribing interviews for his PhD (which became the book *Rationality and Power* 1998). We were both 'finding our way' at the time. Who were we each becoming? Whose work mattered to us? What ideas? More than a decade later, Bent came out with a book that I still require my PhD students to read: *Making Social Science Matter* (2002). I related to that book because it asked key questions about power and knowledge; because it validated a values-driven approach to scholarship; and because it privileged contextual knowledge, in the form of case studies. Some years later we connected again around his work on *phronesis*. He invited me to contribute a chapter to his edited collection developing the argument about 'real social science' being the work of applied *phronesis*, that is, of practical wisdom in the service of public policy (Flyvbjerg, Landman and Schram 2012). By then, Giovanni Attili and I had made a documentary in collaboration with two small First Nations in north central British Columbia, a project that we framed as an intervention into a highly charged political force field, and which seemed like a perfect example of applied phronesis (Sandercock and Attili 2012). But I'm getting ahead of myself now. That is discussed in much more detail in the Commentary to Part III.

Janet Abu-Lughod, a prof at the New School in New York, had been a student in the Planning Program at the University of Chicago with John in the 1950s and was still going strong in the 1990s. She and I had immediate chemistry. We had a sort of mutual admiration club going. She admired my *Property, Politics and Urban Planning* (1990), wrote a generous endorsement for *Towards Cosmopolis*, and was fascinated that I was writing screenplays.[6] In turn, there were two of her books that I thought were impeccable historical scholarship (Abu-Lughod 1980, 1995), and she was now working on what was to become her masterpiece, a historical sociological comparative study, *New York Chicago Los Angeles* (1999). Whenever I was in New York I visited her apartment (barely able to breathe through the fog of her cigarette smoke), and we'd go eat in a neighborhood restaurant and have the best academic gossips ever.

★

What about the 'elephant in the room' up until now, John Friedmann. Did his presence, our relationship, influence my intellectual path? After the publishing of what most people regard as his magnum opus, *Planning in the Public Domain: From Knowledge to Action* (1987), John backed away from teaching planning theory. During this decade, he produced a couple of books, *Life Space and Economic Space* (1988), followed by *Empowerment: A Politics of Alternative Development* (1992), both of which brought some closure to his many years in the field in Latin America. I had nothing to contribute to that work, nor did he to where I was going with planning theory. He was, though, very supportive of the new directions in which I was taking what he called 'the theory enterprise'. He was even more interested in the screenplays I was writing. Every night he'd come home eager to hear what my characters had been up to during the day.

What was deeply enriching about life with John was his love of poetry and music, which I shared, and which was part of the initial attraction between us. John was an occasional poet himself and a serious and brilliant translator of the poetry of Pablo Neruda, Federico Garcia Lorca, Rainer Maria Rilke, and Paul Celan, as well as spending a lot of time with Chinese poets and 20th-century Russian poets. One of the most delightful social aspects of our life together were the poetry parties that John loved to host, inviting mostly grad students and encouraging them to read either their own poems or their favorite poets. Given the multi-national origins of our students, we were treated to poetry in Mandarin, Vietnamese, Spanish, Portuguese, Arabic. We hosted at least one such party every year, everywhere we lived, the final one on the occasion of John's 91st birthday, two months before his death.

John was also a very good amateur cellist. No matter how busy he was with things academic, he always found an hour every day to play his cello. John had grown up in Vienna in a home filled with chamber music, and the single greatest 'learning' I acquired from him was the chamber music repertoire, from Bach through to the late 20th century. In turn, I was able to get him to pay a wee bit of attention to popular culture. We always read each other's writing in progress, and were each other's harshest critics as well as staunchest fans. But I can't say that John had much influence on my work in planning theory.[7]

★

However, my parallel film-school and subsequent screenwriting work definitely influenced my intellectual trajectory in this decade, in at least three ways. When I was writing *Towards Cosmopolis*, I was thinking about what images I might want to use. My film training made me think outside the box in this respect, and I sought to collaborate with a visual artist to create thought-provoking images rather than mere illustrations of some place or event in the text. The Melbourne artist/photographer Peter Lysiottis read my text and came up with brilliant, dissonant images that take the book to another level. A second way I was immediately influenced by my new-found love of storytelling was how I approached my teaching. I now took it as a challenge to teach through story rather than through abstractions, and this liberated me from my previously very uptight, note-beholden teaching style and turned me into a good teacher. In my lectures I incorporated music, video clips, and so on.[8] I also encouraged my students to look for more creative ways to approach their assignments. The resulting work from my UCLA students (and beyond) never ceased to amaze me. I came to realize there's so much creativity that goes untapped because of our academic straightjacket, with its rigid rules about what constitutes acceptable formats and languages of expression. I began to see that creativity could be an enormous asset to planning practitioners, especially in the realm of community engagement, activating other than scientific and technical ways of knowing. *Activating the heart*. I began to enjoy teaching immensely instead of dreading it, and I think the enjoyment was reciprocal.

The third and most important manifestation of the influence of film school is right there in that phrase, 'activating the heart'. *Towards Cosmopolis* began as both an intellectual project and a socio-political one. The intellectual project was threefold: the theoretical restructuring, the radical historiographical critique, and the epistemological exploration (Chapters 1–3 here). The socio-political project was the postmodern utopian dreaming that I call 'the cosmopolis project' (Part 2). I'm no longer comfortable with the descriptor labels 'post-modern utopian'.[9]

At the time, it seemed one had to identify with either modernism or post-modernism. I had rejected modernism in *Towards Cosmopolis*, although I tried to complicate that dismissal in the first chapter of its sequel, *Cosmopolis 2*. I was also uncomfortable with 'post-modern' in so far as it was typically associated with White male French theorists whose work I could not relate to (Lyotard, Baudrillard, etc.). It all seemed devoid of real connection and right relationship to people and place. But since folks whose work I had embraced, such as bell hooks, Cornell West, and others whom I cite extensively in Chapter 3, themselves embraced the postmodern label, I settled for it then. Now I'm uncomfortable with all such labels which divorce us from grounded relationships and practical wisdom. I prefer the terms 'dreaming' or 'envisioning', developed through reflection, connection, thinking, and action.

But *Towards Cosmopolis* was also something else, something even more meaningful to me. I was searching for a new language, or an expanded language for planning: the language of the heart (love, loss, betrayal, anger, fear, desire, connection …) and the language of the soul (spirit, memory, reverence, awe, the sacred …). In the final chapter of that book, I wrote about those things: about recognizing the importance of the City of Desire, the City of Spirit, the City of Memory. This was some of my best writing, and I know that for some readers this was the part of the book that spoke most deeply to them, suggesting *new possibilities for planning*.

Put another way though, my screenwriting life in this decade was a life in itself, an extraordinarily satisfying one from a creative perspective, and I had hopes that this could be my path. It involved a lot of learning: first about the craft of dramatic writing, then about how the business side of things worked, the crucial role of agents, and how to 'pitch' a story to producers. When the first script I wrote after film school was optioned and then produced as an ABC

TV Movie of the Week in 1991, I became a card-carrying member of the Writers Guild of America and was proud of that. The whole experience of having one's script turned into an actual movie, and being invited on location to watch them shoot it, was unforgettable. The story was set in the Pacific Northwest and the filming took place mainly on the west coast of Vancouver Island. I brought my mum over from Australia and she and I went on location and were treated very kindly by the producers and the cast. The four leads in the film were, at the time, quite celebrated TV actors: John Stamos, Chad Lowe, Joanna Kearns and Barry Bostwick. Driving up to the set and seeing the dozens of huge trucks that always accompany a film shoot, I had the delightful realization that my 100-page script was now a job generator for about 100 people. My title for the script had been 'Season of Fear', but the producers in their dubious wisdom changed it to 'Captive'.[10] There was a premiere, involving a red carpet, at some theatre in West Hollywood, before the Sunday night TV airing. It was all a bit unreal. And thrilling. The success of the film, from the producers' point of view, resulted in them giving me a three-year contract to write scripts for them. It also enabled me to get a better agent at a better agency (William Morris Agency). It felt at the time like I had 'made it', albeit in a minor way, and that I might well be able to support myself as a scriptwriter.

It took a few more years for me to realize that the Hollywood world was not my world. One anecdote here will suffice to explain this. On the afternoon of April 29, 1992, I accompanied my producers to a meeting at the Turner TV network, where I was going to pitch one or several story ideas and hope the Turner network would want to finance the development of at least one of them. The story I'd put most work into was based on the actual life of Biddy Mason, a woman born into slavery as part of a Mormon family living in Pennsylvania. That family was part of the great wagon train trek to Salt Lake City in the late 1840s. Biddy bore numerous children, fathered by her slave owner, and in the early 1850s he was directed to move to southern California to establish the first Mormon church in San Bernadino, outside LA. But California was not a slave-owning state, and when Biddy learned of that from a fellow church goer, she (although illiterate) decided to petition for her freedom. Many nefarious twists and turns later, she succeeded in obtaining freedom for herself and her children and set herself up as a midwife. Decades later, she was doing so well that she had bought several properties in what is now downtown LA, where she was known and revered as a philanthropist who was especially interested in helping immigrants.[11] I pitched this story in much more detail than I've provided here, and we waited for the executive's response. His exact words were: 'Hmmm. This is another one of those "triumph of the human spirit" stories. We've done one of those this year. What else have you got?'

The meeting lasted half an hour. As we shuffled out into the lobby, deflated, a bank of TV screens showed that LA was burning. Riots had just erupted in response to the White jury's acquittal of the LA police officers in the brutal beating of Rodney King. I couldn't help but see a connection between the cynicism of the Turner executive and what was now happening in LA. Fury over the acquittal, stoked by years of racial and economic inequality in the city, resulted in five days of rioting, looting, and burning, and ignited a national conversation about racial disparity and police use of force that continues to this day. And that was the day that my personal disillusion about working in the film industry in LA set in.

By 1994, as my scriptwriting contract with Ten Four Productions was coming to an end, I felt I was losing my moral compass, second guessing what producers might want to produce (and pay for) instead of writing what I wanted to write. A gentle observation from my husband in this moment was suggestive. John noticed that on the days I was teaching at UCLA I was animated and outgoing, whereas on the days I was working on my film scripts, or 'pitching' my ideas to some studio executive, I was withdrawn and mildly depressed. He was right. I was immensely enjoying teaching my graduate classes in planning theory, planning history, and race, ethnicity, and the city. John opportunely suggested that I join him at a conference in Norway that summer on Planning Theory, and perhaps give a paper there – testing the academic waters. I had recently co-written with Ann Forsyth, a graduate student at GSAUP, a paper on Gender and Planning Theory (Sandercock and Forsyth 1992). With some trepidation, I decided to present a version of that.

All of the nervousness that had underscored the performance-in-public aspects of my previous academic life returned as I faced a tiered lecture theater of mostly male faces. But there in the middle of the auditorium, when I dared to look up from my notes, was a woman listening so intently that I just pretended she was the only person in the room and that I was speaking only to her. When I was done, she jumped in with such supportive comments and engaging questions that I floated out of that session on a high. That woman turned out to be Patsy Healey, a brilliant planning scholar (still going strong in 2022 with a new book). She was to become a generous long-distance colleague, a friend, and founder in 2000 of the (now leading) journal *Planning Theory and Practice*. Patsy embodies the best of what academia can be. Thanks to her generous engagement that day in 1994, I left Norway resolved to return to academia, somehow, somewhere. That story continues in the Commentary to Part II.

*

Oddly, for quite a few years beyond LA and film school, I stubbornly resisted analyzing what I had come to think of as 'the power of story'.[12] I was afraid of destroying the magic by intellectualizing it. It wasn't until the early 2000s that I began to write about the importance of story and storytelling in planning (Chapter 7). And it was thanks to a serendipitous meeting with Giovanni Attili, an Italian PhD student, at a conference in Florence in 2001 that a path subsequently opened up for me to explore the emotional as well as critical languages of storytelling through the use of film in the planning field.

More about that in the Commentary to, and chapters comprising, Part III.

Notes

1 Pragmatically, the money was important for paying the rent, but I had also cashed in ten years' worth of pension contributions before leaving Australia to pay my way through film school.

2 UCLA is situated on the wealthy west side of LA, across the road from Bel Air. Close to half of our grad students came from the eastside (predominantly Latino/a) or from south central LA (once predominantly Black, but by the 1990s somewhat more mixed ethnically).

3 This experience also made me double my thoughtfulness about the diverse materials I was using in my teaching, such as video and music. I needed to ask how a different positionality might experience these differently.

4 And of course, the opposite is equally true: practice must inform theory, which is why so many of us are drawn to John Forester's approach to planning theory – listening to practitioners.

5 *The Grand Domestic Revolution* (1981) and *The Power of Place* (1995) were breakthrough works, foregrounding the ideas and activism of women in cities and neighborhoods, challenging traditional gender roles and divisions of labor.

6 Janet was not the only visitor to our home who was fascinated by the fact that I was writing film scripts. Pretty much everyone, but especially academics, were always more interested in hearing about my fiction than they were about John's ongoing theoretical work. This was a bit of an irritation for John, but he took it well. There's a deeper point here though, I think, which is the universal one about the appeal of stories (see Commentary, Part III).

7 If the index entry under my name in John's final book, *Insurgencies*, is any indication, I had a considerable influence on his thinking in the last couple of decades.
8 Memorably, in a lecture on the ambivalent legacy of the Enlightenment, I began with excerpts from Beethoven's *Eroica* symphony followed by Pink Floyd's *The Wall*. When I left UCLA, the students organized a farewell party at which they all sang the Pink Floyd 'anthem'. It was the best possible tribute.
9 I thank Elizabeth Walsh for questioning me about these terms, which has led to this clarification.
10 Subsequently the film was acquired for distribution by Disney, and I continued to get royalty cheques for several decades thanks to re-runs on TV.
11 I'd learned all this personally from Dolores Hayden when she was working on what became *The Power of Place* (1995). Dolores had been active in creating a memorial to Biddy on the site of her former downtown properties.
12 Stories also have the power of the dark side, and I address this in Sandercock 2003, which is Chapter 7 here.

References

Abu-Lughod, J. 1980. *Rabat: Urban Apartheid in Morocco* (Princeton: Princeton University Press).
Abu-Lughod, J. 1995. *From Urban Village to East Village: the Battle for New York's Lower East Side* (Oxford: Wiley-Blackwell).
Abu-Lughod, J. 1999. *New York Chicago Los Angeles: America's Global Cities* (Minneapolis: University of Minnesota Press).
Bluestone, B. and Harrison, B. 1982. *The Deindustrialization of America* (New York: Basic Books).
Flyvbjerg, B. 1998. *Rationality and Power: Democracy in Practice* (Chicago: Chicago University Press).
Flyvbjerg, B., Landman, T. and Schram, S. 2012 (Eds.). *Real Social Science* (Cambridge: Cambridge University Press).
Hayden, D. 1981. *The Grand Domestic Revolution* (Cambridge, MA: MIT Press).
Hayden, D. 1995. *The Power of Place* (Cambridge, MA: MIT Press).
Kelly, J. 1984. Women. *History and Theory: the Essays of Joan Kelly* (Chicago: University of Chicago Press).
King, M. 1981. *Chain of Change* (Boston: Southend Press).
Leavitt, J. and Saegert, S. 1990. *From Abandonment to Hope: Community Households in Harlem* (New York: Columbia University Press).
Marris, P. 1987. *Meaning and Action* (London: Routledge).
Marris, P. 1988. *Dreams of General Jerusalem* (London: Bloomsbury Publishing).
Reynolds, H. 1981. *The Other Side of the Frontier: Aboriginal Resistance to the European Invasion of Australia* (Townsville: James Cook) University Press).
Reynolds, H. 1987. *Frontier: Aborigines, Settlers, and Land* (Sydney: Allen and Unwin).
Reynolds, H. 1989. *Dispossession: Black Australia and White Invaders* (Sydney: Allen and Unwin).
Sandercock, L. 1990. *Property, Politics, and Urban Planning: A History of Australian City Planning 1890–1990* (New Brunswick: Transaction Publishers).
Sandercock, L. 1995a. *Introduction, Planning Theory, Special Issue 13(Summer)* (Milan: Franco Angeli).
Sandercock, L. 1995b. Voices from the borderlands: a meditation on a metaphor, *Journal of Planning Education and Research* 14(2): 77–88.
Sandercock, L. and Attili, G. 2012. Unsettling a Settler Society: film, phronesis, and collaborative planning in small town Canada. In: Flyvbjerg, B., Landman, T. and Schram, S. (Eds) *Real Social Science* (Cambridge: Cambridge University Press).
Sandercock, L. and Forsyth, A. 1992. A gender agenda: new directions for planning theory, *Journal of the American Planning Association* 58(1): 49–59.

1

REWRITING PLANNING HISTORY

Official and Insurgent Stories (1998)

> Subversive historiography connects oppositional practices from the past and forms of resistance in the present, thus creating spaces of possibility where the future can be imagined differently – imagined in such a way that we can witness ourselves dreaming, moving forward and beyond the limits of confines of fixed locations.
>
> *bell hooks (1994)*

The Power of History

I read these prophetic sentences of bell hooks as a provocation to those of us engaged in re-thinking and re-writing planning history, demanding a more critical stance towards past practices, and enabling more expansive, inclusive, and imaginative conceptions of the future of the profession. Professions, like nations, keep their shape by molding their members/citizens understanding of the past, causing them to forget those events which do not accord with a righteous image, while keeping alive those memories that do. Novelist Milan Kundera has said that the struggle of people against power is a struggle of memory against forgetting (Appleby, Hunt and Jacob 1994:270). For historians, the struggle of particular memories against particular exercises of omission or suppression also involves power. Stories about the past have power and bestow power. The impulse to tell new stories about the past points up the fact that time itself is a perspective in the construction of histories. Successive generations of scholars revise historical knowledge, investing it with contemporary meaning.

This chapter sets out to revise planning history, to offer not one 'official history' but many histories, to diversify both the themes and the subjects of these histories. If diversity, or difference, is a defining characteristic of multicultural cities, why is that not reflected in our accounts of city-building? For planners there is a further challenge. There are several senses in which history has power and bestows power. In constructing histories of itself, the planning profession is molding its members' understanding of past struggles and triumphs, and simultaneously creating a contemporary professional culture around those stories. And in choosing to tell some stories rather than others, a professional identity is shaped, invested with meaning, and then defended. What are the erasures and exclusions implicit in the process of forging a

DOI: 10.4324/9781003325451-3

professional identity? What are some of the hidden meanings and practices of planning, its *noir* face? In taking the planning dimensions of these historical debates further in this chapter, I concentrate primarily on the United States, for the sake of a single historical and geographical focus. But because my argument is that the line of questioning is internationally pertinent, I will offer two succinct examples of recent work, from Australia and South Africa, which tell radically different stories from the mainstream planning histories of those countries.

I begin with 'the official story' of planning history, the story that we desire to believe about ourselves. It is a heroic story. But is it a true story? Or is it a myth, a legend? At the very least, there is a *noir* side to this story, which also needs to be told. Further, the official story is the story of planning by and through the state, part of a particular tradition of city-building and nation-building. But there are, and always have been, alternative traditions of planning, existing outside the state and sometimes in opposition to it. These *insurgent planning histories* (Sandercock 1995, 1998) challenge our very definition of what constitutes planning. They also provide a foundation for an emerging alternative (to the modernist) paradigm for planning in multicultural cities. In that sense, they provide us with a 'future imaginary'.

The Official Story

In the United States, a course in planning history is a required part of any professionally accredited planning program. But what is it that students read in such courses? The sub-field of planning history has emerged as part of the discipline of planning only in the last 30 years. Since the first major works in the 1960s – J.W. Reps' *The Making of Urban America: A History of City Planning in the United States* (1965) and Mel Scott's *American City Planning since 1890* (1969) – interest in the field has grown and its scope has broadened. There are now many volumes of essays on the subject, the best known and most widely used of which are those edited by Donald Krueckeberg, *The American Planner: Biographies and Recollections* (1983) and *Introduction to Planning History in the United States* (1983); and Daniel Schaffer, *Two Centuries of American Planning* (1988). There is a more recent best-seller by Peter Hall, *Cities of Tomorrow. An Intellectual History of Urban Planning and Design in the Twentieth Century* (1988), the scope of which goes well beyond US planning history. And there are a host of historical case studies of particular pieces of planning history, covering an era, an agency, a city, or a theme. Almost without exception, these studies come from within planning,[1] and are unabashedly modernist in their orientation. What does this mean?

What is planning history? What constitutes its proper field of inquiry? The answer given by the historians identified above is a fairly simple one: to chronicle the rise of the planning profession, its institutionalization, and its achievements. There are various strands to these histories: from the emergence of the profession itself, to accounts of the key ideas and/or people (always great men) shaping the emergence of planning, to histories of specific policies within the field, housing, transportation, garden cities, and so on. All of these works employ a descriptive approach in which the rise of planning is presented as a heroic, progressive narrative, part of the rise of liberal democracy with its belief in progress through science and technology and faith that 'the rational planning of ideal social orders' can achieve equality, liberty, and justice (Harvey 1989:11–13). The choice of individual hero or heroes in these narratives is quite eclectic, with some championing Ebenezer Howard, others Patrick Geddes or Le Corbusier, as the founding fathers of the profession, and most giving prominence to such local heroes as Daniel Burnham, Frederick Olmsted, and Robert Moses. But beyond these individuals, planning itself is the real hero, battling foes from left and right, slaying the dragons of greed and irrationality and, if not always triumphing, at least always noble, on the side of the angels.

In these mainstream, modernist portraits, planning has no fatal flaws. If battles are sometimes, or even often, lost, it is not the fault of the heroic planner but of the evil world in which s/he must operate. Common to these mainstream histories are the following characteristics. The role of planning and of planners is unproblematic. It is assumed that planning is 'a good thing', a progressive practice, and that its opponents are reactionary, irrational, or just plain greedy. It is assumed that planners know what is good for people and possess an expertise that ought to prevail, in a 'rational' society, over politics. It is taken for granted that planners have agency – that what they do and think has autonomy and power. It is seen as natural and right that planning should be 'solution-driven', rather than attentive to the social construction of what are held to be 'urban problems'. There is no scrutiny of the ideology: that is, the class, gender, race, or ethnic origins or biases of planners, or of the class, gender, or ethnic effects of their work. The rise of the profession is a cause for celebration rather than critical scrutiny.

For example, Mel Scott's *American City Planning Since 1890* outlines what have become the traditional themes of US planning historiography: beginning with the attempts to grapple with issues of urban sanitation, slum housing, and population concentration on the part of late-19th-century reformers and settlement house workers; followed by transformations in the city's built environment according to the standards of the City Beautiful campaign in the early part of the 20th century; the development of a 'scientific' foundation for the profession under the crusade of the City Functional movement; the emergence of planning at regional and national scales by mid-century; and finally, a call for a renewed human-centered comprehensiveness. In this sweeping narrative, Scott offers the history of urban planning practice as an almost seamless evolutionary continuum in which ideas take root and mature into legislative proposals, which in turn give birth to planning agencies and institutions, which must then develop procedures of policy implementation. Along the way there are many obstacles, which the hero, with his 'will to plan', must overcome.[2]

Peter Hall's *Cities of Tomorrow* chooses a dozen major themes, rounding up all the usual suspects – slum and sanitation reform, the garden city, the City Beautiful, the birth of regional planning, the Corbusian city of towers, the automobile city, and more – and devotes a chapter to each. His method is to trace these themes to the ideas of a few 'visionaries', most of whom lived and wrote in the decades straddling the dawn of the 20th century, and then to follow the fate of these grand ideas and visions as others – implicitly lesser mortals – seek to implement them. Hall's main theme, which he describes as 'the real interest in history', is individual human agency. He wants to show, in the face of what he calls the economic reductionism of Marxist historians, that individuals can and do make a difference, 'especially the most intelligent and most original among them' (Hall 1988:4–5). Hall's heroes are Ebenezer Howard and Patrick Geddes, the 'fathers' of modern city planning – 'there were, alas, almost no founding mothers' (Hall 1988:7) – and their interpreters in the New World like Lewis Mumford, Clarence Stein, Stuart Chase, Benton MacKaye, Rexford Tugwell, and Frank Lloyd Wright. But there is an elegiac note to his narrative, in his lament over the gap between the visionary quality of the ideas and their diluted impacts 'on the ground', where sometimes these grand ideas are 'almost unrecognizably distorted'. And indeed, after 100 years of planning practice, 'after repeated attempts to put ideas into practice, we find we are almost back to where we started' (Hall 1988:11). What begins as an evolutionary tale, then, ends in a kind of circular finale and lament. But Hall seems unable to offer any satisfactory explanation for this gap between vision and reality, perhaps precisely because he dismisses theory and chooses to focus on individuals rather than on social forces. In his story, therefore, it is the *idea* of planning which is the true hero, rather than the practice.

What is missing from these mainstream/modernist histories? At the most fundamental level, there has been a failure to address two basic questions. What is the object of planning history? And who are its subjects? The boundaries of planning history are not a given. These boundaries shift in relation to the definition of planning (as both ideas and practices) and in relation to the historian's purpose. If we define planning as bounded by the profession, and its objective as city-building, then we generate one set of histories. If we define planning as community-building, we generate another. If we define planning as the regulation of the physicality, sociality, and spatiality of the city, then we produce planning histories that try to make sense of those regulatory practices over time and space. But in emphasizing planning as a regulatory or disciplinary practice, we may miss its *transformative possibilities*, which in turn may be connected to histories of resistance to specific planning practices and regulatory regimes. The writing of histories is not simply a matter of holding a mirror up to the past and reporting on what is reflected back. It is always a re-presentation, a textual reconstruction of the past rather than a direct reflection of it. Mainstream planning historians have typically seen their subject as the profession, and their object as describing and celebrating its emergence. There have been two significant consequences of this approach. One is the absence of diversity in these texts. The other is the absence of any critical/theoretical perspective. These sins of omission are the *noir* side of planning.

The Noir of Planning History

In his critical history of Los Angeles, *City of Quartz*, Mike Davis delineates an earlier tradition of boosterism in the writing about the city that parallels what I describe as the mythologizing of the planning profession in mainstream planning histories. In the absence of a critical tradition of historical writing about the city in Los Angeles from the 1940s through to the 1970s, Davis argues that this city came to understand its past, instead, through a robust fiction genre known as *noir*, in which the image of the city is repainted as a deracinated urban hell. The *noir* novelists (James Cain, Horace McCoy, Nathaniel West, and Raymond Chandler are the best known) created a regional fiction concerned with puncturing the image of Southern California as the golden land of opportunity and the fresh start (Davis 1990:38). Most interesting of all was the brief appearance of 'Black *noir*', embodied in the fiction of writers like Langston Hughes and Chester Himes, who portrayed Los Angeles as a racial hell in which Blacks are destroyed or driven to self-destruction by the capricious and psychotic dynamics of White racism (Davis 1990:43).

What follows is a puncturing or de-mythologizing of the heroic image of planning history by means of injecting a series of critical themes. Perhaps the most glaring omission from the saga of the rise of planning is the absence of all but White professional males as the actors on the historical stage. Were there no women? No African Americans, Mexican Americans, Japanese and Chinese Americans? Were there no gays and lesbians? Where are they, both as subjects – engaged in planning, contributing to city and community building, researching urban problems – and as objects, victims, of planners' neglect or desire to regulate these groups' particular existence, concerns, and needs in cities?

Women?

Peter Hall justifies their absence from his study by the bold assertion that there were no 'fore-mothers of city planning', only forefathers (Hall 1988). That is simply wrong, as the works of feminist scholars like Dolores Hayden (1981), Eugenie Birch (1983), Susan Wirka (1989, 1994), Barbara Hooper (1992), and Gail Dubrow (1998), among others, have clearly shown. Feminist approaches to city planning history range from the chronicling of the 'Great Women' (Jane Addams, Edith Elmer Wood, Mary Simkhovitch, Florence Kelly, Lillian Wald, Catherine Bauer), to the documenting of a whole tradition of feminist home design and community planning (Hayden 1981; Wirka 1989, 1994), to critiquing ways in which women's contributions have been memorialized (Dubrow 1992). More recently, these White feminist historians have acknowledged the absence of women of color from earlier accounts, and have begun to address that absence (Dubrow 1998, Hayden 1995). Some feminist historiographers are challenging the traditional periodizations of planning history (Sandercock 1990); others are doing new textual readings of existing accounts in order to explore new themes – like the social control or disciplinary elements of planning practice (Wilson 1992; Hooper 1992, 1998). Wirka has argued persuasively for a re-definition of planning to include the City Social as well as the City Practical, in order to appreciate the contributions of women social reformers and community builders (Wirka 1989, 1994; Marcuse 1980).

In the absence of such a (re)definition of planning, mainstream historians have failed to appreciate women's contributions. Mel Scott's work has two references to Addams and Simkhovitch, each no longer than half a sentence. Addams is noted as the founder of Chicago's Hull House, the first Settlement House in the US, and as someone who, along with Jacob Riis, had an advanced insight into the social needs of the community (Scott 1969:72). But it is Riis to whom Scott devotes a paragraph, attributing Riis' insight into New York's housing problems to his recent immigrant status and empathy for the plight of poor immigrants. Yet it was Addams who, through Hull House, had been working with poor immigrants on the south side of Chicago since 1889 – well before Riis' study – and had pioneered social survey research among them. And what of Mary Simkhovitch, whom Scott also mentions, but only in passing, as an outstanding Settlement House worker and as a member of the 1907 Committee on Congestion of Population, and who pops up again, in 1931, as president of the National Public Housing Conference? Despite her obvious longevity in the planning and housing movements, Simkhovitch never rates more than these fleeting mentions, while Benjamin Marsh is given all the credit for the work of the Committee on Congestion of Population. As Susan Wirka's research has shown, Mary Simkhovitch not only wrote extensively on housing and social planning issues, but also worked tirelessly as a public activist on these issues, was the first to outline a comprehensive vision of neighborhood planning, and was the first to locate such a planning in its broader metropolitan context (Wirka 1989, 1994). The work of recovering the contributions of individual women to mainstream planning continues, as does the task of reconceptualizing women's work in urban and social reform issues and in community development as another kind of planning, albeit at grass-roots level rather than through state agencies – planning as community-building.

Invisible Minorities?

What of the absence of African Americans and other ethnic minorities from mainstream accounts? There is the unspoken assumption here that there are no African/Mexican/Asian

American forefathers or mothers of city planning. There is another implicit assumption in mainstream narratives – that planning has been race-neutral in its practices, rather than supportive of a White power structure's policies of segregation and discrimination. Joan Fitzgerald and William Howard have addressed the first assumption, making the case that there is indeed a Black planning history, and that Blacks were involved in planning around urban problems long before the civil rights era. They focus on the activist research of W.E.B. DuBois, beginning with his monumental study, *The Philadelphia Negro*, in 1898, and continuing in his investigations reported in the Atlanta University Publications, which provided a comprehensive portrait of urban African Americans. Through these publications, DuBois 'made a great contribution to urban research and community development planning, especially as such planning related to the Black community' (Fitzgerald and Howard 1993:10–11).

Along with the work of the Urban League, Black churches, and Black women working in their communities (Gilkes 1988), there is a body of research, political action, and urban social services that collectively represents a distinctively African American urban planning and community development tradition. Cheryl Gilkes (1988) and Gail Dubrow (1992) are among a growing group of researchers who are documenting the role of Black women in community-building. If we redefine planning to include the community-building tradition – planning 'from below' – then we must include these narratives in our planning histories.

According to Fitzgerald and Howard, there are two reasons why this tradition has been ignored both by the emerging planning profession and by mainstream histories. First, the researches of DuBois and of the Urban League drew attention to histories of racial tension and strife in American cities. But in the emerging profession of city planning in the first half of the 20th century, the matter of White racism was a taboo subject, and this continued to be the case at least until the challenges of the civil rights era. Second, the planning tradition that came to dominate the profession was based on shaping the physical environment – the city-building tradition – while the focus of the African American tradition was on employment and economic concerns, social work and urban service delivery, and collective political action – a community-building tradition.

The silence of mainstream historians on the subject of the racism that has been so prominent in American cities has led to an avoidance of the ways in which planning practice has often worked to reinforce racial segregation and discrimination. One needs to go to the work of the legal scholar C.E. Vose for a systematic study of the ways in which Whites used the planning tool of restrictive covenants to exclude Blacks (and Jews, and Mexicans) from their neighborhoods for the first half of the 20th century, until the NAACP and the ACLU took the matter to the courts (Vose 1967). And we need to go to the 'new ethnic histories' (Hayden 1995), like Ricardo Romo's *East Los Angeles: History of a Barrio* (1983) to get an account of the multiple ways in which minorities have been spatially excluded from large parts of American cities.

This theme of the racist consequences of urban and regional planning schemes receives full engagement in recent works by June Manning Thomas (1994, 1998) and Clyde Woods (1998) and the edited collection by Thomas and Marsha Ritzdorf, *Urban Planning and the African American Community* (1997). Thomas (1994) argues for a more racially conscious perspective in planning history, one that is 'more sensitive to the history of African American urbanization'. She suggests a new, four-part periodization for city planning history to bring it into line with Black urban experience. This would begin with the era during and immediately after World War One, which saw the first substantial migration of southern Blacks to northern cities and, not coincidentally, the first major race riots of the 20th century. This migration created industrial, civic, housing, and religious issues for city officials. Their planning response was the creation of residential controls: zoning for social segregation by race, and restrictive covenants built into land titles (Thomas 1994). The second stage in this story is the era of public housing after the passage of the Wagner-Steagall Act in 1937, and including World War Two housing and post-war urban renewal. This was a period in which residential segregation was reinforced and ghetto boundaries consolidated as local politicians and planners, responding to overt displays of White racism, used the planning system to keep Black housing projects out of White neighborhoods. This was also the time of the second great migration of southern rural Blacks to the north and west, and of the clearances of 'Black slums' for freeways. The third stage is the era of civil rights and civil rebellions, an era in which the planning profession developed a consciousness of and conscience about race and racism and when social planning and advocacy planning were responses to this new awareness (Thomas 1994). In the era that Thomas describes as the 'racially separate metropolis', since the 1970s, the Black community has experienced both an increase in political power (at least as measured by the number of Black elected officials) and disastrous economic decline thanks to processes of de-industrialization and economic restructuring. Thomas describes how planning affects, and is affected by, race and racism, as well as showing the historical linkages between urban development and racial oppression of Blacks by Whites.

In a more recent essay, Thomas (1998) argues for making the issues of race and racism central to the history of urban planning in the United States. Criticizing existing histories for their dismissive treatment of African Americans as either invisible or as passive victims of urban policies, or worse, as the carriers of social pathologies, Thomas gives us a fine-grained account of the historical links between race, racism, and planning. While not avoiding the extent to which planners have been complicit with oppressive policies in the past, she also gives a subtle reading of the difficulties of urban improvement efforts in a social context of racial oppression. She describes the persistent struggle by African Americans to plan their own communities, emphasizing how these stories of struggle can empower present and future generations.

Similar studies are needed for other minorities, from the exclusionary 19th-century zoning actions against Chinese immigrants (Kayden and Haar 1989), to the restrictive covenants against Mexicans and Jews through the first half of the 20th century (Romo 1983), to the whole colonial system of 'planned reservations' for Native Americans, as part of a broader reinterpretation of the work of planning as social control of (certain) bodies in space – women, 'minorities', the poor, Indigenous Peoples. This is indeed a *noir* side of planning history, something which planners would rather not acknowledge in their collective professional past. And yet, unless we discuss these hidden intentions and consequences of planning practices, we will continue to perpetuate them, as both Thomas (1998), and Beauregard (1998) – talking about South African planning – have argued.

Diversity in the multicultural city has many faces. Alongside the struggles for recognition and place-claiming by racial and ethnic 'minorities' are those of gays and lesbians, a minority with a long history of oppression and of resistance. Like other groups, gays and lesbians have stories to tell about the ways in which their lives in cities and neighborhoods have been and are impinged on by planning's social and spatial policies, and how they, in turn, have contested certain policies. Making these stories an integral part of planning history requires us to address questions such as the following: have gays, as individuals or as couples, been excluded from particular housing developments and certain neighborhoods? How have planning and housing policies created or reinforced such barriers? How have policies affecting the design and use of public spaces impacted the ability of gays and lesbians to live openly and without fear of perpetual harassment? What assumptions about the 'normal' family/household are built into suburban planning codes and how does this discriminate against gay couples/households? How do public housing agencies and private landlords treat gays who are parents? How can planners help to create safer streets and neighborhoods for gays and lesbians? How have gays and lesbians acted to create and protect their urban places? How have they made certain parts of cities their own? How have they interacted with planners in this pursuit?

Gay urban politics and social movements inevitably spill over into planning issues and into questions of who controls city councils and planning agencies. Gays and lesbians have become involved in electoral struggles precisely to influence the kinds of neighborhoods in which they live and to provide services specific to their economic, recreational, and health needs. Just as planning policies are neither gender- nor race-neutral (Sandercock and Forsyth 1992; Grigsby 1994; Mier 1993; Thomas 1994), nor are they neutral with respect to gay and lesbian communities.[3] We need to understand how planning policies, historically, have affected the quality of urban life of gays and lesbians, and indeed how they may have reinforced their broader societal oppression. Scholars like Moira Kenney (1995, 2001) and Eric Reyes (1993) have done pioneering work in these areas, employing cognitive mapping and other techniques to make visible the histories of gays and lesbians in American cities, their place-claiming, and the ways in which they have transformed the contemporary city.

In recent years, a wealth of scholarship about the history of gays and lesbians in the city has revealed, on the one hand, extensive and systematic practices of discrimination and oppression targeting them as 'threats' to the urban social order; and on the other, a century of resistance to this oppression (Chauncey 1994; Faderman 1991; D'Emilio 1983). Kenney (1995, 2001) asks planning historians to acknowledge the importance and relevance of this research as part of a broader shift away from the focus on institutional responses to urban inequalities, and towards the collective, street-level responses of those who are targets of discrimination. She also reminds us of the fluidity of the boundaries we have drawn around other divisions in the city – class, race, gender, age – and of how the gay and lesbian struggle crosses these other boundaries and compounds many of the difficult social and cultural questions we have only recently begun to consider. Kenney's work is important in yet another way, in drawing attention to the complexity of roles that marginalized groups have played in the development of American landscapes, as both subjects and objects. Our planning histories have yet to grapple with these complexities.

History and Theory

There is a fundamental critique embedded in drawing attention to some of the glaring absences in mainstream accounts of planning history. These absences are not innocent. They are systematic exclusions which emerge from the authors' epistemological positions concerning the proper subject and object of planning, concerning the writing of history, and concerning the relationship of planning to power and the power of systems of thought. In order to understand these systematic exclusions, we need theory. As a discipline, historians have acknowledged the importance of theory at least since the inception of journals like *History and Theory* (1961), *Radical History Review* (1974), and *Marxist Perspectives* (1978). Over the past two decades, there has been a proliferation of 'new histories': feminist, post-colonial, ethnic, queer, cultural, and more. The very titles of some of these works – *Remaking History* (Kruger and Mariani 1989), *The New Cultural History* (Hunt 1989), *Selected Subaltern Studies* (Guha and Spivak 1988), and *Telling the Truth About History* (Appleby, Hunt, and Jacob 1994) – indicate the challenges across many fronts to 'traditional histories', not only of 'dates and greats' but of masculinist, White, and Eurocentric accounts. These new histories begin with the recovery of neglected, repressed, and forgotten cultures, the recuperation of names and faces erased from past accounts. This process of 'recovery' – what Joan Kelly (1984) has called 'compensatory histories' – is essential in disrupting mainstream accounts of planning history, and is also an important factor in the contemporary politics of identity. Recovering one's history is a first step in constructing or reconstructing both individual and group identity. But the process of recovery is not the end of the story. There are further levels of excavation and analysis. In the 1990s, new work influenced by a range of critical theories, from Marx to Foucault, from feminism to post-colonialism, post-structuralism, and queer theory, began to re-present planning's histories (see Sandercock 1995, 1998; Yiftachel 1992).

Insurgent Planning Histories

It is the argument of this book that planning in the multicultural cities of the 21st century requires a very different approach than that of the modernist paradigm. In order to imagine the future differently, we need to start with history, with a reconsideration of the stories we tell ourselves about planning's role in the modern and postmodern city. In telling new stories about our past, our intention is to re-shape our future. If we can uncouple planning history from its obsession with the celebratory story of the rise of the planning profession, and demonstrate its multiple and insurgent histories, we may be able to link it to a new set of public issues – those connected with the challenge of planning for a future of multicultural cities and regions. The future multicultural city – *cosmopolis* – cannot be imagined without an acknowledgement of the politics of difference (which insurgent planning histories embody), a belief in inclusive democracy, and the diversity of social justice claims of the disempowered communities in our existing cities. If we want to work towards a politics of inclusion, then we had better have a good understanding of the exclusionary effects of planning's past practices. And if we want to plan in the future for multiple publics, acknowledging and nurturing the full diversity of the many social groups in the multicultural city, then we need to develop a new kind of multicultural literacy. An essential part of that literacy is familiarity with the *multiple histories* of urban communities, especially as those histories intersect with struggles over space and place-claiming, with planning policies and resistances to them, with traditions of Indigenous planning, and with questions of belonging and identity and acceptance of difference. What follows are some examples of insurgent planning histories: insurgent in their theoretical approach to understanding what constitutes planning history as well as in the particular stories they have to tell about Native Americans, African Americans, women, and gays and lesbians.

There is a difference between rewriting history by adding the forgotten or repressed contributions of particular groups, and reconceptualizing planning history by using gender and race as categories of analysis. In a landmark essay, historian Joan Kelly argued that women's history has shaken the foundations of historical study by making problematic three basic concerns of historical thought – periodization, the categories of social analysis, and theories of social change (Kelly 1984). One of the themes of feminist scholarship has been the issue of women's status, that is, the roles and positions women hold in society by comparison with those of men. In historical research, this means we look at epochs or movements of great social change in terms of their liberation or repression of women's potential. Once we do this, the period, or set of events, may take on a wholly different meaning from the normally accepted one.

> If we apply Fourier's famous dictum – that the emancipation of women is an index of the general emancipation of an age – our notions of so-called progressive developments, such as classical Athenian civilization, the Renaissance, and the French Revolution, undergo a startling re-evaluation.
>
> *(Kelly 1984:5)*

Kelly's own work on the question 'Was there a Renaissance for women?' provides the substantive evidence for her theoretical argument. With her question in mind, as to whether significant turning points in history have the same impact for women as they do for men, we can turn to planning histories and examine their periodizations from very different points of view. Instead of celebrating every milestone in the evolution of planning as a profession, we need to ask whether women (and other minorities) were part of this emerging profession, and what effect each milestone had on the lives of women living in cities. My own periodization of Australian planning history fell into three eras: an ideas phase (late 19th century through to World War One); a legislative phase (between the two World Wars and immediately after World War Two); and an implementation phase, primarily post-World War Two (Sandercock 1990). Applying a feminist framework to this periodization, a number of questions arise. With respect to the early formative ideas, were women active in writing or propagating any of these ideas? Were women organizing around urban issues at this time? How did the ideas of the 'great men' or 'founding fathers' potentially affect the lives of women in cities? What male–female relationships are implied in ideas about 'the good city' and good planning? If women played only a marginal role in the emerging *profession*, why was that the case, and is that the end of the story, or should we broaden our horizon to include women's work in housing reform, social work, the playgrounds movement, and community organizing? In the United States, we might want to focus on the split between the housing reform movement and the planning profession, analyzing the role of gender in this split. Similar questions can be asked of the phases of legislation and implementation. Were women involved as actors in these processes? What were the consequences of particular pieces of legislation on women's lives in the city? Has any planning legislation ever tried to broaden the opportunities for women to participate in public life? What assumptions about relations (of equality or inequality) between the sexes are built into planning legislation?

Similarly to Kelly's second challenge – the categories of social analysis. If gender is a category as fundamental to our analysis of the social order as other classifications such as class, and if we regard the relations between the sexes as socially rather than naturally constituted, then this would lead a planning historian to ask, of any set of ideas or practices, 'What are the male–female relationships implied here?' For example, what roles are being assigned to women when we design houses and neighborhoods and transportation systems?

Kelly's third challenge is based on the second. If the relation of the sexes is as necessary to an understanding of human history as the social relationship of classes, then a theory of social change that incorporates the relation of the sexes has to consider how general changes in production affect and shape production in the household and, thereby, the respective roles of men and women. This in turn requires us to address the relations between the domestic and the public orders, and to look to the organization of the productive forces of society to understand the domestic order to which women, historically, have been primarily consigned (although not all women, and not for all of history). Conversely, women's history also views women as agents and the family as a social and productive force, and asks us to analyze the positive aspects of women's lives within household and community, producing use values and providing the caring and attachment without which both individual and social life is the poorer. Theories of social change that are derived from gender-neutral approaches to history assume that when 'things' change for the better, those things are better for women too. A feminist approach to planning history needs to demonstrate that this has not always been the case and is not always a reasonable assumption. Planners who want to create a better world for both men and women need a theory of social transformation which has at its heart a consideration of gender relations, how those relations are shaped by and in the built environment, and how that built environment is socially produced in accordance with pre-existing notions of what constitutes appropriate gender relations.

The feminist work discussed above has as a starting point an understanding of gender as a social construction. There is a group of French feminists whose work begins from a rather different starting point, that of the body as a social construction. When this work is linked with that of Foucault on disciplining the self and on the social creation of 'docile bodies', an interesting new line of inquiry emerges around regimes of regulation and discourses about regulatory regimes. This comes together in the planning history field around discourses of bodies, cities, and social order.

Bodies, Cities, and Social Order

Paul Rabinow's *French Modern* (1989) was the first work to emerge from that intellectual milieu with a direct relevance to our understanding and re-presentation of planning. *French Modern* is a study of the construction of norms and the search for forms adequate to understand and regulate modern society. While planning is not the focus of his study, Rabinow describes how, beginning in the 19th century, the 'corrective sociology' of positivist social and natural sciences began to define and valorize norms – along the binaries healthy/pathological, normal/abnormal, productive/non-productive, and so on – and to provide these norms with architectural and urban forms. He explores the ways in which the rising professions and professional experts created disciplinary practices which served to control and regulate people in modern societies – essentially to regulate bodies in space. Regrettably, Rabinow does not pick up on the gender and other implications of applying Foucault's work to planning practices. Such an inquiry would ask how planning, as a profession, has functioned as a regulatory regime, specifically in its discourse on and rules concerning the 'appropriate' (or 'normal') place of women, or gays, or people of color in the city.

We find a tantalizing beginning to such a project in Elizabeth Wilson's study of cities and planning, *The Sphinx in the City* (1991), and in an essay in *New Left Review*:

> With the intensification of the public-private divide in the industrial period, the presence of women on the streets and in public places of entertainment caused enormous anxiety, and was the occasion for any number of moral and regulatory discourses.
>
> *(Wilson 1992:90)*

Wilson argues that during the 19th century there was an increasing preoccupation among urban social reformers, politicians, and writers with the subject of urban disorder. One of the causes of this growing disorder was believed to be the presence of women in the new industrial towns. She further suggests that 'the androgenous woman, the lesbian, the prostitute, the childless woman ... all aroused fears and created anxieties concerning the eroticization of life in the metropolis' (Wilson 1992:106). She argues that the emerging planning profession was preoccupied with controlling menacing sexualities in the city, thereby posing planners as early enemies of gay lifestyles. These ideas demand further historical research and re-reading of classic planning texts and plans in order to reveal such agendas.

On this theme, George Chauncey's *Gay New York: Gender, Urban Culture, and the Making of the Modern Gay World, 1890–1940* is an important work for planning historians. Chauncey argues that the thinking about the city that ultimately coalesced into urban planning theory identified 'sexual deviants' as one of the causes of social disorganization:

> Some theorists in the first generation of American urban sociologists, who echoed many of the concerns of the reformers with whom they often worked, expressed similar anxieties about the enhanced possibilities for the development of a secret homosexual life that urban conditions created. Urbanization, they warned, resulted in the breakdown of family and social ties that kept an individual's behavior under control in smaller, more tightly organized and regulated towns. The resulting 'personal disorganization', the sociologist Walter Reckless wrote in 1926, led to the release of 'impulses and desires ... from the socially approved channels', and could result 'not merely in prostitution, but also in perversion'.
>
> *(Chauncey 1994:132)*

Chauncey's research places gays and lesbians at the center of this debate over the threats to urban moral order. He argues that the attraction of the city was not merely the anonymity, but 'an organized, multi-layered, and self-conscious gay subculture, with its own meeting places, language, folklore, and moral codes. What sociologists and reformers called the social disorganization of the city might more properly be regarded as a social reorganization' (Chauncey, quoted in Kenney 1995:79). Kenney's work provides a crucial link between the research of gay and lesbian historians and the issues that ought to be of interest to planning historians. She looks at three aspects of systematic discrimination against gays and lesbians: harassment in public places, housing regulations like exclusionary zoning laws, and public accommodation statutes, which govern access to the whole range of public spaces from streets to commercial establishments (Kenney 1995). Her study of the processes of place-making and place-claiming in the gay community in Los Angeles, *Mapping Gay L.A.* (2001), reinforces Chauncey's argument that what is happening here is a social reorganization, a new process of community-building.

The next two examples of insurgent planning histories are concerned with action, change, and empowerment. 'If modernist planning relies on and builds up the state, then its necessary counter-agent is a mode of planning which addresses the formations of insurgent citizenship', argues anthropologist James Holston (1998:47), who urges planning theorists and historians to study the grass-roots mobilizations and everyday practices which subvert state agendas. The research of Clyde Woods and of Theodore Jojola answers this call.

Recoveries (1): The Blues Epistemology

Clyde Woods' account of the failed effort at regional economic development planning embodied in the short life history of the Lower Mississippi Delta Development Commission (LMDDC) forges a new historiography of regional planning (Woods 1998). His project is the recovery of the memory, voices, and visions of the African American community of the Delta. Drawing on Foucault's archaeological/genealogical method, he 'excavates' a repressed African American tradition of resistance to the hegemonic definitions of planning and development in the Delta. Woods makes visible the existence of a regional epistemology grounded in African American experience – what he calls the 'Blues epistemology' – that, he argues, embodies an alternative theory of social, economic, and cultural development and change. Woods describes how the Blues operated to instill pride in a people facing daily denigration, as well as channeling folk wisdom, descriptions of life and labor, and critiques of bosses and sheriffs, planters, and the plantation regime. He argues that the Blues operated as a self-referential system of social explanation – as an epistemology. This new epistemology proceeded from the assumption that the indigenously developed folk culture, its orature, its ethics, its tradition of social explanation, and its prescriptions for social action, were the basic representational grid of working-class African American consciousness. Woods also explains why those voices, that tradition, were systematically marginalized in the interests of a hegemonic modernization of which planning was and is a complicit partner.

Within the oral and written record of these 'arrested agendas and movements' rests the regionally Indigenous knowledge upon which to construct new relationships and new regional structures of equality and democracy. Woods is concerned as much with an empowering vision of the future for the region as he is with the region's past. Re-presenting the region's past, however, is the first step in imagining a better future. Reacting to the appalling consequences of the modernization project for the African American population of the Mississippi Delta, Woods' insurgent historiography embodies an activist social imagination that draws on what already exists in the history of African American resistance to hegemonic definitions of planning and development in the Delta.

Recoveries (2): The All-Indian Pueblo Council

There could be no more glaring absence from the pages of the planning history of the United States than that of its original inhabitants, Native Americans, who were forcibly removed from their lands in order for most American cities to be built and farms to be established. But, once we have acknowledged that tribal communities have been victims of colonization and urbanization, have we put a swift end to their inclusion in the planning narrative – allowing them an appearance on stage only briefly, as victims?

The most important contribution to an understanding of this hitherto invisible part of planning history is the work of Native American/Pueblo scholar and activist Theodore Jojola (1998). In his essay 'Indigenous Planning: Clans, Intertribal Confederations, and the History of the All Indian Pueblo Council', Jojola acknowledges that tribal communities have been victims of colonization, but the story he tells about this is twofold. He argues that tribal communities themselves have a history of planning traditions that have been rendered invisible by virtue of not being seen as part of the dominant narrative of modernization and nation-building. And he outlines a story of hundreds of years of resistance and survival despite overwhelming odds, a story of selective interaction and adaptation. He argues that this survival was possible precisely because of Indigenous planning traditions, specifically the role of clans in community development and the tradition of consensus-seeking among inter-tribal confederations. Together, these constitute traditions of community and regional planning with a much longer history than their parallels in mainstream Euro-American planning. While Jojola's essay focuses on the history of these traditions among the Pueblo Indian nations of the Southwest, he notes that there are variations on these same models throughout what is known as Native America.

Jojola likens the clan to the neighborhood unit in planning theory. And when clans of various cultural communities came together to form regionally based groups, these groups – or tribal confederations – became the means by which Native Americans organized their own political community. Jojola equates this with a regional planning model, arguing that the confederations established an interdependent economic system that, in turn, was tied to managing and sustaining a shared ecosystem. One of the surviving tribal confederations is the All Indian Pueblo Council of the Southwest. Jojola describes the unified actions of the Pueblos in resisting first, Spanish intrusion and exploitation; continuing regional resistance after the 1821 transfer of Spanish authority to the Mexican government; and again, after the American territorial government was established in New Mexico. With the induction of New Mexico into statehood in 1912, the Pueblos continued to mobilize against the unjust practices of local authorities.

Jojola's long march through the centuries can in fact be described as a success story – albeit with tragic overtones. In the long view, in spite of decimation, assimilation, and alienation from their lands, tribal communities have survived and are being revitalized, and the Indigenous planning traditions that have made this survival possible are being reasserted. In keeping with the notion of the past as future – with history-writing directed toward present and future policy concerns –Jojola outlines the expanded policy role of the All Indian Pueblo Council today in matters of community and regional development and national policy, and advocates the formation and revitalization of other confederations which existed at the time of first European contact.

Contesting the Spaces of Historic Representation

Traditional planning histories have presented planning as the voice of reason in modern society, as the carrier of the Enlightenment mission of material progress through scientific rationality. The insurgent planning histories discussed above both implicitly and explicitly challenge the accuracy of this representation of planning. The 'rewritings' of Thomas, Woods, Jojola, Chauncey, and Kenney represent 'subaltern consciousness' (of Blacks, tribal communities, gays and lesbians) as a rationality of its own, which has been subjected to erasure by those in power. Each has captured/recovered subaltern consciousness through mixtures of oral history, political economy, cognitive mapping, critical theory, and textual analysis. And each underscores ways in which, through resistance and acquiescence, the subaltern confronts the world and may transform it. The variety of insurgent planning histories outlined here all challenge the mainstream approach of planning history as synonymous with the history of the profession. They also, implicitly, challenge the very idea of '*the*' history of planning, emphasizing instead plural histories of planning practices and plural readings of planning texts, according to the theoretical lens in use.

There is no single theoretical framework that informs this new work, nor is there a dominant theme. The goal is not to produce a new, unified, postmodern or feminist or post-colonial interpretation of planning history, the new 'official story'. I began by criticizing the inadequacies of mainstream or modernist planning histories, what they have rendered invisible, their *noir* side. Beyond this deconstruction, I have outlined new ways of thinking about and doing planning history, drawing on the works of scholars of the past decade. In part, this involves a reconceptualization of the subject – of mainstream definitions of planning and the role of planners as well as disputing dominant definitions of who is and is not a planner. We still need to see planning as, in part, a state-directed activity, a branch of city-building, but this need not be seen as an exclusively top-down, expert-driven process. There have always been oppositional movements within and outside of mainstream planning, from the City Social tradition dominated by women urban reformers to the community-building traditions documented by Woods, Jojola, Kenney, Fitzgerald and Howard, and Thomas for African American, Native American, and gay and lesbian communities, as well as other White working-class and ethnic communities. Stories of resistance to 'planning by the state' are as important a part of the historical narrative as the more familiar heroic stories of master plans and master planners, of planning legislation and state planning agencies. There is a tradition of community resistance, from people whose life space is at stake in the urban and regional development process, and of community planning, which needs to be incorporated as a counterpoint to the modernist narrative.

All of these forms of rewriting of American planning history could be equally well applied to other national planning cultures. Oren Yiftachel has for the past decade been critiquing the history of Israeli planning practices as discriminatory (Yiftachel 1992, 2000). In post-apartheid South Africa, scholars have been engaged in documenting the ways in which the planning system was central to the implementation of racial segregation (Mabin and Smit 1997). Australian scholar Sue Jackson has broken new ground with her study of planning's colonial history. Her case studies of the role of planning in the northern Australian towns of Darwin and Broome expose a story of domination over Indigenous socio-spatial relations by using the environmental and racial discourses of settler colonialism. In her narrative, planning is a spatial technology of cultural dominance, repressing and/or marginalizing Indigenous peoples' perspectives and concepts of place. She makes a persuasive case as to why contemporary Australian planning must begin to address the special nature of Aboriginal relationship to place and to develop a procedural equity rather than the current emphasis on management plans. She also shows how apparently well-meaning planning legislation such as the Heritage Act, in the context of Broome, is yet another means of containing and disempowering Aboriginal society, and yet also illustrates how Aboriginal groups have used such legislation in an insurgent way, as their own *de facto* planning tool (Jackson 1998).

What this chapter has demonstrated is that planning by the state is, and always has been, only one part of the story. Today, not all professionally trained planners work in state agencies. Increasingly, planners work for diverse local communities, in a new dialogic relationship that goes far beyond advocacy. Planning as a profession has spread its wings to include environmental policy, historic preservation, community development, anti-poverty planning, and so on. There are many strands of planning, loosely held together through a common education and a professional culture that includes the teaching of some version of planning history. But is the subject presented in its full richness and diversity, its *noir* and insurgent as well as its triumphal versions? The diversifying of the profession over the past three decades – the influx of women, people of color, activist gays, and lesbians – has brought to the forefront the need to re-present planning history as part of a contemporary project of rethinking planning's future. *We have moved from planning history – the official story – to planning's contested and multiple histories.*

Notes

1 Mel Scott's book was, literally, an official history, in that it was commissioned by the American Institute of Planners on the occasion of the 50th anniversary of the Institute.
2 My own history of Australian city planning from the 1890s to the 1970s, *Cities for Sale. Property, Politics and Urban Planning* (1975), followed this same pattern and adopted more or less the same sets of assumptions, which I subsequently critiqued in the Introduction to the second edition (Sandercock 1990).
3 The notion of planning 'as a heterosexist project' has been well argued by Frisch (2002).

References

Appleby, J., Hunt, L. and Jacob, M. 1994. *Telling the Truth about History* (New York: W.W. Norton).

Beauregard, R. 1998. Subversive histories: texts from South Africa. In: Sandercock, L. (Ed.) *Making the Invisible Visible* (Berkeley: University of California Press).

Birch, E. 1983. From Civic worker to city planner: women and planning, 1890–1980. In: Krueckeberg, D. (Ed.) *The American Planner* (New York: Methuen).

Chauncey, G. 1994. *Gay New York: Gender, Urban Culture, and the Making of the Gay Male World 1890–1940* (New York: Basic Books).

Davis, M. 1990. *City of Quartz* (New York: Verso).

D'Emilio, J. 1983. *Sexual Politics/Sexual Communities* (Chicago: University of Chicago Press).

Dubrow, G. 1992. Claiming public space for women's history in Boston, *Frontiers: a Journal of Women's Studies* 13(1): 111–148.

Dubrow, G. 1998. Feminist and multicultural perspectives on preservation planning. In: Sandercock, L. (Ed.) *Making the Invisible Visible* (Berkeley: University of California Press).

Faderman, L. 1991. *Odd Girls and Twilight Lovers: A History of Lesbian Life in Twentieth Century America* (New York: Penguin).

Fitzgerald, J. and Howard, W. 1993. Discovering an African American planning history, unpublished paper, presented to Association of Collegiate Schools of Planning Conference, Philadelphia.

Gilkes, C. 1988. Building in many places: multiple commitments and ideologies in black women's community work. In: Bookman, S. and Morgen, E. (Eds) *Women and the Politics of Empowerment* (Philadelphia: Temple University Press).

Grigsby, J. E. 1994. In planning there is no such thing as a "race neutral" policy, *Journal of the American Planning Association* 60(2): 240–241.

Guha, R. and Spivak, G. C. 1988. *Selected Subaltern Studies* (New York: Oxford University Press).

Hall, P. 1988. *Cities of Tomorrow. An Intellectual History of Urban Planning and Design in the Twentieth Century* (London: Blackwell Publishing).

Harvey, D. 1989. *The Condition of Postmodernity* (London: Blackwell Publishing).

Hayden, D. 1981. *The Grand Domestic Revolution* (Cambridge, MA: MIT Press).

Hayden, D. 1995. *The Power of Place* (Cambridge, MA: MIT Press).

Holston, J. 1998. Spaces of insurgent citizenship. In: Sandercock, L. (Ed.) *Making the Invisible Visible* (Berkeley: University of California Press).

hooks, b. 1994. House, *Assemblage* 24: 22–29.

Hooper, B. 1992. Split at the roots: a critique of the philosophical and political sources of modern planning doctrine, *Frontiers* 13(1): 45–80.

Hooper, B. 1998. The poem of male desires: female bodies, modernity, and "Paris: Capital of the nineteenth century". In: Sandercock, L. (Ed.) *Making the Invisible Visible* (Berkeley: University of California Press).

Hunt, L. ed. 1989. *The New Cultural History* (Berkeley: University of California Press).

Jackson, S. 1998. Geographies of Co-existence: native title, cultural difference and the decolonization of planning in North Australia (Unpublished PhD dissertation (Sydney: School of Earth Sciences, Macquarie University)).

Jojola, T. 1998. Indigenous planning: clans, intertribal confederations and the history of the all Indian pueblo council. In: Sandercock, L. (Ed.) *Making the Invisible Visible* (Berkeley: University of California Press).

Kayden, J. and Haar, C. 1989. (eds) *Zoning and the American Dream* (Chicago: Planners Press).

Kelly, J. 1984. *Women, History, and Theory* (Chicago: University of Chicago Press).

Kenney, M. 1995. Remember, stonewall was a riot: understanding gay and lesbian experience in the City, *Planning Theory* 13: 73–88.

Kenney, M. 2001. *Mapping Gay LA: the Intersection of Place and Politics* (Philadelphia: Temple University Press).

Kruger, B. and Mariani, P. eds. 1989. *Remaking History* (Seattle: Bay Press).

Mabin, A. and Smit, D. 1997. Reconstructing South Africa's cities? The making of urban planning 1900–2000, *Planning Perspectives* 12(2): 193–223.

Marcuse, P. 1980. Housing in early city planning, *Journal of Urban History* 6(2): 153–176.

Mier, R. 1993. *Social Justice and Local Development Policy* (Newbury Park: Sage).

Reyes, E. 1993. *Queer Spaces: the Spaces of Lesbians and Gay Men of Color in Los Angeles* (Unpublished Master's thesis, Graduate School of Architecture and Urban Planning, UCLA).

Romo, R. 1983. *East Los Angeles: History of a Barrio* (Austin: University of Texas Press).

Sandercock, L. 1990. *Property, Politics and Urban Planning: a History of Australian City Planning* (2nd edn.) (Brunswick: Transaction Publishers), 1890–1990.

Sandercock, L. 1995. Introduction: new historiographies for planning. *Planning Theory* 13(Summer): 9–34.

Sandercock, L. 1998. *Towards Cosmopolis: Planning for Multicultural Cities* (London: John Wiley and Sons).

Sandercock, L. and Forsyth, A. 1992. A gender agenda: new directions for planning theory, *Journal of the American Planning Association* 58(1): 49–59.

Scott, M. 1969. *American City Planning Since 1890* (Berkeley: University of California Press).

Thomas, J. M. 1994. Planning and the black urban experience: linkages and contemporary implications, *Journal of Planning Education and Research* 14(1): 1–11.

Thomas, J. M. 1998. Race and empowerment: necessary theoretical constructs for understanding US planning history. In: Sandercock, L. (Ed.) *Making the Invisible Visible* (Berkeley: University of California Press).

Vose, C. 1967. *Caucasians Only: the Supreme Court, the NAACP, and the Restrictive Covenant Cases* (Berkeley: University of California Press).

Wilson, E. 1992. The invisible flaneur, *New Left Review* 191: 90–110.

Wirka, S. 1989. A Foremother of Planning: Mary Kingsbury Simkhovitch and the First national Conference on City Planning, 1899–1909 (Unpublished Master's thesis, Graduate School of Architecture and Urban Planning, UCLA).

Wirka, S. 1994. Introduction to Mary Simkhovitch's housing chapter. In: Krueckeberg, D. (Ed.) *The American Planner* (2nd edn.) (New Brunswick: Center for Urban Policy Research).

Woods, C. 1998. *Development Arrested: the Blues and Plantation Power in the Mississippi Delta* (New York: Verso).

Yiftachel, O. 1992. *Planning a Mixed Region in Israel: the Political Geography of Arab-Jewish Relations in the Galilee* (Aldershot: Avesbury).

Yiftachel, O. 2000. Social control, urban planning and Ethno-class relations: Mizrahim in Israel's development towns, *International Journal of Urban and Regional Research* 24(2): 417–434.

2

WHO KNOWS?

Exploring Planning's Knowledges (2003)

> Country is underneath us all the time, but it's all covered up and we in our minds are all covered up. So when we walk in the land, we can't see anything for a while. We got all our possessions with us, and through these things we look at the land. Do you feel the sand you walk on? Are you aware of where your feet step? Are you aware of the trees you just passed, the birds that just landed? How much do you see? That has to shift and as soon as it does, we get a shift in mind which drops down to feeling. Then we wake up to feeling, what we call le-an here, and we become more alive, we start feeling, we become more sensitive. You start to read the country. Then all of a sudden there's an opening down there. Before there was only a wall, but now that tree has meaning, now that rock has meaning and all of a sudden that thing takes you. You just follow. … See, you are that land, and the land is you. … You wake up, and you see a lot of things and the country starts living for you. Everything is based on that feeling ***le-an***, seeing through that feeling.
>
> *(Sinatra and Murphy 1997, 15)*

'How much do you see?' asks Frans, a White man from Holland who has been living in the remote north-western region of Australia for the past two decades. Frans has been learning the importance and meaning of land from an Aboriginal perspective. His teacher is Paddy, an Aboriginal elder who is the custodian of the Lurujarri (coastal dunes) Dreaming Trail. The Lurujarri Dreaming Trail, an 80-mile coastal strip near the town of Broome, is part of the great song cycle of the continent according to Aboriginal cosmology. Thinking politically, Paddy has taught Frans tribal knowledge of the land (as reflected in his words in the opening quotation) to enable Frans to act as spokesperson and mediator in Paddy's efforts to protect the living landscape. Paddy teaching Frans was the first step in creating cross-cultural appreciation of the meaning of the land, as a way of protecting it. Paddy invites people to walk with his family members along the Lurujarri Trail so they can appreciate the importance of local beliefs by experiencing this cultural and spiritual landscape. Together, Paddy and Frans' teachings provide an opportunity for people to see the beauty of the landscape, not in the Western sense of a Kodacolor-registered beautiful landscape, but through understanding the

DOI: 10.4324/9781003325451-4

power that land has to sustain life. It is Paddy's hope that the people who walk the Trail will return home to become caretakers of their own 'country'.[1]

None of what Paddy teaches is written down. Aboriginal law/lore and cosmology are passed down through storytelling traditions, through song and dance and ceremony. Maintained within this oral tradition are laws about how to live in harmony with everything that makes up the living country. Frans talks about 'a shift in mind, which drops down to feeling', what the local people call 'le-an'. 'I mean, you're trained one way or other and you actually look through that upbringing at the land. You project through your training process the reading in the land. And all of a sudden it doesn't fit anything ...' Frans is talking about a shift in cognitive reality, discovering new ways of knowing. 'Everything is based on that feeling ***le-an***, seeing through that feeling' (Sinatra and Murphy 1997, 15)

It may appear odd to place a discussion of Aboriginal perceptions of land in the far north-west of Australia at the beginning of a chapter on knowledge in the planning profession. It is intended to appear so. The intent is to ask what is missing from accounts of knowledge in planning, ways of knowing cities and regions that ignore a rich variety of ways of seeing/knowing, by privileging only one way – the epistemology of the Enlightenment. In fact, the Lurujarri Dreaming Trail is not so remote from the concerns of planners as this opening seems to suggest. The Trail begins near the fishing port and town of Broome. Aboriginal people make up 40% of the population of the Shire of Broome, and yet in the 1993 Broome Planning Strategy, there is almost no demographic or social, let alone cultural, information about the Aboriginal population (Jackson 1998). In and around Broome there is land that is in the hands of Aboriginal custodians, and recent Native Title legislation makes it possible for Aboriginal tribes to make further land claims. How can a planning strategy claim to be speaking for 'the community' when it systematically ignores 40% of that community, and when it operates in a professional language that is the equivalent of a foreign language to the Aboriginal community?

Why Epistemology Matters

The story of Paddy and Frans, of a Black man and a White walking a Dreaming Trail together, is a graphic way into the subject of epistemology in planning. Why is or should epistemology be important to planners? Why should we care about the philosophy of knowledge? A short answer: legitimacy and authority. The questions at the heart of a planning epistemology are: What do I (think I) know? How do I know that I know? What are my sources of knowledge? How is knowledge produced in planning? What level of uncertainty or ambiguity can I tolerate? What forms of knowledge offer me most security? How adequate is my knowledge for the purpose at hand? How can I improve the knowledge base of my (and others') actions? What rights does my knowledge confer on me as a planner? What responsibilities do I assume for the application of what I claim to know? What is valid knowledge in planning? Who decides that? What is the relationship between knowledge and power? And who possesses knowledge that is relevant to planning? If these are the questions, the answers to them must surely constitute the very heart, or core, of planning.

How do we know what we know? How do we arrive at truth (a debated concept, as I shall discuss later) or certainty, or some adequate foundation for action? There have been at least half a dozen answers since the time of the ancient Greeks: Greek rationalism, mysticism, empiricism, dialectical materialism, pragmatism, and theoretical anarchism. Plato and Aristotle certainly didn't agree on an answer. Aristotle distinguished three kinds of knowledge: *episteme*, scientific or theoretical knowledge; *techne*, technical applications of theory; and *phronesis*, judgment or practical wisdom, which allowed a significant role for intuition, imagination, and emotion. For rational deliberation and choice in personal and public life, Aristotle advocated *phronesis*. For Plato, on the other hand, the idea that rational deliberation might draw on, even be guided by these elements (emotion, imagination, intuition) was a conceptual impossibility, since he defined the rational in opposition to these 'irrational' parts of the soul. Plato insisted that emotion and appetite (desire) were corrupting influences, and that the intellect needs 'to go off by itself' in order to make practical deliberations (Nussbaum 1990:76). Plato rejected sensuous cognition as part of his general rejection of the bodily. Imagination was rejected as being wedded to particularity and the recognition of incommensurables, and therefore a threat to the impartial assessment of facts and probabilities.[2]

Mysticism, through the ages, by contrast, has suggested that that which is most worth knowing is achievable through contemplation and meditation, and communicable only through poetic image and metaphor. Despite the attacks on mysticism since the Scientific Revolution, beginning with Isaac Newton in the late 17th century and continuing during the Enlightenment of the 18th century, mysticism appears in many arenas today, from New Age environmentalism and deep ecology to the descriptions by some Nobel Prize winning scientists of their actual methods of work. Albert Einstein, for example, wrote that 'only intuition, resting on sympathetic understanding, can lead to discovery; the daily effort comes from no deliberate intention or program, but straight from the heart'. And geneticist Barbara McClintock argued that 'reason is not by itself adequate to describe the vast complexity, indeed mystery, of living forms' (Keller 1993:199). Perhaps this is also the 'seeing through feeling', the *le-an* of Aboriginal culture.

The Scientific Revolution of the 17th century ushered in what has come to be called the Age of Reason, and the scientific method of empiricism became the dominant way of knowing. Observation, hypothesis, experiment; the search for mathematically based laws of nature; and a sharp distinction between reason and emotion: these became the defining characteristics of the empirical method which has dominated Western approaches to truth/knowledge since the Enlightenment, and out of which developed the social sciences in the 19th century, earnestly trying to replicate the methods of the physical sciences. Apart from the voices of the romantic poets, who embraced a pre-Enlightenment mysticism and reverence for nature that was in stark contrast to the Age of Reason's confidence in conquering and dominating nature, the most important source of opposition to Enlightenment empiricism in the 19th century were the critiques of Karl Marx and Friedrich Engels.

Marx made the radical argument that all knowledge is a reflection of specific material conditions, and that the so-called objective scientific model of the Age of Reason was in fact the knowledge that served the emerging bourgeoisie, whose industrial wealth was created by the technical applications of the Scientific Revolution. Marx dared to argue that knowledge is class-based and reflects class interests. Bourgeois social science was, in his words, mere ideology. He proposed to replace this ideology with a 'truly' scientific method, which he called dialectical materialism. Based on German philosopher Hegel's dialectics, Marx and Engels developed a method in which thought proceeds by contradiction and reconciliation of contradiction, to a new synthesis. Marxist social science was intended to expose the ideological foundations of 'bourgeois' social science, and to establish universal (dialectical) laws of society, nature, and thought. These dialectical laws, according to Engels, revealed that both nature and society are in a continuous process of evolutionary, though conflict-laden, development. Interestingly, Marxist doctrine is clearly based on the same Enlightenment belief in objectivity and reason, only claiming that its method actually possesses those characteristics, while 'bourgeois' social science does not.

The New World had been very receptive to Enlightenment thought in the 18th and 19th centuries. By the late 19th century, America was producing its own responses to empiricism (that is, to observation-based scientific method). The philosophy of pragmatism was very much an American invention of this era. Seeking action-relevant knowledge, philosophers like John Dewey criticized what they called the 'spectator theory of knowledge', arguing that the aim of knowledge was to make a difference in the world. What was important, therefore, was to assess the value of ideas by their outcomes, rather than by their (ideal) intentions. Dewey's writings gave an emphasis to social experimentation, to the reciprocity of theory and praxis, knowledge and action, facts and values. Out of this philosophical tradition came an epistemology of practice that we might call 'learning by doing', or social learning – a tradition that has developed a strong following in the planning literature (Friedmann 1987). But in the West, the most dominant epistemology since the 17th century has been based on empiricism, the verification of the physical world through scientific observation. It is customary to refer to this as the epistemology of the Enlightenment, and it is this epistemology that has underpinned the history of planning thought.

Planning's Enlightenment Epistemology

Modernist planning is a child of the Enlightenment, that period stretching from the late 17th to the late 18th century (roughly from the time of the English revolution to that of the French revolution) which began with scientific breakthroughs in our understanding of nature and of our place in the heliocentric universe and ended not only with the technical applications of those scientific breakthroughs transforming the entire system of production (the Industrial Revolution) but also with challenges to the entire social and political order represented by the absolute power of Church and monarchy, and a feudal class structure. In his 1784 essay 'What is Enlightenment?' the German philosopher Immanuel Kant described it as that point in history when humanity put its own reason to use, without subjecting itself to any authority, particularly ecclesiastical authority. 'It is humanity's passage to adult status'.

Historians today use the word 'Enlightenment' also as an adjective, as in 'the Enlightenment tradition', denoting a specific attitude of mind that gradually gained ascendancy among European intellectuals and bourgeoisie during this 100-year period. We might characterize this attitude of mind as dedication to human reason, science, and education as the best means of building a stable society of *free men* on earth. Enlightenment thinkers were hostile to tradition, opposed to any authority based on custom and faith alone, and some (particularly the French *philosophes*) were anticlerical as well. This was an era that was intoxicated with the idea of progress through reason, of perfecting the good life on earth guided solely by the light of reason, which would free *men's* minds from the bonds of superstition and ignorance. Liberty, equality, and fraternity were its slogans. In epistemology, the claim was that the world could be known objectively. The practice and logic of the physical sciences involved hypothesis-testing through experiment and careful measurement. Rigorously applying this 'scientific method' would result in the discovery of universal laws about the world of material facts.

While the history of planning thought contains many streams (Friedmann 1987), all of them come from the same headwaters of Enlightenment epistemology, flowing through the social sciences, to produce what I called in Chapter 1 the 'heroic model' of modernist planning. As planning emerged as a profession in the years after World War Two, it consolidated itself around a social science core that emphasized its claims to rationality and objectivity. In such classics as Harvey Perloff's *Education for Planning: City, State, and Regional* (1957) or Andreas Faludi's *Critical Rationalism and Planning Methodology* (1986), the ideal planner appears as rational, detached, a-political; a servant of a benign state; confident in *his* own expertise, and in *his* ability to develop 'general principles applicable not in one situation alone, but, with modifications, in any similar one' (Perloff 1957:140); confident, that is, in the universality of planning principles.

> Planning involves the careful elaboration and integration of a series of projected actions to attain the desired goals. Planning thus centers on the making of decisions and scheduled effectuation of policies. It takes form in a number of closely integrated steps, from the analysis of problems, the setting of broad objectives and the survey of available resources, to the establishment of specific operating targets; and through various succeeding stages until the results can be checked against the targets established and needed adjustments proposed.
>
> *(Perloff 1957: 142)*

Here speaks the voice of reason, in a detached, dispassionate tone. According to Perloff, planning is a kind of decision-technology, the science of society. In other words, planning was captured by a particular view of knowledge that privileges technical rationality and instrumental problem-solving ability over an array of what I will argue are equally important alternatives, especially value rationality. But before I set out these alternatives, I want to listen to some recent voices that have been raised in opposition to what Herbert Marcuse, in a book of the same title, called so succinctly '*one dimensional man*' – the man of technical reason, Rational Man.

Friedmann's Mutual Learning

Critiques of Enlightenment epistemology have been swelling over the past two decades. Within planning, there are at least three internal critiques that must be acknowledged as chipping away at the heroic model of Rational Man that the profession has aspired to. In *Retracking America* (1973), John Friedmann diagnosed a dual crisis in post-industrial society: a crisis of values and a crisis of knowing. Describing the latter, he talked of the growing polarity between experts and actors (or planners and people), with experts confident in their science-based, professional knowledge. Actors, on the other hand, possess a great deal of experiential knowledge that is not acknowledged as having any validity in the planning process. Further, experts tend to formulate problems in a language that actors often don't understand, thereby widening the gap. In Friedmann's view, neither camp has all the answers necessary for problem-solving, and the way forward is to bring the two together in a process of *mutual learning*. To bridge this gap between experts and client-actors, Friedmann urged the adoption of a *transactive style* of planning which would place more emphasis on interpersonal relationships and skills. He described the new world of transactive planning as revolving around the life of dialogue, with its emphasis on human worth and reciprocity. He wrote of accepting the authenticity of the other person; of a necessary fusion of thinking, moral judgment, feeling, and empathy; recognition of the importance of the non-verbal; and the acceptance of conflict as inevitable and something that must be dealt with rather than denied.

Schön's Reflective Practitioner

Ten years later, in *The Reflective Practitioner* (1983), Donald Schön also identified a 'crisis of confidence in professional knowledge', a crisis which he attributed in part to the way in which universities as institutions are committed to a particular epistemology, 'a view of knowledge that fosters selective inattention to practical competence and professional artistry' (Schön 1983: vii). He contrasted the 'hard knowledge of science and scholarship' with the 'soft knowledge of artistry' and set out to inquire into the epistemology of practice, based on a close examination of what some practitioners actually do. Practitioners, he argued, usually know more than they can say, possessing a complex 'knowledge-in-practice', most of which remains unarticulated. Schön's book takes apart the 'dominant epistemology of practice', the model of technical rationality, a model of instrumental problem-solving made rigorous by the application of scientific theory and technique. With positivism ascendant in the social sciences` since the mid-19th century, technical rationality became 'the positivist epistemology of practice' (Schön 1983:40).

For Schön, the main problem with this model is that it obscures or denies the difference between problem-solving and problem-setting. For problem-solving, technical rationality is often the most appropriate model, but before a problem can be solved, *it has to be defined*. Problem-setting is the process by which we define the decision that is to be made, the ends to be achieved, and the relevant or appropriate means. In order to convert a problematic situation to a problem, Schön explained, a practitioner must do a certain kind of work. She must make sense of an uncertain situation that initially makes no sense. For example, when a planner considers what road to build, s/he faces a complex and ill-defined situation in which geographic, social, political, topological, and financial issues are all mixed. Deciding what road to build, where, is so much harder than deciding how best to build it. It is the difference between *phronesis* and *techne*. And it is this sort of situation that is central to planning practice. Problem-setting, which is a necessary condition for technical problem-solving, is not itself a technical problem. When we set the problem, we impose upon it a coherence that allows us to say what is wrong and what needs changing. 'Problem-setting is a process in which, *interactively*, we name the things to which we will attend and frame the context in which we'll attend to them' (Schön 1983:40).

The persistence today of the idea of technical rationality is obvious in the hunger of students for technical skills. Perhaps one of the most difficult tasks of planning educators is to teach the limits of technical rationality, to demonstrate that the scope of technical expertise is limited by situations of great uncertainty, instability, uniqueness, and conflict. According to Schön, this means that the only real claim to expertise that a professional/planner can make is the claim to be well-prepared to *reflect in action*. 'Professionals are best seen as participants in a larger societal conversation' (Schön 1983:353). The reflective planner participates in these societal conversations, and in so doing, helps to construct the problem to be solved. Schön's work makes a very strong statement about the efficacy of technical and quantitative analyses. They are tools, no more and no less, frequently valuable, but incomplete without the richer study of human ends that they themselves cannot perform.

Forester's Talking and Listening

The notion of a societal conversation leads us to the work of John Forester and others (Healey 1992: Innes 1995) who have been attracted to the work of the contemporary German social philosopher and political theorist Jürgen Habermas. Forester has spent a lot of time observing planners *in situ* and talking with them about what it is that they are really doing. He concludes that planning is, more than anything, an interactive, communicative activity. As an emerging paradigm (Innes 1995), the idea of planning as communicative action turns its back on the model of technical rationality in favor of a more qualitative and interpretive mode of inquiry, seeking to understand the unique and the contextual rather than arriving at general rules for practice. These 'communicative' thinkers are also attracted to the storytelling mode, and to seeking insights from such stories about practice (Throgmorton 1996; Mandelbaum 1991). Forester's *Planning in the Face of Power* (1989), and *The Deliberative Practitioner* (1999) have become the 'classic' texts in this emerging field. Surely influenced by Schön's lead, Forester defines the planner's key activities as focusing and shaping attention, talking, and listening. Drawing on Habermas' work, particularly the *Theory of Communicative Action* (1984) and his concept of communicative rationality, Forester and his colleagues who are shaping this new paradigm are proposing a new method of knowing. One element is self-reflection, designed to identify one's own rationalizations and uncover what is hidden in the self. A second is that emancipatory knowledge comes from discourse and dialectic. A third way of knowing comes through praxis, through action in the world, through experience and practical know-how (Innes 1995:186). Finally, Forester emphasizes the political nature of all planning activity, in which relations of power are always involved and systemic inequalities influence outcomes. 'To be rational, be political', he advises (Forester 1989: 25). Be aware of systemic inequalities and work to redress them. Pay attention to imbalances of information, to lack of representation. Make sure all the major points of view are heard, and not only those of the most articulate or powerful.

Interestingly, the combined works of Friedmann, Schön, and Forester, which constitute a clear challenge to the hitherto hegemonic Enlightenment epistemology, can also be seen as drawing on a much older tradition, that of Aristotelian thinking about practical wisdom. This link has been made consciously in the recent work of Danish scholar Bent Flyvbjerg (2001), who proposes a synthesis of Aristotle, Nietzsche, and Foucault that he calls 'progressive *phronesis*' and 'the science of the concrete'. What is particularly important about Flyvbjerg's reworking of Aristotle's original conception of *phronesis* is that he foregrounds the issues of power, values, and conflict. Flyvbjerg has produced a deep epistemological critique of the social sciences, arguing that their strength is not in attempting to imitate the natural sciences in explanatory and predictive theory, but rather in recognizing that their *forte* is the 'reflexive analysis of values and interests, which is the prerequisite for an enlightened political, economic, and cultural development in any society' (Flyvbjerg 2001:3).

Since there is now an array of individual challenges from within planning to its foundational epistemology, it may be appropriate to summarize some of the recent systematic critiques of that Enlightenment epistemology – feminist, postmodern, and post-colonial – before drawing all these arguments together and outlining a more eclectic epistemology for planning.

Feminist Critiques

An epistemology is both a theory of knowledge and a justificatory strategy. It frames questions such as who can be a knower? Can women, can non-experts? What tests must beliefs pass in order to be validated as knowledge: tests from observation, experiment, calculation, only? Or tests based on experience, perception, imagination? What kinds of things can be known? Can subjective truths count as knowledge? Is there such a thing as objectivity? Does it require the absence of a point of view? What is the appropriate relationship between the researcher and her subject/s? Must, can, the researcher be detached, dispassionate? What should be the purpose of the pursuit of knowledge?

From the early 1970s, feminists' interest in epistemology was provoked by the apparent exclusion of women from the epistemology that has dominated the social sciences since their emergence in the 19th century – that of positivism. Feminists argue that positivist epistemology excludes women as 'knowers' and excludes women's life experiences as valid foci of study; that the voice of science is a masculine one; and that history has been written only from the point of view of men of the dominant class and race. Since the late 1970s, feminists have been proposing alternative theories of knowledge that legitimate women's claims as knowers (Harding and Hintikka 1983; Jaggar and Bordo 1989; Nielsen 1990) and raise the possibility of 'women's ways of knowing' (Belenky et al. 1986). Most of the earlier feminist work began by wanting to make a better science by 'adding women' to traditional analyses and by acknowledging that women, historically, had been excluded from the institutions where knowledge is produced and validated, particularly with respect to scientific knowledge (Keller 1993; Wertheim 1995).

Subsequently, feminists began to challenge the privileging of scientific and technical knowledge at the expense of other ways of knowing, such as knowledge based on experience, intuition, and imagination, and to outline a distinctively feminist epistemology, elements of which include the notions of 'connected knowing' (Belenky et al. 1986), passionate knowledge, 'maternal thinking' (Ruddick 1989), the importance of dialogue, and the validity of emotions in the pursuit of knowledge and understanding (Collins 1990). Although the idea of a 'connected knowing' (the head and the heart, reason and emotion) suggests an attempt to transcend the dualism at the heart of positivist epistemology, the very idea of a *feminist* epistemology reinforces all those Enlightenment dualisms, from male/female to rational/irrational, mind/body, culture/nature. Challenges to the very notion of a feminist epistemology have come from three directions: from women of color, from non-Western feminists, and from postmodern feminist philosophy. These will be further discussed in the next sections on post-colonial and postmodern critiques. Before I move on to these critiques, however, it is important to summarize broad areas of feminist agreement with respect to theories of knowledge.

Feminists insist on the importance of discussing the politics of theory and method and the origins and implications of our theoretical hierarchies. We insist on paying attention to the political content of knowledge creation, to the ways in which knowledges are institutionalized, and to who benefits from the production of which knowledges. We argue that knowledge is not, has never been, gender-neutral and that knowledge in planning, for example, is loaded with assumptions about appropriate gender relations (of subordination and domination) as well as with assumptions about who is a legitimate knower. We argue that knowledge is, above all, a social construction and as such will always be an unfinished business.

Postmodern Critiques

The most familiar names in the postmodern pantheon, Lyotard, Derrida, Foucault, have each mounted formidable attacks on Enlightenment epistemology, which have in turn come under scrutiny from feminists and people of color writing from a post-colonial perspective. Postmodern critiques begin with an attack on the very idea of a possible theory of knowledge (or justice or beauty), arguing that the pursuit of such theories rests upon the modernist conception of a transcendent reason, a reason able to separate itself not only from historical time and place but also from the body. 'There is no Reason; only reasons' (Lyotard 1984, in Yeatman 1994). The critique of the Enlightenment concept of rationality and its unitary definition of truth forms the basis of postmodern thought. Postmodern thinkers claim that the Enlightenment privileged rational discourse by identifying it as the sole avenue to truth, and by defining that discourse in terms of its abstraction from social context. Postmodernism rejects this privileging of rational discourse, arguing that there cannot be one such privileged discourse.

Foucault explored the ways in which particular discourses create their own definitions of truth. In drawing attention to the more sinister connections between knowledge and power, Foucault exposed a dark side of Enlightenment thought which has necessarily transformed our understanding of the role of institutionalized knowledges. Foucault provided a critique of the way modern societies control and discipline their populations by sanctioning the knowledge claims and practices of the human sciences – medicine, psychiatry, psychology, criminology, sociology. His analysis makes us critical of the presumed rationality of our discourses and practices, taking us behind the facade of universality and objectivity to reveal the operation of modern techniques of domination – of which the modern, self-examining, self-policing, self-disciplining, 'normal' individual is a product. He asks us to look carefully at how the familiar objects of our everyday experience, from the self and our bodies to our social institutions and scientific norms, are 'objects produced in historically variable relations of power' – that is, neither 'natural' nor 'scientific'. In Foucault's analysis, modern power is, above all, disciplinary. Its goal is normalization and the production of docile and useful bodies. The particular form modern power takes is without a center; it is capillary. Rather than located in the state or in capital, power is a moving substratum of force relations, local and unstable. And the agents of this distinctively modern form of normalizing or disciplinary power include social workers, teachers, doctors, social scientists, and ordinary citizens who internalize the categories and values of the dominant regime of power.

Foucault's analysis of the role of the professions in power regimes should surely be extended to the planning profession. If the rise of the professions created the possibility for new forms of domination, then each new profession carries within it the possibility and techniques for controlling others. I have already noted, in Chapter 1, how Foucault's work provides a new basis for exploration and interpretation of planning's history, an unveiling of its disciplinary and oppressive aspects. But there is also a liberatory side of Foucault's thinking which is potentially useful to planners. For if power is anchored in the micro-practices of everyday life, then that is also where oppositional politics needs to begin, with a deconstruction of the power relationships built into everyday practices, and a reconstructed, political planning.

The preceding discussion of postmodern thinking has deliberately focused on the names most commonly associated with the discourse. But, as noted at the outset, these writers have been challenged both by feminists and by those who do not accept the privileging of Western thought and thinkers. In particular, women of color, lesbians, and non-Western women have disputed the assumption that the category 'women' is or can be a unifying category capable of transcending differences in class, race, ethnicity, and sexual preference. They remind us that feminist theory has for the most part been written by White, western, middle-class, straight women, unconscious of their own position and special interests as members of a dominant culture (Collins 1990; Anzaldua 1990; hooks 1989; Narayan 1989). Chapter 3 explores in some depth the ways of knowing and acting, living and loving, embraced by these 'voices from the borderlands'.

Post-colonial Critiques

The post-colonial critique of Enlightenment epistemology begins with an epistemological politics of representation in which two strategic questions are posed: whose representations of reality prevail? And who has the authority to represent reality? Who must be silenced for these representations to prevail? Whose voice is deprived of authority so that they may prevail? 'This is a politics of representation which insists on the material effects of discursive power ... and which refuses legitimacy to the consensual community of rational scientists which both Karl Popper and Jürgen Habermas invoke in their respective conceptions of science and rational discourse' (Yeatman 1994:31).

Post-colonial discourse is clearly attempting to open up contested epistemological spaces, as bell hooks explains in a much-quoted passage: 'Spaces can be real and imagined. Spaces can tell stories and unfold histories ... The appropriation and use of space are political acts' (hooks 1990:152). And Barbara Christian identifies in the Black narrative tradition a unique form of theoretical resistance to power:

> our theorizing ... is often in narrative forms, in the stories we create, in riddles and proverbs, in play with language ... And women, at least the women I grew up around, continuously speculated about the nature of life through pithy language that unmasked the power relations of their world.
>
> *(Christian 1988:68)*

Clyde Woods has made a similar argument with respect to the importance of the Blues in African American culture and its history of resistance. He identifies a 'Blues epistemology' – a system of social explanation that has informed daily life, social institutions, and social movements, affirming the value of the culture and its ability to deal with chaos and repression. 'The essential motive behind the best blues songs is the acquisition of insight, wisdom' (Larry Neal, in Woods 1998:84).

The works of Edward Said, Homi Bhabha, Guyatri Spivak, Trinh Minh-ha, Chinua Achebe, and Ngugi wa Thiong'o, along with Christian, Woods, Anzaldúa, and hooks, are challenging not only Enlightenment epistemology but Eurocentrism itself, in so far as Western thought has worked as a colonizer of other cultures and sought to impose its own rationality and language. 'The bullet was the means of physical subjugation. Language was the means of the spiritual subjugation' (Thiong'o 1986). The voices of post-colonial critics offer and defend new/old ways of knowing, including the right to speak one's own language.[3]

*

If we accept what is common to all of the above critiques – that all knowledge is embodied, historically situated, shaped by language, and embedded in power relations, institutionalized or not – then the very idea of the expert planner able to arrive at an understanding of 'the public interest' through rational deliberation will have to be revised in favor of a notion of planning for multiple publics, based on an epistemology of multiplicity. It is toward that new epistemology for planning that we are moving in the final sections of this chapter.

Towards an Epistemology of Multiplicity for Planning Practice

The hunger of students for skills, about which Donald Schön has written so well, is reflected in the pride of place given to courses on quantitative methods in most planning programs. These courses reinforce the centrality of Enlightenment epistemology in planning education at the expense of equally important alternatives – experiential, intuitive, and local knowledges; knowledges based on practices of talking, listening, seeing, contemplating, sharing; and knowledges expressed in visual and other symbolic, ritual, and artistic ways. *Without discarding these scientific and technical ways of knowing*, we need to acknowledge, as well, the many other ways of knowing that exist; to understand their importance to culturally diverse populations; and to discern which ways of knowing are most useful in what circumstances. Such an *epistemology of multiplicity* for planning would consist of at least six different ways of knowing, in addition to what is usually taught in planning schools: knowing through dialogue; from experience; through seeking out local knowledge of the specific and concrete; through learning to read symbolic, non-verbal evidence; through contemplation; and through action-planning.

Knowing through Dialogue

Earlier in this chapter I touched on the emergence of a new paradigm of communicative action developed by John Forester, Judith Innes, Larry Susskind, and others in the US, and Patsy Healey in the UK. Best elaborated by Forester (1989, 1999) and Healey (1997), this paradigm defines planners' key activity as focusing and shaping attention, and their most important skills as talking and listening. In a wonderful chapter on 'Listening', which he describes as 'the social policy of everyday life', Forester begins with the acknowledgement that in planning practice, reason and emotion, fact and feeling, are usually tightly entwined, and anger and fear are always close at hand because people have large stakes, emotional and financial, in the built environment. In many situations, planners are dealing as much with people's passions as with their own earnest technical predictions. But how to deal with passion? 'Planners must not only be able to hear words; they also must be able to listen to others carefully and critically. Such careful listening requires sensitivity, self-possession, and judgment'. Forester talks about the 'work of listening', not only in planning but in everyday life as well, noting that how we respond, or fail to do so, makes up the politics of our everyday lives.

> We can hear words, but miss what is meant. We can hear what is intended, but miss what is important. We can hear what is important, but neglect the person speaking. As we listen, though, we can learn and nurture relationships as well. Listening is an act of being attentive, *a way of being* in a moral world. We can make a difference by listening or failing to do so, and we can be held responsible as a result.
>
> *(Forester 1989: 108, my italics)*

To listen well also means to be able to ask good questions about deeper interests, about what others care about, hope for, and fear. Listening necessarily addresses questions of possible actions, of 'what can be done?' If planners do not listen carefully to members of the public, they lose any reputation for responsiveness or fairness. They send signals to the community that what they (the community) have to say is not really important, or valid. 'Listening well, we can act to nurture dialogue and criticism, to make genuine presence possible, to question and explore all that we may yet do and yet become' (Forester 1989:118).

But to whom and to what should we listen? The advocates of the communicative action paradigm would answer the first part of the question – to whom should we listen – with: the relevant stakeholders. Identifying the relevant stakeholders is itself a deeply political question, which can only be answered in the particulars of a situation, but clearly implies something like 'all those affected', and not only those affected financially but also those affected emotionally, like about-to-be-displaced tenants. To be even more explicit regarding the ever-present power relations in planning, we should make an effort to listen to those who are least powerful. To what should we listen? Beyond the obvious structured situations of formal dialogue between planners and communities of interest, I suggest that we need to tune in to oral traditions, such as storytelling (Part 3).

The work of planners in multicultural societies is taking place in situations of increasing cultural diversity in which some groups, historically and/or culturally, have relied exclusively or semi-exclusively on oral traditions. The most important, sometimes the only way of acquiring knowledge, for some groups is through stories and the act of storytelling. African tribes, Indigenous Peoples, are obvious examples. As these groups become a more acknowledged part of planning processes and decision-making, it would seem that planners must be able to listen to their stories if they really want to understand people's cares, hopes, and fears, and to learn from them – to tap into local knowledges. To return to the story with which I opened this chapter, how can the White planners for the community of Broome in the remote Northwest of Australia imagine that they are planning for the whole community, 40% of whom are Aboriginal people, unless they listen to and understand the songlines of the local people? African American writer Zora Neale Hurston collected stories from poor Black folks for her novel *Their Eyes Were Watching God* (1937), a classic tale of empowerment. Might not planners' attention to peoples' stories be equally empowering, in terms of validating their knowledge and moving forward to a situation of mutual learning through respect for that knowledge?

Knowing from Experience

Tacit knowledge can be defined as that which people know but cannot say. As the philosopher Michael Polanyi (1958) observed, people usually know more than they can say. When that is so in a planning situation, it must be the responsibility of the planner to try to tap into people's tacit knowing. Particularly when they are working with disempowered communities who have always received messages about their ignorance and/or inferiority, planners need to begin the process of communication by helping people to articulate what they already know. This can be as simple as sitting at someone's kitchen table and saying 'Tell me about your street/village/neighborhood'. A wealth of information and understanding is likely to tumble out, though not in a form that planners are used to digesting. In more structured participatory situations, planners can adopt the same approach. A social planning consultant who has been hired by a local council or planning agency to bring local residents into the discussion about a controversial development proposal might call a public meeting, to which 500 people show up. How such a meeting is run determines whether it is an empty public relations exercise or a genuine attempt to work with the community to find a solution.

Intuitive knowledge is harder to describe. To intuit that a problem exists usually means that we are using our senses to interpret signs, in the environment, in people's behavior, in a situation. The intuition comes not out of the blue, but precisely because we are immersed in something and are paying close attention to detail (like Barbara McClintock 'listening' to the ears of corn that were the subject of her research, as described in Keller 1993). Good intuitive skills are also an important part of good inter-personal skills, which are now being recognized as a vital part of a planner's repertoire.

Tacit and intuitive knowledge are each part of something larger that we describe as experiential knowing, a knowing that comes from living, from simply being, as well as from doing. Planners who have been around for a while are constantly drawing on their fund of experiential knowledge, often without consciously acknowledging that they are doing so. But planners could draw more consciously on the experiences of others. How better to learn about the social and urban impacts of, say, immigration policy, than to ask those who are living it, both immigrants and host communities? How better to design a playground than by working with children, or a senior citizens' center than by working with seniors? The very idea of an apprenticeship is an acknowledgement of the value – to an aspiring doctor, sculptor, planner, chef – of learning at the side of someone who has more experience. But in urging planners to be more receptive to experiential knowledge, I have in mind something much broader than professional experience. And that is the individual and collective experiential knowledge of those who dwell in the place in question – whether street, village, city or region, ethnic enclave, or red-light district – those who possess local knowledge.

Learning from Local Knowledge

Wisdom, as an African proverb has it, comes out of an ant heap. Practical wisdom, in Aristotelian thought, is concerned with the concrete, with particulars. To know a city is to know its streets, we might say. And who knows those streets better than those who live in them and use them? Who knows the needs of a village better than the villagers? A planner from Washington DC flies into Quito, Ecuador, with a solution to 'underdevelopment' in her briefcase, a solution which is in 'the public interest' because it was generated by consultation with objective data files and through familiarity with economists' models of economic growth. Why be surprised if the solution is rejected by either the local politicians or the people of the villages, or both? Should planners be surprised if, despite substantial financial backing, the plan falters in implementation?

If it seems so obvious today that we need to draw on local knowledge in the planning process, whether the locale is south central Los Angeles or an Andean village, why then is it still the exception rather than the rule? In part, it is because planners believe that 'uncertified' people can't understand 'the complexities'. It is also because of the lingering belief – the all-too-pervasive influence of Enlightenment epistemology – that local knowledge is 'tainted' by self-interest: that is, by the passions, whether greed, love, attachment, anger, faith, power, prestige, beauty. I suggest that many planners, because of their positivist training, are afraid of the presence of these emotions in the community, do not know-how to deal with their eruption in the midst of what is meant to be rational deliberation, and therefore choose to hide behind the apparent safety and alleged objectivity of data. We are a profession in a state of arrested emotional development.

Learning to Read Symbolic and Non-verbal Evidence

There are many forms in which knowledge can be expressed. I have already mentioned stories and storytelling as ways of knowing that might be explored by planners in particular cultural and class settings. We might also pay attention to music, painting, poetry, and theatre.

We might think of music as including rap, folk songs, the blues, reggae, and more; of painting as including murals, graffiti, Aboriginal bark paintings, chalk drawings on pavements; of theater as street demonstrations, performance art, rituals of all sorts. Once we pay attention to these forms of expression, and learn how to 'read' them, we can learn from them about what's on people's minds, their hopes and anxieties. So, tuning in, getting a better understanding of particular groups (like youth) or marginalized peoples (Aboriginal, Native American), is one reason for paying attention to symbolic forms. Another is that we may be able to use such alternative ways of communicating in the planning process, as part of community consultation, or as a tool of community organizing. Two examples follow.

In the Watts neighborhood of Los Angeles, African American performance artist James Woods uses chalk drawings on a pavement as part of his community organizing and educating work. He gathers people together, young and old, supplies chalk, occupies public space, and asks people to draw what's on their mind concerning a particular topic. Woods' approach is part of a broader interest in the way in which the arts can be used as part of community organizing and place-making.

In the township of Hastings on Westernport Bay, 70 km south-east of Melbourne (Australia), industrial development, recreational boating, tourism, and real-estate investment were, by the 1980s, beginning to threaten an area of immense ecological value, including a landscape of mudflats, salt marsh, and white mangroves. Dredging, swamp draining, the human waste created by large settlements on the Bay, and the industrial waste from the steel mills and gas and oil refineries that were the area's large employers were devastating the delicate environment. Residents became concerned, and arguments raged with industry and a pro-development council about the future of the area. A group called the Westernport and Peninsula Protection Council (WPPC) was formed and decided to raise local awareness by entering a float in the annual Hastings parade. Using his theatre and puppetry skills, one of the local residents, Ian Cuming, constructed an elaborate display – The Seagrass Story – which pointed to the unexplained disappearance from the bay of large tracts of seagrass, the basis of the food chain. The sensational appearance of this float in the parade precipitated a three-year campaign that involved many parts of the community (Winikoff 1995:69).

This community participated in workshops on biological research, oral history, puppetry, costume-making, acting, and music. The result was the Seagrass Project, a three-year-long environmental theatre project which developed annual dramatizations of the story of the Bay in all its environmental and industrial complexity, and which brought together councilors, artists, scientists, business people, conservationists, and residents. The theatre performances, which were held at twilight on a park adjacent to the bay that had formerly been a garbage tip, attracted huge audiences. The Seagrass Project helped people to see the problem from a different perspective, helped them to connect emotionally to a particular environment, and to come together in a non-confrontational way to explore solutions to a complex set of problems.

Learning through Contemplative or Appreciative Knowledge

In the Enlightenment tradition, the truth of the world as it emerges from scientific inquiries is validated by becoming the basis for the mastery of the world. Learning the secrets of nature, we learn to fly, to harness the sun's energy, to transmit signals over vast distances. The social validation of knowledge through mastery of the world puts the emphasis on manipulative knowledge. But knowledge can also serve another purpose: the construction of satisfying and meaningful images of the world. Such knowledge, which we pursue primarily for the worldview it opens up, may be called appreciative knowledge. Contemplation, and the creation of symbolic forms expressing that contemplation, continue to be pursued as ways of knowing about the world, but because they are not immediately useful, they are not validated or respected socially.

This kind of appreciative, contemplative knowledge is very much at odds with the instrumental knowledge of the Enlightenment tradition. That instrumental knowledge has contributed to what Max Weber called the 'disenchantment of the world', a feeling of profound alienation. Perhaps, if we are interested in re-enchanting the world, in re-enchanting our life spaces in cities and regions rather than having to escape from them in order to 'experience nature' or spirituality, then the way forward is to re-valorize contemplative knowledge and to follow where it takes us.

Learning by Doing, or, Action-Planning

Acting in the world, and continuously and critically reflecting on that action, is the only workable way of knowing, according to John Dewey and the philosophy of pragmatism, and a number of other sources that constitute this 'social learning tradition' (Friedmann 1987). From Jane Addams' founding of the Settlement House movement to Miles Horton and the Highlander School's approach to popular education, the argument is that our knowledge comes from our practice, and returns to it. The social learning approach has, historically, been strongest in the world of community organizing, particularly among practitioners whose goal is community empowerment. Those ideas have in turn infiltrated into planning through the community-based action research approaches practiced by King (1981), Leavitt and Saegert (1990), and Heskin (1991), in which the planner works with, and from the perspective of, the poor and the disempowered rather than from the perspective of state-directed or expert-centered planning practices. As opposed to the advocacy model, in which professionals work *on behalf of* poor communities, in the empowerment model, the role of the professional planner is to enable communities to do things for themselves.

★

In this chapter I have asked what constitutes valid knowledge in planning, and who is recognized as a legitimate knower. I have examined mainstream planning's answer to these questions by exploring the Enlightenment epistemology on which planning has been founded. That epistemology has been shown to have serious weaknesses, as reflected in the range of critiques discussed. I have argued that we need to revise our dominant epistemology to better reflect these new understandings of knowledge. And, finally, I have proposed an *epistemology of multiplicity*, which recognizes at least six different ways of knowing. In principle, there should be no hierarchy between the ways of knowing that I have outlined. They are all potentially valuable. A good planner will be sensitive to them all, without *a priori* privileging any particular one. And part of the skill of the good planner will be her perception of when to use which way, or ways, of knowing, and to see them all as context-dependent. This is the artistry of planning.

Notes

1 'Country' is an Aboriginal English term that refers to the collective identity shared by a group of people, their land (and sea) estate, and all the natural and supranatural phenomena contained within that estate (Porter 2002:16).
2 For a brilliant refutation of this position, and an argument for the 'intelligence of the emotions', see Nussbaum (2001).
3 I began this chapter with a quotation about an Indigenous way of 'seeing', which I came across thanks to a colleague at the time, Jim Sinatra, but when I wrote this in 1997, the now rich literature on Indigenous ways of knowing and traditional ecological knowledge had yet to emerge (see Archibald 2008, Kovach 2010, Wilson 2008). Part IV of this book hopefully redresses this absence.

References

Anzaldua, G. (Eds.) 1990. *Making Face, Making Soul* (San Francisco: Aunt Lute Books).

Archibald, J. 2008. *Indigenous Storywork* (Vancouver: UBC Press).

Belenky, M., Clinchy, B., Goldberger, N. and Tarule, J. 1986. *Women's Ways of Knowing* (New York: Basic Books).

Christian, B. 1988. The race for theory, *Feminist Studies* 14(1), 67–9.

Collins, P. H. 1990. *Black Feminist Thought: Knowledge, Consciousness and the Politics of Empowerment* (Boston: Unwin Hyman).

Flyvbjerg, B. 2001. *Making Social Science Matter* (Cambridge: Cambridge University Press).

Forester, J. 1989. *Planning in the Face of Power* (Berkeley: University of California Press).

Forester, J. 1999. *The Deliberative Practitioner* (Cambridge, MA: MIT Press).

Friedmann, J. 1973. *Retracking America* (New York: Doubleday Anchor).

Friedmann, J. 1987. *Planning in the Public Domain: from Knowledge to Action* (Princeton: Princeton University Press).

Harding, S. and Hintikka, M. (Eds.) 1983. *Discovering Reality: Feminist Perspectives on Epistemology, Metaphysics, Methodology, and Philosophy of Science* (Dortrecht: Reidel).

Healey, P. 1992. A planner's day: knowledge and action in communicative practice, *Journal of the American Planning Association* 58(1): 9–20.

Healey, P. 1997. *Collaborative Planning* (London: Macmillan).

Heskin, A. 1991. *The Struggle for Community* (Boulder: Westview Press).

hooks, b. 1989. *Talking Back* (Boston: South End Press).

hooks, b. 1990. *Yearning: Race, Gender, and Cultural Politics* (Boston: South End Press).

Innes, J. E. 1995. Planning theory's emerging paradigm: communicative action and interactive practice, *Journal of Planning Education and Research* 14(3): 183–9.

Jackson, S. 1998. Geographies of Co-existence: native title, cultural difference, and the decolonization of planning in North Australia (Unpublished PhD dissertation (School of Earth Sciences, Macquarie University)).

Jaggar, A. and Bordo, S. (Eds.) 1989. *Gender/Body/Knowledge* (New Brunswick, NJ: Rutgers University Press).

Keller, E.F. 1993. *A Feeling for the Organism* (New York: Freeman).

King, M. 1981. *Chain of Change* (Boston: South End Press).

Kovach, M. 2010. *Indigenous Methodologies* (Toronto: University of Toronto Press).

Leavitt, J. and Saegert, S. 1990. *From Abandonment to Hope: Community Households in Harlem* (New York: Columbia University Press).

Lyotard, J. F. 1984. *The Postmodern Condition* (Manchester: Manchester University Press).

Mandelbaum, S. J. 1991. Telling stories, *Journal of Planning Education and Research* 10(3): 209–14.

Narayan, U. 1989. The project of feminist epistemology: perspectives from a non-western feminist. In: Jaggar, A. and Bordo, S. (Eds) *Gender/Body/Knowledge* (New Brunswick, NJ: Rutgers University Press).

Nielsen, J. (Eds.) 1990. *Feminist Research Methods* (Boulder, CO: Westview Press).
Nussbaum, M. 1990. *Loves Knowledge: Essays on Philosophy and Literature* (New York: Oxford University Press).
Nussbaum, M. 2001. *Upheavals of Thought: the Intelligence of Emotions* (Cambridge: Cambridge University Press).
Perloff, H. 1957. *Education for Planning: City, State and Regional* (Baltimore: The Johns Hopkins Press).
Polanyi, M. 1958. *Personal Knowledge: Towards a Post-critical Philosophy* (New York: Harper and Rowe).
Porter, E. 2002. *Indigenous Planning Knowledges*, paper presented to ACSP Conference, Baltimore.
Ruddick, S. 1989. *Maternal Thinking* (New York: Ballantine Book Company).
Schon, D. 1983. *The Reflective Practitioner* (New York: Basic Books).
Sinatra, J. and Murphy, P. (Eds.) 1997. *Listen to the People, Listen to the Land* (Melbourne: Melbourne University Press).
Thiong'o, N. 1986. *Decolonising the Mind: the Politics of Language in African Literature* (London: James Currey Publishers).
Throgmorton, J. 1996. *Planning as Persuasive Storytelling* (Chicago: University of Chicago Press).
Wertheim, M. 1995. *Pythagoras' Trousers. God, the New Physics, and the Gender Wars* (New York: Times Books).
Wilson, S. 2008. *Research Is Ceremony: Indigenous Research Methods* (Halifax: Fernwood Publishing).
Winikoff, T. (Eds.) 1995. *Places Not Spaces. Placemaking in Australia* (Sydney: Envirobooks).
Woods, C. 1998. *Development Arrested. The Blues and Plantation Power in the Mississippi Delta* (New York: Verso).
Yeatman, A. 1994. *Postmodern Revisionings of the Political* (London: Routledge).

3

VOICES FROM THE BORDERLANDS

A Meditation on a Metaphor (1995)

> Spaces can be real and imagined. Spaces can tell stories and unfold histories ... The appropriation and use of space are political acts
>
> *(hooks 1990)*

> Because we are not allowed to enter discourse, because we are often disqualified or excluded from it, because what passes for theory these days is forbidden territory for us, it is VITAL that we occupy theorizing space, that we not allow White men and women solely to occupy it. By bringing in our own approaches and methodologies, we transform that theorizing space.
>
> *(Anzaldua 1990)*

'Borderlands' are both real and imagined spaces. A common understanding of the space of the borderlands is of somewhere or someone that is marginalized, on the outside, somehow 'less than', not belonging. But both bell hooks and Gloria Anzaldua, whose ancestries are African American and Mexican respectively, are seeking to disrupt this understanding in their theory and their practice as scholars and activists. In this essay, I take inspiration from their writings in seeking to disrupt the taken-for-granted gender- and race-neutral categories of planning theory. My project here is to diversify existing approaches to planning theory by arguing for the relevance of including these and other (self-identified) 'voices from the borderlands'.

The small community of scholars actively working at planning theory is open to a remarkable diversity of concerns – so diverse, in fact, that there is considerable disagreement as to what constitutes the subject of planning theory. My project in this essay is to add fuel to that disagreement by advocating the expansion of an already quite catholic discourse. I am interested in connecting planning theory with other theoretical discourses – specifically debates around marginality, identity and difference, and justice, debates which empower groups whose voices are not often heard by planners. My purpose is to introduce a group of authors – the 'voices from the borderlands' of my title – whom, I suspect, many planning theorists have never read, or, if they have read them, don't think of them as colleagues in the theorizing enterprise.

DOI: 10.4324/9781003325451-5

This is a very personal take on planning theory, based on my experiences teaching the subject over the last five years to a very diverse student body in the planning program at UCLA. It is personal in another way also. I entered the planning world 20 years ago from a background in the humanities (literature and history), and was made to feel like an outsider because of this background. My colleagues were plainly puzzled at what I could possibly contribute to the field, given this 'soft' intellectual background. Twenty years on, although we have not exactly come full circle, the wheel has certainly begun to turn. There is renewed interest in the humanities among planners, evident in such themes as the importance of storytelling and the growing popularity of ethnographic approaches in planning research.

There is now a perceptible shifting of focus in planning education. Although nobody has yet advocated outright abandonment of the dominant social science paradigm, there is a refreshing groundswell of openness to the possible contributions of cultural studies, ethnography, cultural anthropology, literature, and history to the planning field. This essay adds its grain of sand to this shifting dune by arguing for the inclusion of new voices into the discourse of planning theory. These voices belong to people of color, and they could be called postmodern, although this label carries such confusing and off-putting baggage that I am almost reluctant to use it.

In his landmark essay 'Between Modernity and Postmodernity: The Ambiguous Position of US Planning', Robert Beauregard (1989) outlines how the modernist planning project is being challenged by the political and economic manifestations of postmodernity and its corresponding cultural practices. Suspended between modernity and postmodernity, modernist planning is, he argues, 'in the grip of a postmodern helplessness' (Beauregard 1989, 389). Beauregard's essay is the most thoughtful and thorough exploration of the significance of the postmodern critique for the field of planning to appear in the past decade. He describes, in some detail, the centrifugal disintegration of planning practice and shows us how the deconstruction of modernist planning represents a clash not only between new and old forms of urban political economies, but also between new and old forms of social thought. And he notes the paradoxical failure of the planning theory literature to address the challenges of the postmodern critique, scolding planning theorists for their avoidance of current intellectual debates and their ignorance of the foundations of modern social thought.

The first response in the planning literature to this challenge appeared in the symposium 'Into Postmodern Weightlessness' in the *Journal of Planning Education and Research* in the summer of 1991 (Milroy 1991; Beauregard 1991; Liggett 1991; and Mandelbaum 1991). In this essay I want to take Beauregard's challenge in a rather different direction than any of the contributors to that symposium, and in a different direction from Beauregard (1989, 393) himself, who concludes by reasserting the modernist core of the planning profession, urging planners to re-dedicate themselves to city-building processes and analyses.

For the past five years in my planning theory class I have been discussing the various streams of postmodernism: the critiques of positivism, the political deficiencies of technical expertise, the demise of metatheory (or the master narrative), debates around identity and difference, the belief in the possibility of critical distance and the authority of objective knowledge, deconstruction, anti-essentialism, the problems with dualistic thinking, and, more broadly, with the epistemology of the Enlightenment.[1] While much of this critique can be exhilarating and potentially liberating, it can also be experienced by students as paralyzing, producing severe cases of Cartesian Anxiety (Bernstein 1985) and leading to political acquiescence and hopelessness. As educators of young people who come to us wanting to make a difference in the world (by subscribing to the modernist faith in planning as social reform), I believe we have a responsibility to bridge this postmodern abyss. We have to find a meaningful way forward, without denying the validity of the critiques noted above.

Perhaps one way to do this is to acknowledge that postmodernism is not a metanarrative awaiting its historical moment to take over from the modernist metanarrative. Rather, it is a *multiplicity* of critical, deconstructive, and oppositional voices hovering over the corpse of modernism and primarily preoccupied with postmortems. It has largely been a dismantling exercise directed against the failures and disastrous consequences of the project of modernity, rather than a search for new ways to perceive, think about, and act (Garcia 1993). But within this multiplicity of voices there are radical as well as conservative tendencies. We need to build on the former if we are to advance a progressive planning practice into the 21st century – a practice built on a politics of hope, a concern for economic and social justice and equality, a new moral vision or consciousness, and an 'epistemology of multiplicity' (Gomez-Pena 1993; Sandercock 1998). But where do we find such voices?

Interestingly, for the argument I want to develop here, all of the postmodern authors whom Beauregard (1989) cites are homogeneous with respect to race and gender. They are all White and male and European. And for the most part, their political stance is cynical and nihilistic. But if we broaden the planning theory literature to include the works of feminists and people of color, we find writers who are addressing the condition of postmodernity in constructive and progressive ways, within a revised radical democratic tradition. In this revised democratic tradition, the analysis of injustice and inequality does not begin and end with class, but is broadened to an understanding of the systematic and intersectional inequalities which are based on race, gender, sexual preference, and ethnicity (Lorde 1984; West 1993; Jordan 1992; Anzaldua.1990; hooks and West 1991). And grand revolutionary metanarratives (such as Marx's dynamics of class struggle) have been discarded.

•

Twenty years ago, a political economy perspective made its presence felt in the literature of urban and regional planning as an oppositional voice to the dominant paradigm of rational comprehensive planning (Harvey 1973; Castells 1977). Many of us embraced this perspective as the liberatory voice of its time, only much later coming to realize the exclusions of a purely class-based perspective and the paralyzing effects of structuralism. The demographic and economic restructuring of our cities and regions over the past 20 years (and reflected in the changing composition of our student bodies) calls for a further radical shift in perspective. We are being challenged in the city and in the academy by frontiers of difference. We must listen to these voices, for they are telling us not only what is wrong with our cities, but also what is wrong with our way of looking at the world, and providing clues as to what might be better ways of dealing with both. In other words, we are faced with a challenge to both our theory and our practice by these voices from the borderlands.

This essay takes the form of a meditation on a metaphor – the metaphor of frontiers/borderlands/margins – and a reflection on the relevance of this metaphor for planning theory and practice. I have been struck in recent years by the recurrence of metaphors of frontiers/La Frontera, borders/borderlands, margins, and boundaries in the writings of women of color in North America. It seems to be a theme or preoccupation that is shared by Chicanas (Gloria Anzaldua, Cherry Moraga), African Americans (bell hooks, June Jordan, Toni Morrison, Audre

Lorde, Patricia Williams), Native Americans (Leslie Marmon Silko), and Vietnamese, Chinese, and Japanese Americans (Trinh T. Minh-ha, Maxine Hong Kingston, Mitsuye Yamada).

The metaphor itself seems an evocative part – and also a condition – of the landscape of postmodernity – certainly not the exclusive property of women of color, but rather reflecting global changes in economies, cultures, and demographies, all of which problematize borders, boundaries, and territories. This recurring metaphor of the borderlands/margins/frontier and its significance is my starting point for this essay. I argue that these voices from the borderlands are significant for planning theory, and that they raise questions for planning practice that are arguably the most difficult we face in this and coming decades.

We do need to distinguish different meanings and uses of the term *theory*. If planning theory is seen as reflection on planning practice, then the writers I'm about to discuss don't have anything directly to contribute. But if we think of theory in planning as being receptive to and applying the intellectual currents within the humanities and social sciences where they seem relevant to planning concerns, then the writings of these radical postmodernists can be seen as addressing issues of central importance to us. And if theory is seen as synonymous with practical philosophy – that is, thought crafted to guide action – we can see even greater relevance.

The range of themes that recur in the writings of women/people of color in North America include identity and difference; multiple voices; the importance of storytelling; the search for new ways of being alongside new and old ways of knowing; new spaces of opposition, resistance, and consciousness; and new ways of theorizing. In their work, the condition of postmodernity is a way of being and knowing, acting, and loving that provides us, as planners, with a way forward in the face of uncertainty and in spite of the death of grand theories about social change. It is in their work that we find the most textured attempt to come to terms with a world divided along lines not only of gender or of class, but also of race, ethnicity, and sexual preference. The writings of women/people of color, inhabiting the margins/borderlands –psychologically, epistemologically, ontologically, and methodologically – challenge us, showing us a new way, a new space, a space that is not comfortable but risky, not rigid but flexible, but above all, *a space of radical possibility* – a way forward.

What is the nature of the challenge? And what can we learn from this work?

⋆

I focus on the works of two writers, Gloria Anzaldua and bell hooks, with more limited references to the other writers, simply because of space limits. But I also draw on Cornel West and Guillermo Gomez-Pena because I do not mean to imply that these insights are gendered in some essential way. Historically, women have been marginalized, denied access to the public sphere and to knowledge and power, relegated to the domestic domain, while men have controlled public space and public affairs. This is now such a familiar theme in feminist literature that I don't want to dwell on it (see Spain 1992). Rather, I want to point to a very significant shift within feminism concerning thinking about this marginal status. While the predominantly liberal White feminist movement of the 1960s and 1970s demanded that women take center stage, come out of their homes, and participate in public life, there has been a subtle but crucial shift in the language and program of women of color. As their voices have emerged over the past decade, they have not only challenged the exclusiveness of White feminism, but have also started altogether new debates about identity and difference in which the notions of 'choosing the margin' (hooks 1990, 145) and shaping a poetic vision out of the shadowy space of the borderlands (Anzaldua 1987) strike me as crucial and crucially different. All of them celebrate rather than lament, choose rather than feel exiled to the margin or the borderlands – a part physical, part metaphorical territory which is described as the new frontier or the new space of radical resistance. While they are certainly not celebrating the history of power relations that subjugated them and confined them to the margins, they are celebrating this position as conferring on them a unique vantage point. They are exploring the political, cultural, intellectual, and psychological dimensions of living on/in the borderlands. They are imagining new cartographies. Implicitly or explicitly, they offer powerful critiques of enlightenment epistemology, which must affect the way we think about planning theory and practice. If planning is theorized normatively as an emancipatory practice, as some theorists have recently argued it ought to be (Friedmann 1987; Soja 1989; Leavitt and Saegert 1990; Heskin 1991; Sandercock and Forsyth 1992), then perhaps we have much to learn from these writings.

I suggest that their importance lies in their attention to the themes of identity and difference, as well as more inclusive ways of thinking about justice; their concern about the ways knowledges are invented and institutionalized; their insistence on the importance of storytelling; and, more broadly, their project of rewriting history using race, class, gender, and ethnicity as categories of analysis. They are searching for a new consciousness, or, as Anzaldua (1987, 81) says, 'a new story, to explain the world and our participation in it, a new value system with images and symbols that connect us to each other and to the planet'. A challenge, that is, to the past 200 years of liberal individualism. And a significant part of this new consciousness is what Cornel West (1993, 15) calls a 'politics of audacious hope' and bell hooks (1990, 152) calls a new 'politics of location … marginality as a site of resistance. Enter that space. Let us meet there'. This space of resistance, of counter-hegemonic cultural practice, is found not just in words but in habits of being. Perhaps above all else it is this latter theme, this ethical concern with 'habits of being and the way one lives' (hooks 1990, 149), that is compelling about this work, the mapping of ways of living, knowing, and acting in the postmodern world that take us beyond the stalemate of postmodern despair, a practical philosophy which counters fashionable *fin-de-siècle* cynicism.

We begin, then, with poet/storyteller/theorist/writer Gloria Anzaldua, whose work explores the notion of a new consciousness, moving beyond the assertion of racial pride and separate identity. In the central chapter of her book *Borderlands/La Frontera* (Anzaldua 1987), in a poetic essay titled 'La conciencia de la mestiza' (the consciousness of the woman of mixed blood), she describes the emergence of a new mestiza consciousness, 'una conciencia de mujer' (a woman's consciousness). It is a consciousness of the borderlands, a consciousness shaped by cultural collision, forged out of mental and emotional states of perplexity and confusion. The new mestiza copes by developing a tolerance for ambiguity. She has a plural personality. Her being is engaged in working out a new synthesis because the future depends on the breaking down of paradigms; it depends on the straddling of two or more cultures. 'The work of mestiza consciousness is to break down the subject–object duality that keeps her a prisoner and to show in the flesh and in the images in her work how the duality is transcended' (Anzaldua 1987, 80).

The style of *Borderlands* is consistent with the author's desire to break down the subject–object duality. She moves between prose and poetry, personal story, folklore, and analysis as she discusses questions of epistemology, ontology, and ecology. She returns to her roots in the Lower Rio Grande Valley ('how I love this tragic valley of South Texas' [Anzaldua 1987, 90]), that borderland between the Nueces and the Rio Grande where some of the poorest people in the US live. She evokes the valley, with its small towns with chicken pens and goats picketed to mesquite shrubs, junk cars lining the front yards of hot-pink and lavender trimmed houses, fields of aloe vera and red pepper, rows of sugar cane, corn hanging on stalks – and the despair and the hostility. 'I still feel the old despair when I look at the unpainted, dilapidated, scrap lumber houses consisting mostly of corrugated aluminum. I remember how the White teachers used to punish us for being Mexican' (Anzaldua 1987, 89).

Her family is still there. She talks with her brother about falling corn prices, helps her mother prune roses, and remembers getting off the school bus, changing into work clothes, and bending to the ground to plant watermelon seeds. Emphatically, she reminds us (Anzaldua 1987, 91):

This land was Mexican once
was Indian always
and is.
And will be again.

It is in her poetry that Anzaldua most poignantly captures the struggle of the mestiza consciousness, the attachment to and identification with more than one culture, more than one way of life, and the need to keep moving on, to explore, to go beyond. The poem 'Letting Go' describes the necessity, and yet the terror, pain, and ongoing tension, of belonging to more than one culture (Anzaldua 1987, 165):

Meet the dragon's open face and let the terror swallow you.
You dissolve in its saliva
no one recognizes you as a puddle
no one misses you
you aren't even remembered
and the maze isn't even
of your own making.

You've crossed over. All around you space.
Alone. With nothingness.

While the new mestiza can rejoice in her roots, return home, and help her mother prune the roses, she now lives in another kind of borderland. In the introduction to her most recent book, the edited collection *Making Face, Making Soul*, Anzaldua (1990) asks what it means to be a thinking subject, an intellectual, for women of color from working-class origins. She answers that it means 'not fulfilling our parents' expectations, it means often going against their expectations by exceeding them. It means being in alien territory and suspicious of the laws and walls. It means being internal exiles' (Anzaldua 1990, xxv).

In *Making Face, Making Soul*, the focus is on the wounds of racism, ways of combatting sexism and racism, the silencing strategies of the privileged, the painful passage from psychic numbness to learning to 'sing our songs' (Anzaldua 1990, xvii), working with other ethnic groups and marginalized Whites, making critical theories and theories of consciousness, and exploring other ways of knowing, including traditional ethnic modes.

Anzaldua invites White readers to listen in to women of color talking to each other, and sometimes talking to and against Whites. She and her contributors are exploring how identities are simultaneously mediated, marked, and influenced by race, class, gender, and vocation. They refuse to talk about identity in the singular, insisting that we are all constituted of multiple identities. Anzaldua (1987, 77) presents herself as not only inhabiting borderlands, but also as perpetually traveling, traversing the boundaries/frontiers between classes, races, cultures:

> Because I, a *mestiza*,
> continually walk out of one culture and into another,
> because I am in all cultures at the same time …

Because of this, she is willing to share, to make herself vulnerable to foreign ways of seeing and thinking. She surrenders all notions of safety, of the familiar. 'Like corn, the mestiza is a product of cross-breeding. She holds tight to the earth. She will survive the crossroads' (Anzaldua 1987, 82–83). Out of all this confusion around identity and belonging, Anzaldua chooses to blur the boundaries, to cross over and cross back. Rather than identify as Chicana or lesbian or Indian or Spanish or intellectual, she identifies as all of those, and acknowledges the possible tensions between them.

Similarly, Mexican border artist Guillermo Gomez-Pena, now living in Los Angeles, explains how, in his fractured reality, there cohabit two histories, languages, cosmologies, artistic traditions, and political systems, which are dramatically counterposed. His response to this conflict is to opt for 'borderness'. Eight years after his departure from Mexico, he describes the border as his 'house, laboratory and ministry of culture'. His identity possesses 'multiple repertoires' (Gomez-Pena 1993, 37)

> I am Mexican but I am also Chicano and Latin American. At the border they call me *chilango* or *mexiquillo*; in Mexico City it's *pocho* or *norteno*; and in Europe it's *sudaca*. The Anglos call me 'Hispanic' or 'Latino', and the Germans have, on more than one occasion, confused me with Turks or Italians. I walk amid the rubble of the Tower of Babel of my American postmodernity.
>
> *(Gomez-Pena 1993, 37)*

As a result of this process, Gomez-Pena (1993, 37–38) calls himself a 'cultural topographer, border-crosser, [a member of] the migrant tribe of fiery pupils'. He talks of what has been lost by the weary travelers, the dislocated – the loss of country, class, the language and the literary culture of his native tongue, the loss of ideological meta-horizons and of metaphysical certainty. And yet he celebrates what has been won – a vision of a more experimental culture, a multi-focal and tolerant one.

> Going beyond nationalisms, we established cultural alliances with other places, and we won a true political conscience … as well as new options in social, sexual, spiritual and aesthetic behavior. Our artistic product presents hybrid realities and colliding visions within coalition. We practice the epistemology of multiplicity and a border semiotics.
>
> *(Gomez-Pena 1993, 38)*

When he talks of creating alternative cartographies, he is talking, like Anzaldua and bell hooks, of both material and metaphysical alternative spaces. And his proposal for new creative languages echoes theirs. He celebrates 'the funeral of modernity and the birth of a new culture', despite the pain of the crisis of identity and of the destruction of borders (Gomez-Pena 1993, 39).

Each of these authors is also engaged in another kind of crossing over or border activity: the attempt to move between and bridge the worlds of the academy and the community. In their roles as theorists and social critics, each shares common concerns. Anzaldua makes a number of interesting arguments about the making of theory. She talks about the need to occupy theorizing space, 'that we not allow White men and women solely to occupy it'. She talks of the need to transform that theorizing space, since what is considered theory in the dominant academic community is not necessarily what counts as theory for women of color. She emphasizes the need to rewrite history using race, class, gender, and ethnicity as categories of analysis; to de-academicize theory; to connect the community to the academy; to give up the notion that there is a correct way to write theory. Her search for 'theories that cross borders, that blur boundaries' is part of a broader movement among women of color who are in the process of trying to formulate marginal theories that are partially outside and partially inside the Western frame of reference, theories that overlap many worlds (Anzaldua 1990, xxv).

> We are articulating new positions in these in-between, Borderland worlds of ethnic communities and academies, feminist and job worlds. In our literature, social issues such as race, class and sexual difference are intertwined with the narrative and poetic elements of a text, elements in which theory is embedded.
>
> *(Anzaldua 1990, xxvi)*

One of the key aspects of this new theoretical space is the refusal to separate theory from lived experience, and the assertion that people of color have always done theory, but in forms sometimes quite different from the Western form of abstract logic. Barbara Christian and Patricia Hill Collins have each articulated this. Collins (1990) outlined an Afrocentric feminist epistemology which emphasizes concrete experience as a criterion of meaning; the use of dialogue in assessing knowledge claims; an ethic of caring (which emphasizes the appropriateness of emotions in dialogues); and an ethic of personal accountability. Barbara Christian (1988, 68) identifies in the Black narrative tradition a unique form of theoretical resistance to power:

> our theorizing ... is often in narrative forms, in the stories we create, in riddles and proverbs, in the play with language. And women, at least the women I grew up around, continuously speculated about the nature of life through pithy language that unmasked the power relations of their world.

There would seem to be many similarities between Anzaldua's project and that of bell hooks, the prolific African American writer whose work defies disciplinary boundaries as she enters debates on postmodern theory, cultural criticism, film, and the politics of race, class, and gender. Anzaldua talks of being loosened and empowered by mestizaje theories. hooks identifies marginality as a space of resistance. Anzaldua (1990, xxv) describes the pain of crossing over into or inhabiting the intellectual space of Anglo culture, of exceeding parents' expectations, of finding oneself in alien territory, of making that new space one's own by 'continually challenging institutionalized discourses'. hooks similarly describes growing up in an old-school, southern Black world where children were meant to be seen and not heard, where talking back was punished with silence, or worse. 'I had to leave that space I called home to move beyond boundaries' (hooks 1990, 148). She reflects on that painful journey, on the need to escape the censorship of a southern Black community, 'to struggle and resist to emerge from that context', to leave her class origins, her family's folk traditions, journeying to the city, the university, to read theory, eventually to write theory. But at some point, 'I need to return there', which forces her to explore the meaning of home. At times, she writes, 'home is nowhere'. There is only extreme estrangement and alienation. Then home becomes, of necessity, a variety of locations. 'Home is that place which enables and promotes varied and ever-changing perspectives, a place where one discovers new ways of seeing reality, frontiers of difference' (hooks 1990, 148).

What an intriguing and illuminating description of the ideal of home – as more than a physical and emotional space, as an intellectual space where one is free to explore, to discover, to move, to grow, where one is not required to have a rigid, unchanging identity. She is writing about the experience of coming from a poor Black community into a university or other privileged cultural setting, and the 'need to create spaces within that culture of domination if we are to survive whole, our souls intact'. In an extraordinarily powerful passage, hooks describes the pain and fear and challenge of being between two worlds:

> Back in those spaces where we come from, we kill ourselves in despair, drowning in nihilism, caught in poverty, addiction, in every postmodern mode of dying that can be named. Yet when we few remain in that 'other' space, we are often too isolated, too alone. We die there, too. Those of us who live, who 'make it', passionately holding on to aspects of that 'downhome' life we do not intend to lose while simultaneously seeking new knowledge and experience, invent spaces of radical openness.
>
> Without such spaces we would not survive.
>
> *(hooks 1990, 148)*

The quest for a human connectedness and new moral vision that transcends race, class, and gender divisions is a common thread in the writings of bell hooks, Gloria Anzaldua, and Cornel West. hooks writes of a 'space of radical openness' where more than one identity can be embraced, where one can experiment, redefine home, rethink boundaries, and forge new coalitions. 'For me this space of radical openness is a margin – a profound edge. Locating oneself there is difficult but necessary. It is not a "safe" place. One is always at risk' (hooks 1990, 149). The language in which hooks talks about this space of radical openness is crucial. *The margin. The edge.*

Is hooks' margin the same as Anzaldua's borderlands/La Frontera? I think so. Both of them arrive at these locations after reflection on and appreciation of their home communities. Both have had to leave those communities to further their own development, to escape the rigidities, the censorship of those communities. Both have moved into Anglo-dominated cultural institutions without being prepared to be depicted as the exotic Other, refusing, in Bob Marley's words, to 'be what you want us to be', trying instead to forge new, multiple, flexible identities (quoted in hooks 1990, 150). Both writers accept the importance of returning to those home communities, yet know that they are returning with their own voices, which cannot be silenced. They have changed, and yet they would not have been able to change without the unique perspective that their upbringing on the margin gave them.

> Living as we did – on the edge – we developed a particular way of seeing reality. We looked both from the outside in and from the inside out. We focused our attention on the center as well as on the margin. We understood both.
>
> *(hooks 1984, iv)*

It was here, on the wrong side of the tracks, that an 'oppositional worldview' developed, a different way of seeing that sustained them in their struggle to transcend poverty and despair, and strengthened their sense of self and of solidarity. Thus, hooks describes marginality as much more than a site of deprivation. It is also 'the site of radical possibility, a space of resistance' (hooks 1990, 149). Both are determined, having left that space in the margins, to keep alive those ways of knowing reality which affirm the primacy of resistance and the necessity of a resistance that is sustained by remembrance of the past, including 'recollections of broken tongues giving us ways to speak that decolonize our minds, our very beings' (hooks 1990, 150).

For hooks, talking Black talk and storytelling are critical to her purpose. And for Anzaldua, writing in a mixture of English, Spanish, and texmex and writing flexible combinations of prose and poetry, anecdote and folklore are essential in her effort to describe the new consciousness of the borderlands. In this insistence on other ways of knowing, and the connection between ways of knowing and ways of writing and in the theme of hybrid identities, hooks and Anzaldua share a remarkable common ground with women of color from other cultural backgrounds. I think of Leslie Marmon Silko (1977, 1981), a Laguna Pueblo writer and storyteller, whose beautiful novel *Ceremony* deals with the same pain of the mestizo existence about which Anzaldua writes. Or I think of Maxine Hong Kingston (1975), whose classic *The Woman Warrior* is a celebration of the Chinese custom of talk story at the same time as it is a scream of pain from a daughter caught between two cultures, inhabiting the borderlands, searching for a new way.

June Jordan's (1992) essays discuss her own multiple, and sometimes conflicting, identities as poet, mother, teacher, African American, activist, and middle-class cosmopolitan world traveler. Exploring the limitations of the notion of roots, she recalls Buckminster Fuller's words 'a man is not a tree,' and comments.

> The whole world will become a home to all of us, or none of us can hope to live on it, peacefully. But much of the American dream mistakenly supposes that, like a tree, we will grow and flourish standing in one place, where we murmur doomed declarations about our roots, about finding our roots, or putting down roots.
>
> *(Jordan 1992, 22)*

Jordan writes of the paralysis of identity politics and suggests that the way out of the paralysis is a moral attachment to a concept beyond gender and race – the concept of justice.

> There is another realm of possibility, political unity and human community based upon concepts that underlie or supersede relatively immutable factors of race, class, and gender: the concept of justice, the concept of equality, the concept of tenderness.
>
> *(Jordan 1992, 164)*

Trinh T. Minh-ha (1989) is a Vietnamese American writer, filmmaker, composer, and theorist, also working at the intersection of several fields and challenging the cultural and theoretical establishment. Her concerns are the processes of cultural displacement and cultural hybridization and the resulting decentered realities, fragmented selves, and multiple identities which are hooks' and Anzaldua's main themes. Like them, she is preoccupied with marginal voices and with the language of rupture. She incorporates the poetic in the analytic and stays away from the old way of theorizing through systematic dissection, combination, and recapitulation. Trinh discusses questions of language and writing in relation to notions of ethnicity and femininity; of identity, authenticity, and difference; and of storytelling as one of the oldest forms of building historical consciousness.

> The storyteller, besides being a great mother, a teacher, a poetess, a warrior, a musician, a historian, a fairy, and a witch, is a healer and a protectress. Her chanting or telling of stories ... has the power of bringing us together.
>
> *(Trinh 1989, 140)*

This theme of bringing us together is elaborated forcefully in the work of Bernice Reagon Johnson, a singer, writer, activist, theorist, and creator of the African American women's vocal ensemble Sweet Honey in the Rock. The widely published transcript of a talk she gave in 1981 at a women's music festival, 'Coalition Politics: Turning the Century', is a primer in multicultural feminism. Reagon (1981, 357) asserted then,

> We've pretty much come to the end of a time when you can have a space that is 'yours only' – just for people you want to be there ... we've finished with that kind of isolating. There is no hiding place. There is nowhere you can go and only be with people who are like you. It's over. Give it up.

Her deconstruction of the notion of home, her insistence on coalition-building across the boundaries of race and gender and class, and her ability to translate/communicate across these boundaries – musically, poetically, and theoretically – positions her, along with the other writers discussed above, as mapping new cartographies of loving and living, knowing and acting.

The voices from the borderlands to which we've been listening belong to people who dwell in cultures of displacement and transplantation, and/or cultures with a long history of oppression. They challenge familiar notions of ethnicity and identity because they speak from within unresolved, and perhaps unresolvable, historical dialogues between continuity and disruption, essence and positionality, home and journeying, old roots and new. But the postmodern condition which they describe is the condition of life for many people in today's Los Angeles, Chicago, New York, Washington D.C., Miami, San Diego, Houston, and all the towns and cities along the border with Mexico. We need to hear their stories because they are telling us something about what dwelling (or home) means today, something about the importance of memory, and about the possibility of forging a new vision from/based on the borderlands/frontier experience as a way of life. And they are telling us that this marginal space, this place of unease and discomfort, can be a place of strength, a way forward. But forward to what? And what does all of this have to do with planning education and planning theory?

*

This essay was prompted by Robert Beauregard's (1989) challenge to planners to engage in current intellectual debates and to address the challenges of postmodern critiques. I agree with Beauregard that much of the postmodern critique undermines the very foundations of planning theory and practice, but, unlike him, I do not believe that this leaves us in the grip of a postmodern helplessness. I have argued that there are multiple voices within postmodern theorizing, some of which encourage a nihilistic passivity, others of which are searching for new political practices, new ways of being and knowing. These latter seem to me to be the voices to which we ought to be paying attention. I have defined these as voices from the borderlands. The voices belong to people of color, to folks who have been marginalized but are now insurgent and who are turning their very marginality into a creative space for theorizing. These are the voices that I bring into my planning theory classes and which I would like to see included in the broader planning discourse because they are addressing the condition of postmodernity in constructive and progressive ways. They are formulating new ways of being and alternative ways of knowing. They are finding new ways of acting and creating a new oppositional politics, all of which are relevant to the present dilemmas of planning theory and practice, which are dilemmas of meaning and action.

The postmodern condition as defined by Michael Dear (1986) involves questions of style (aesthetics, architecture), of epoch (understanding the post-Fordist political economy of global capitalism), and of method (deconstruction). As defined by Beth Moore Milroy (1991), it is more a matter of positions along a continuum. Do we want to revise modernism, displace some of its major tenets, or replace it altogether? But for the authors I have been discussing, the postmodern condition is more a way of being and knowing, loving, and acting. As a way of being, they describe the state of living on/in the borderlands, living in between, living on the margins. They accept the necessity of living with uncertainty, living without universals, without the panaceas of the 'revolution' or of 'progress'. But they do not live without hope or without meaning. They embrace uncertainty as a potential space of radical openness which nourishes the vision of a more experimental culture, a more tolerant and multi-focal one. They accept the challenge of continually crossing boundaries, perpetually traveling between classes, races, and cultures. They reject the notion of a static and frozen identity in search of a safe place, a homogeneous community, and a clearly defined border between us and them. They talk of (our individual) multiple identities and the need to understand and accommodate the resulting tensions and conflicts. They embrace borderness, or marginality, and accept its risks. They reject the very idea of a space that is yours only (Reagon 1981), because of the separatism and isolating implicit in that desire. They describe a new consciousness, a consciousness of the borderlands, shaped by cultural collision, forged out of mental and emotional states of perplexity and confusion, a consciousness in which dualities are transcended and we make ourselves vulnerable to 'foreign ways of seeing and thinking' (Anzaldua 1987, 82).

How Does This Translate into the Domain of Planners and Planning?

These authors show us that it is possible to live with uncertainty, and essential to acknowledge our multiple identities as active subjects. They encourage us to reach out, to take risks, to explore the borderlands and transcend differences between classes, races, ethnicities, genders. And they warn us to give up the search for a safe place, a homogeneous community, and to embrace difference and diversity. Beyond these ontological issues, all of the writers I have discussed are concerned with epistemological questions, with how knowledge gets produced, by whom, for what purposes, and to what effect. They are concerned with who is included in and who is excluded from theoretical discourse and with how theory is defined. They argue that there are different ways of doing theory, that the voice of Western abstract logic and reason is not the only possible theoretical voice. They refuse the critical distance of enlightenment epistemology, the split between subject and object, the very notion of knowledge without a knowing subject. They argue instead that theory is always embodied. They stress the importance of contextualizing knowledge rather than abstracting it.

They celebrate the value of experiential knowledge, and of theorizing that grows from that knowledge. They value alternative methods of knowing, learning, discovering, including traditional ethnic modes – methods like storytelling, the importance of listening, the skills of interpreting and understanding, and searching for meaning rather than for truth. And alternative, more flexible ways of writing are essential: styles that move between prose and poetry, between anecdote and analysis, between vernacular and standard English, storytelling styles, conversational styles, writing that is accessible, that can bridge the gap between the academy and the community. Gomez-Pena (1993, 38) calls this an 'epistemology of multiplicity'. How does it speak to planners? It tells us of the limits and distortions of scientific and technical knowledge. It validates experiential, embodied, contextualized knowledge as opposed to objective, disembodied, abstract systems of reasoning. It is dialogic. It emphasizes listening and interpreting, developing skills that are sensitive to, able to pick up on, everyday ways of knowing. It suggests a whole different practice, in which communicative skills, openness, empathy, and sensitivity are crucial; in which we respect class, gender, and ethnic differences in ways of knowing; and in which we actively try to learn and practice those ways in order to practice a truly inclusive and democratic planning.

Finally, we are presented with ways of acting in spite of, or in the face of, the postmodern abyss. All of these writers challenge us to acknowledge and respect difference, or diversity, without being paralyzed by it, without succumbing to the quicksands of identity politics. Instead, they outline a new cultural politics of resistance and opposition, a politics that transcends (yet builds on) the concern with individual identity and becomes a collective struggle for justice and equality. These writers have not lost sight of certain key principles of the enlightenment – justice and equality – but they have seen through the exclusivity involved in the application of those principles. They are calling for a new, more inclusive, moral vision shaped by the consciousness of the borderlands, informed about racism and sexism, xenophobia and homophobia, homelessness and poverty. They are offering a theory of social transformation based on an oppositional worldview and a powerful critique of the dominant culture, rooted in communities of resistance, informed by principles of justice and equality, caring, tenderness, and love.

Gomez-Pena (1993, 41) talks of 'new models of communication and association'. Every one of these authors advocates the practice of coalition politics – forging coalitions that force us to move out of that safe place with 'our own people', and to build bridges, to acknowledge the humanity and the needs of 'the other'. Reagon (1981) gives advice about 'how to cross cultures and not kill yourself'. She gives hands-on lessons in the kind of postmodern feminism that is wary of structures and theories of wholeness. In multicultural, multiracial coalitions, she argues, fictions of sameness, reflected in the 'code word, *woman*', must give way to the articulation of other sources, other issues. We must move beyond the 'womb stage' (of identity politics). Anzaldua and Moraga (1983, iv) insist on the connectedness and interdependence of our struggles, 'White Black straight queer male female'. June Jordan dreams of political unity and human community. bell hooks imagines alliances of progressives, irrespective of their race, class, sexuality, ethnicity.

How does this moral vision and political practice translate to the domain of planning? Most obviously, we need to develop ways of planning {and of theorizing planning) which acknowledge and respect difference and reflect diversity. We could call this the need to plan for multiple publics and for diversity, rather than for the (myth of the) public interest and homogeneity. We could critique our own practices and history for their racist and sexist policies and assumptions (see Mier 1994; Grigsby 1994; Cordova 1994). We could scrutinize the internal culture of planning (that is, what goes on in planning agencies every day) for its reproduction of a patriarchal and racist and homophobic culture (Sandercock and Forsyth 1992). We could look around the room when we are in planning meetings and ask whether the faces present actually represent the diversity of the population. We could ask whether the readings and sources we use in the classroom offer more than the White Western male pantheon. We could deconstruct our plans (planning documents, regulations, legislation) to see who is the subject; who the object; who the knower, the author; and what interaction there has been between author and community. And we could ask how we might harness the energy of these mobilized voices and communities in our social planning.

I have argued for the inclusion of new voices into the domain of planning theory – voices from the borderlands, folks who would not think of themselves as 'planners'. The reason for including this group of writers is a very practical one. They are addressing issues of importance to the work of planners. The demographic and economic restructuring of our cities and regions over the past 20 years – a shift that is reflected in the increased diversity of our student bodies – requires a complementary restructuring of our thinking. We need to be asking not only what is wrong with our cities, but also what is wrong with our way of looking at the world – that is, what is wrong with our theoretical tools.

The global changes in economies, cultures, and demographies that characterize the 'landscape of postmodernity' (Jameson 1984; Soja 1989; Zukin 1991) are presenting us with challenges of meaning and of action (see Marris 1987). The writers I have discussed are all thinking constructively and progressively about these challenges. They all embody the notion of postmodernism as a form of theoretical restructuring. They are all producing new ways of constructing knowledge, new ways of being in the world and, most importantly, new ways of acting. They deserve to be read, both by planning students and by those of us writing and teaching planning theory.

Note

1 Many, though not all, of my sources for these discussions are from feminist literature. See Spelman 1988; Pateman 1988; Trinh 1989; Anzaldua 1990; hooks 1990; Flax 1990; Nicholson 1990; Collins 1990; Hekman 1990; Code 1991; Morrison 1992; Harding 1993. But also Kuhn 1962; Feyerabend 1975; Foucault 1980; Bernstein 1985; West 1992; Garcia 1993.

References

Anzaldua, G. 1987. *Borderlands/La Frontera. The New Mestiza* (San Francisco: Aunt Lute Books).

Anzaldua, G. (Eds.) 1990. *Making Face, Making Soul. Hacienda Caras* (San Francisco: Aunt Lute Books).

Anzaldua, G. and Moraga, C. (Eds.) 1983. *This Bridge Called My Back. Writings by Radical Women of Color* (2nd edn.) (New York: Kitchen Table, Women of Color Press).

Beauregard, R.A. 1989. Between modernity and postmodernity: the ambiguous position of US planning. *Environment and Planning D* 7(4), 381–395.

Beauregard, R.A. 1991. Without a net: modernist planning and the postmodern abyss. *Journal of Planning Education and Research* 10(3), 189–194.

Bernstein, R. 1985. *Beyond Objectivism and Relativism. Science, Hermeneutics, and Praxis* (Philadelphia: University of Pennsylvania Press).

Castells, M. 1977. *The Urban Question* (London: Edward Arnold).

Christian, B. 1988. The race for theory. *Feminist Studies* 14(1), 67–69.

Code, L. 1991. *What Can She Know? Feminist Theory and the Construction of Knowledge* (Ithaca: Cornell University Press).

Collins, P. H. 1990. *Black Feminist Thought: Knowledge, Consciousness, and the Politics of Empowerment* (Boston: Unwin Hyman).

Cordova, T. 1994. Refusing to appropriate: the emerging discourse on race and planning. *Journal of the American Planning Association* 60(2), 242–243.

Dear, M. J. 1986. Postmodernism and planning. *Environment and Planning D* 4(3), 367–384.

Feyerabend, P. 1975. *Against Method: Outline of an Anarchistic Theory of Knowledge* (London: New Left Books).

Flax, J. 1990. *Thinking Fragments., Psychoanalysis, Feminism and Postmodernism in the Contemporary West* (Berkeley: University of California Press).

Foucault, M. 1980. *Power/Knowledge. Selected Interviews and Other Writings, 1972–1977*, Gordon, C. (Ed.) (New York: Pantheon).

Friedmann, J. 1987. *Planning in the Public Domain* (Princeton: Princeton University Press).

Garcia, R. B. 1993. *Changing Paradigms of Professional Practice, Education and Research in Academe: A History of Planning Education in the United States* (Ann Arbor: UMI Dissertation Services).

Gomez-Pena, G. 1993. *Warrior for Gringostroika.* Saint Paul: Greywolf Press.

Grigsby, J. E. 1994. In planning there is no such thing as a "race neutral" policy. *Journal of the American Planning Association* 60(2), 240–241.

Harding, S. (Eds.) 1993. *The Racial Economy of Science* (Bloomington: Indiana University Press).

Harvey, D. 1973. *Social Justice and the City* (London: Edward Arnold).

Hekman, S. 1990. *Gender and Knowledge* (Boston, MA: Northeastern University Press).

Heskin, A. 1991. *The Struggle for Community* (Boulder: Westview Press).

hooks, b. 1984. *Feminist Theory: From Margin to Center* (Boston: South End Press).

hooks, b. 1990. *Yearning. Race, Gender, and Cultural Politics* (Boston: South End Press).

hooks, b. and C 1991. Breaking bread. *West* (Boston: South End Press).

Jameson, F. 1984. Postmodernism, or the cultural logic of late capitalism. *New Left Review* 146, 53–92.

Jordan, J. 1992.. Technical difficulties. *African American Notes on the State of the Union* (New York: Pantheon).

Kingston, M. H. 1975. *The Woman Warrior* (New York: Random House).

Kuhn, T. 1962. *The Structure of Scientific Revolutions* (Chicago: University of Chicago Press).

Leavitt, J. and Saegert, S. 1990. *From Abandonment to Hope. Community Households in Harlem* (New York: Columbia University Press).

Liggett, H. 1991. Where they don't have to take you. In: The representation of homelessness in public policy. *Journal of Planning Education and Research* 1(0), 201–208.

Lorde, A. 1984. *Sister Outsider* (New York: Crossing Press).

Mandelbaum, S. 1991. Telling stories. *Journal of Planning Education and Research* 10(1): 209–214.

Marris, P. 1987. *Meaning and Action: Community Planning and Conceptions of Change* (2nd edn.) (London: Routledge).

Mier, R. 1994. Some observations on race in planning. *Journal of the American Planning Association* 60(2), 235–240.

Milroy, B. M. 1991. Into postmodern weightlessness. *Journal of Planning Education and Research* 10(3), 181–187.

Morrison, T. (Eds.) 1992. *Race-Ing Justice, En-Gendering Power. Essays on Anita Hill, Clarence Thomas, and the Construction of Social Reality* (New York: Pantheon).

Nicholson, L. (Eds.) 1990. *Feminism/Postmodernism* (New York: Routledge).
Pateman, C. 1988. *The Sexual Contract* (Stanford: Stanford University Press).
Reagon, B. 1981. Coalition politics: turning the century. In: *Home Girls: A Black Feminist Anthology* Smith, B. (Ed.) (New York: Kitchen Table, Women of Color Press).
Sandercock, L. 1998. *Towards Cosmopolis: Planning for Multicultural Cities* (Chichester: John Wiley and Sons).
Sandercock, L. and Forsyth, A. 1992. A gender agenda. New directions for planning theory. *Journal of the American Planning Association* 58(1), 49–59.
Silko, L. M. 1977. *Ceremony* (New York: Viking Penguin).
Silko, L. M. 1981. *Storyteller* (New York: Arcade Publishing).
Soja, E. 1989. *Postmodern Geographies: the Assertion of Space in Critical Social Theory* (New York: Verso).
Spain, D. 1992. *Gendered Spaces* (Chapel Hill: University of North Carolina Press).
Spelman, E. 1988. *Inessential Woman* (Boston: Beacon Press).
West, C. 1993. *Race Matters* (New York: Random House).
Trinh, T. M. 1989. *Woman Native Other* (Bloomington: Indiana University Press).
West, C. 1992. The postmodern crisis of Black intellectuals. In: *Cultural Studies* Grossberg, L., Nelson, C. and Treichler, P. (Eds) (New York: Routledge).
Zukin, S. 1991. *Landscapes of Power. From Detroit to Disneyworld* (Berkeley: University of California Press).

PART II

Imagining Cities of Difference

FIGURE 4.1 'Chaos', artist Susan Zox. Photo, Leonie Sandercock

Commentary: The Cosmopolis Project: From Theory to Practice, 1992–2006

'Imagine the world we will birth when we see no stranger' (Kaur 2020, 299).

It feels accurate to label this historical moment in which I write as uncertain and chaotic, volatile, and complex, spinning wildly out of control. That's why I chose the image opposite, 'Chaos', a collage by my friend, artist Susan Zox (Figure 4.1). When I conceived the 'cosmopolis project' 25 years ago, I wrote that we were entering into a 'new world disorder' characterized by fear of 'the Other', connected to an age of migration which itself had complex causes: political, economic, and ecological. In the 25 years since I wrote *Towards Cosmopolis* (1998), this descent into chaos has only accelerated. Writer and activist Margaret Wheatley describes where we are now as 'an age of retreat': 'from one another, from values that held us together, from ideas and practices that encouraged inclusion, from faith in leaders, from belief in basic human goodness (Wheatley 2017, 6). In this Commentary, I reflect back on the concept of 'Cosmopolis' as an inclusionary project, asking whether it is still worth fighting for, or whether it is fatally flawed.

*

> Radical postmodernism calls attention to those shared sensibilities which cross boundaries of class, race, gender, etc., that could be fertile ground for the construction of empathy – ties that would promote recognition of common commitments, and serve as a base for solidarity and coalition.
>
> *(bell hooks 1990, 27)*

It seems that bell hooks and I were dreaming a similar dream in the 1990s. And Valarie Kaur (2020) continues the fight for this dream today: a world in which we 'see no stranger'. Back then, I called my dream 'cosmopolis'. And while I insisted that cosmopolis was a construction site of the mind, a search for a postmodern utopia that must always be in the making, always fluid, adaptable, I did also spell out a set of guiding principles for this culturally diverse, democratically pluralist project in *Towards Cosmopolis* (1998). That was where all of the theoretical restructuring provoked by the Los Angeles uprising came together (Commentary, Part 1). For the next decade I worked on grounding this theory through my research and my practice.

The praxis I was discovering while still in LA was a praxis of insurgent planning and rainbow coalition building. But I was relying on interviews with practitioners and on secondary sources. I never got my hands dirty while I lived in LA; that is, I never got out into the city and got involved. In part, that was because only half of me was invested in this work; the other half was, until 1995, writing screenplays and trying to sell them. But now I think there's another reason. LA was not my town. I never felt 'at home' there in the way that I'd felt at home in both Melbourne and Sydney. It was all too big and too daunting. My 'home', my community, really was the UCLA environment, and my praxis was *mere* theorizing. I say that with no disrespect to theorists, but it was never enough for me. I'd been very active in Australia, at many levels, from federal government advisory committees to working with trade unions and community organizations (Introduction). I understood how and where to get involved. I understood the scene; I had a network.

Perhaps it's not surprising then, that once I had resolved my moral dilemma – disillusionment with screenwriting in Hollywood (Commentary, Part 1) – and decided to return to academia, I ended up back in Australia. Several things made this possible, serendipity among them, in the shape of an old friend, my Sydney surfing buddy David Wilmoth. David happened to be visiting LA, so we invited him to lunch. He mentioned that Royal Melbourne Institute of Technology University (RMIT), where he was a deputy Vice Chancellor, was looking for a new head of their Planning program. Would I be interested?

This was 1995. At that time, I had never entertained the possibility of returning to Australia because John was so involved at UCLA. Ten years earlier, the prospect of him moving to Australia was not even discussable. If we were to be together, I'd have to move to LA, and that didn't seem like any kind of sacrifice to me at the time, or ever after – quite the contrary, as I have described in the Commentary to Part 1. When David raised the possibility over lunch, I looked sideways at John, expecting a negative reaction. Instead, he suggested that I apply. They flew me to Melbourne for interviews and offered me the job on the spot. I phoned John, animated. He heard my excitement, and said 'let's do it'. John had decided to retire from UCLA by mid-1996 and was, in principle, prepared then to move wherever I wanted to go. Now he was ready for an Australian adventure. This was an unexpected turn of events. I told RMIT that I couldn't take up the position for another year, because I wanted to honor John's final year at UCLA. I was already co-organizing, with Marty Wachs, a farewell event for John: a weekend-long event that combined invited paper presentations on the theme of 'Civil Society' with music and dancing and feasting.[1]

Several things excited me about returning to Australia. First there'd been the recent High Court's 'Mabo decision', recognizing Native Title and opening the door to more extensive Aboriginal land claims. In addition, the old White Australia immigration policy had been replaced by what seemed (at face value) a very progressive approach to immigration and integration. I felt instinctively that I could get involved again in these arenas of public policy. And at RMIT, I was quite beguiled by the dean of the faculty I was joining: a White, South African–born and trained architect, gay, cultured, well-read. I fell for his flattery and believed in his promises.

Two years into my new role as head of department, I resigned over what I experienced as a complete betrayal by the Dean. In a nutshell: in the intervening year between accepting the position and actually moving to Australia, the federal Labor government had been defeated and the country was now in the hands of conservatives, led by John Howard, a small-minded, back-to-the-family-values-of-the-1950s kind of guy. In the first budget his government delivered, in early 1996, funding to the university sector was cut by a whopping 50%. It was the old anti-intellectual strain of Australian culture at its worst: 'Who needs those eggheads?' Thus, the morning after our arrival in Melbourne for our new life in Australia, mid-1996, at breakfast with the Dean, I was told that 'my' Department had a budget crisis, a deficit of $300,000, and I'd have to solve this as well as figuring out how to replace 50% of the previous revenue stream from the federal government. I was given three years to solve this, but I had to produce a 'Business Plan' *within 6 weeks,* explaining how I would do this! The irony was that, when interviewed for this position and asked about my weaknesses, I'd said quite clearly that I was not a fund raiser or money manager. I'm an intellectual leader. And the Dean had said intellectual leadership was exactly what they were looking for.

I gave it my best shot. I don't think I had time to read a book in that first two years, and certainly didn't write anything of note. Copying what every other department head was doing, I travelled to East and Southeast Asia recruiting students who would pay high international fees. I never felt good about that, but I didn't have any better ideas for bringing in money. But two years into this nightmare, while I was in Taiwan on one of these recruiting trips, the Dean pulled the rug out from under me. He dismantled the department I'd worked so hard to unite, by bringing urban planning, environmental policy, and landscape architecture into one degree program. They'd previously been three separate programs that never spoke to each other. Now the Dean carved off Landscape Architecture (his pet) and moved it into Architecture, and relegated the rest, the social science stuff, to another faculty on an outer suburban campus, while creating a small research unit that I could run. I suppose he thought I'd be thrilled about this. I wasn't. Perhaps if he'd consulted me instead of presenting a *fait accompli* I might have considered it, or felt differently. Instead, I experienced his coup as a massive betrayal. I put my resignation letter on the Dean's desk first thing Monday morning, after discussing the situation all weekend with John. By lunchtime, the Dean had called in security and tried to lock me out of my office. I had submitted a legally correct resignation, giving 6 months' notice, so I knew he couldn't lock me out, but he evidently didn't know that. We never spoke again. I served out my 6 months then walked away, unemployed.

Looking back, I think there's a more generous interpretation of the Dean's surgical restructuring. There were many 'personnel' problems in the department I'd inherited. In the previous decade, the second tier of post-secondary institutions (Teachers Colleges and Institutes of Technology) had all been given status as universities, so now their faculty were meant to suddenly become researchers: not part of their original employment contract. There were a number of faculty in my department who had not published anything in 15–20 years and I'd had to suggest to them that they might need to reinvent themselves if they wanted to stay employed. Being good Aussie blokes, their response to me was, literally, 'Go f★★★ yourself!' I think perhaps the Dean was trying to 'remove the dead wood', make it another Dean's problem, and create space for me to do what I was good at: research and writing. Maybe. I didn't see it that way at the time. I was wedded to the efforts I'd made to reform the curriculum, and I wanted to teach in that new degree program I'd created, not just do research in some privileged little bubble. But now there was no program.

I had resigned in anger, and felt brain-dead after two years fighting these institutional fights. But I have never had a moment's regret about that decision. However, the lesson I still had to learn was that I was devoid of conventional leadership skills. Just as I had never received any training on how to 'present' my work to audiences or speak in public, nor had there ever been any leadership training. These things were just missing from the training of academics in Australia. Nor had it occurred to me to seek out such training, off-campus. Instead of soul-searching and self-reflection, it was easier to place the blame on an evil Dean and a dysfunctional system of financing tertiary education. For all of my supposed intelligence, I had yet to develop the qualities of a reflective practitioner. While I have never regretted walking away from that position, I do regret not asking myself what I needed to learn about leadership from this experience.[2]

I thought about looking for positions back in the USA, but we loved our life in Melbourne. John had started his own Pacific Rim research network and was organizing workshops and beginning to start some new research himself on Chinese urbanization. I had tapped back in to the local circle of friends I'd made back in the late 1970s, and John and I hosted monthly Sunday evening 'salons', inviting these friends (poets, painters, politicians, the odd professor) to talk about their work, accompanied by plentiful local wines and cheeses. John had found a very sympatico cello teacher and was working his way through the Bach cello suites. He had been an unfailingly patient listener to my nightly rants about all that was going wrong with my life at RMIT, distracting me with home cooking, hiking, and Sunday trips to our favorite wineries on Mornington Peninsula. We were both prepared to hang out in Melbourne for a bit and see what happened.

A few months later, I was offered a position in the Faculty of Architecture, Building, and Planning at the University of Melbourne, just down the road, and for the next two years found time to research and write as well as teach, and I did not have to worry about budget crises. My new Dean was taking care of all that. My main intellectual focus continued to be the 'cosmopolis project'. I had orchestrated some empirical research on municipal planning in the culturally diverse outer suburbs of Melbourne with an RMIT grad student, Bev Kliger (Sandercock and Kliger 1998a,b), and used those results as part of a major think piece about 'Managing Cities of Difference' (Sandercock 2000), which appears here as Chapter 5. It was my first attempt at articulating the different ways in which multi-ethnic, multi-racial cities are a challenge to planning systems, policies, practices, and education, and it identified four possible ways of responding to these challenges.

In that same year, I was invited by Jean Hillier (then at Curtin Institute of Technology in Perth) to give a keynote address at an event focused on Pierre Bourdieu's concept of *habitus*, with Bourdieu himself making an appearance via teleconference. In that paper I began to explore the nature of fear in the city, especially the fear of 'strangers', that is, immigrants from countries with cultures different from the dominant Anglo-European, since this had clearly become a pressing political issue across Europe and the UK, as well as in Australia (Sandercock 2002). I was influenced by the work of Ghassan Hage, a Lebanese-Australian sociologist, whose book *White Nation* shook my cosmopolis dreams to their foundation. Hage (1998) argued that in Australia, White racists and White multiculturalists share a conception of themselves as nationalists and as managing the national space, a space that is structured around a White culture and in which Aboriginal people and non-White 'ethnics' are merely objects to be moved or removed, according to a White national will. Through interviews, Hage had talked with people who were deeply unsettled by an immigrant presence in their neighborhood, specifically the presence of those whom he describes as 'Third World–Looking People' like himself, a dark-skinned Lebanese. Fears are generated; loss is experienced. 'The other', the stranger, is thought to be taking over, or as having already taken over. I was disturbed and intrigued by this line of inquiry and it led me in a new direction: into the political economy and social psychology of fear of strangers. I needed to understand this fear and the challenge it presented to the 'cosmopolis project'. This line of inquiry became so significant for me, in fact, that I decided to revisit *Towards Cosmopolis*, writing a sequel addressing the nature of these fears and how they affected planning in multi-ethnic, multi-racial cities. That's how *Cosmopolis 2: Mongrel Cities of the 21st Century* (2003) came into being. Looking back, it's significant that it was a racialized immigrant (Ghassan Hage) who provided this critical lens on Australia's version of multiculturalism: the lens of ongoing White supremacy that has always and still defines liberal democracies. I had not been able to see this with my own White eyes. The privilege of (my own) Whiteness was still invisible to me. I will return to this theme at the end of this Commentary, and again, in relation to my subsequent work with Indigenous communities, in the Commentary to Part 4.

In *Cosmopolis 2*, I also developed several other themes that had been simmering inchoately for a few years. I first came up with the term *therapeutic planning* in an essay in *Planning Theory and Practice* in 2000 (Chapter 5) when I was reflecting on the work of the award-winning social planner, Wendy Sarkissian. There I discussed a case study involving conflict over land use in an inner Sydney neighborhood in the late 1990s between Aboriginal and non-Aboriginal residents. Wendy had been brought in to defuse and 'manage' that conflict. But she did much more. In an extremely careful, yet creative and experimental way, Wendy designed a 'space', an occasion, in which all parties could *speak out* their anger, resentment, fear, and hope, without interruption. This was not yet a dialogical space, because the groups who were in conflict were not ready to be in dialogue. (Wendy's wisdom was in understanding that.) Rather, they needed to express their fears, frustrations, histories, and aspirations, and to have those emotions and stories heard by those with whom they were in conflict. It was this practitioner's story[3] that led me to develop my thoughts about the potential of therapeutic planning in *Cosmopolis 2*, by which time I was living in Vancouver. Soon, I would begin to see its application in working with Indigenous communities in Canada (Chapter 9).

The second emerging theme was the power of story in planning. I'd been invited by Jim Throgmorton and Barbara Eckstein to the Iowa Writers Workshop in 2000. They were exploring 'Story and Sustainability', the title of the book produced from the papers presented at this stimulating workshop. It was there that I first met Ken Reardon and learned of his path-breaking community–university partnership work in East St. Louis. Ken dramatized that story in a mesmerizing one-hour lecture that was more like theater to me, such was its power (Reardon 2003). My own paper also included many stories about planning practice and concluded that 'stories and storytelling – in their various incarnations, as film, as mural, as rap or folk music, as dance, as witness, and yes, even as research reports – have a profound place in the work of the city-building professions as we try to build more sustainable cities for the 21st century. Stories are telling' (Sandercock 2003). And there, written in 2000 but not published until 2003, was the spark for another stream of work I would do for the next 20 years, (although I didn't know it at the time), and which is the subject matter for Part 3 of this book, 'the storytelling project'.

My two years of peace at Melbourne Uni had enabled me to write two essays (Chapter 5, and Sandercock 2003) that began to set my agenda for the next couple of decades. It was energizing to reconnect with my curiosity and creativity, and I was grateful for the space I'd been given at Melbourne Uni that enabled this rejuvenation. But there was also a dark side: teaching planning to undergraduates.[4]

I have never been a fan of undergraduate professional planning programs. I believe that undergraduates need a broader education (in the humanities, social sciences, environmental sciences, and beyond) and some work/life experience before choosing planning as a vocation. But in Australia, planning had always been taught at undergraduate level. And now that I'd had the experience of teaching in one of the top graduate programs in the USA, I knew what was missing from this education. Compounding this limitation was the nature of planning systems in Australia: predominantly regulatory, with little to no room for innovation, creativity. It was depressing (arguably even futile and irresponsible) to be filling students' heads with grand ideas of how they could contribute to making the world a better place, only to have them confront the on-the-ground realities of working on development permit applications in municipal planning. The burn-out rate of young planners at that time was a testament to this reality. I started to think about returning to the USA, and John was again supportive, in spite of enjoying life in Melbourne. He and I had submitted a joint application to the Australian Research Council for an ambitious international project examining how planning agencies in a range of cities, from east Asia to Europe and North America, were responding to their increasingly culturally diverse populations. When we got word that our application was not going to be funded, not a cent, it cemented the decision to look for opportunities back in North America.

Timing is everything, as I discuss in the Commentary to Part 3, where I explore 'serendipity'. I was invited by the University of Manitoba to give a keynote on the occasion of the 50th anniversary of their planning program in 2000. To cover my airfare from Australia and an honorarium, they organized a lecture tour of planning programs across Canada. It turned out that people were reading *Towards Cosmopolis* and wanted to talk about it. I travelled from Montreal to Toronto, then Winnipeg, followed by Calgary, ending up at UBC in Vancouver. My hosts were generous everywhere, but none more so than at UBC, where Tony Dorcey was then director of the School of Community and Regional Planning (SCARP). He met me at the airport, took me to meet the co-director of Planning at the City of Vancouver, Larry Beasley, and on campus arranged a chat with the vice chancellor, Martha Piper. He had no idea I was on the job market. But over coffee, just before leaving for my flight to Seattle, I mentioned to Tony that I was being interviewed for a position with the University of Washington. His eyes lit up. SCARP had a position open. Would I be interested?

A few months later, John came with me to Vancouver for the interviews at UBC and we were seduced, by the city and by the university. There was a very appealing communicative and collaborative energy about Tony Dorcey as director, and about what the City was doing: its inner city revitalization planning, its participatory neighborhood planning, and what subsequently became known as 'Vancouverism', a densification alternative to suburban sprawl. It seemed like an excellent two-year Masters program. I'd met with the students and was more than impressed with their work, their excitement about the field, and their connection with my work. Six months later we moved into our townhouse in south False Creek, and the next stage of my professional life began.

*

International moves can be very disruptive of the flow of one's work, and life. Fortunately, I was blessed with a life partner who wanted to support me on my path. John embraced the move with energy and enthusiasm, and when asked by Tony Dorcey if he'd like to have some role in SCARP, John volunteered his services, literally. He became an 'honorary professor', co-taught the Advanced Planning Theory seminar for PhD students with me, took on PhD supervision of a handful of students, and continued with his own writing projects. When I say we 'co-taught' that PhD seminar, though, I should clarify. We actually divided the course into two semesters and took one semester each, so that we were never in the classroom together. This was a very deliberate choice on my part. Back in the UCLA days, I had co-taught a class with John and was shocked at his teaching style, which I can best describe as 'old school'. I didn't consider myself to be a good teacher at that time, but at least I was kind. John could be harsh in his critical responses to students. To him, it was never personal, but to the poor students it felt like it was. When I took on the task of chairing the PhD program at SCARP, this became one of the rare conflicts in our domestic life. John could reduce PhD students to tears, and they'd then come to me, as chair, complaining. It actually took some 'couples therapy' for us to work through this, and related, conflicts. (Just to be clear, this was *after* I'd started writing about the possibilities of therapeutic planning, not the cause of ....)

One thing John and I did do very well together during this first decade at UBC was to inaugurate what we called the Biennial PhD Jamboree. It struck us both that students in a relatively small PhD program, like SCARPs, with 10–20 students spread across half a dozen or more years and with no shared work spaces, were missing an experience of building their own intellectual community. (UCLA's PhD program had around 40 students, who all had work space in one large room, and there was always a sense of energy and activity.) So we convened a week-long workshop, inviting up to 40 participants, initially from North America but subsequently from across the globe, along with a couple of'name' keynote speakers for each event: luminaries such as Patsy Healey, Bent Flyvbjerg, Marty Wachs, Margaret Crawford, Mike Douglass, Libby Porter, and Rebecca Abers. SCARP colleagues offered sessions on their specializations, and students organized themselves into thematic groups and found their kindred spirits. We began each workshop with a welcome dinner at UBC, and closed the week with walking tours in the city, followed by a bang-up party at our home. After each event, we invited written feedback from the participants, so that we could 'refine' the workshop design over the years, and we did hear back wonderful reports of how 'life-saving' the event was for many participants who had hitherto felt so isolated. In part, it had been memories of my own isolation as a doctoral student that had made me throw myself into this summer event, and it turned out to be a hugely satisfying project for John and I to do together. And perhaps, just as I had been inspired by the loving relationship between Hugh and Patsy Stretton, and Peter Marris and Dolores Hayden, decades earlier, perhaps some of the participants in our Jamboree were similarly encouraged by the way John and I worked and played together, to believe that life could be built around love *and* work.

On the research front, I quickly established a local continuity that fueled my dream of cosmopolis. I'd begun writing *Cosmopolis 2: Mongrel Cities of the 21st Century* in Melbourne and was eager to finish it before getting completely immersed at UBC. Tony Dorcey generously gave me a semester free of teaching in that first year, and I did finish the manuscript. Then a timely connection happened. I took on supervision of Steven Dang, a Masters student who was enamored of *Towards Cosmopolis* and knee-deep in multicultural theory. He had a fascinating case study for his thesis. The Collingwood Neighbourhood House (CNH) was a local institution established in 1986 in what had rapidly become the most culturally diverse neighborhood on the eastern edge of the City of Vancouver. I was intrigued by what Steven told me about this place and asked him to take me there. And that was the beginning of many beautiful friendships, and a love affair with CNH itself, if one can fall in love with an institution – actually, not an institution but a place of welcome. I tell this story in more detail in the Commentary to Part 3, and in Chapter 8, where I discuss the subsequent making of the documentary about CNH, *Where Strangers Become Neighbors* (2006).

That film, with Giovanni Attili, was both the start of my new academic life in storytelling through film, and also, in a symbolic way, the joyful closure of the Cosmopolis Project. I believed that CNH embodied the spirit of my postmodern utopia, my cosmopolis in microcosm. The film was a love song, a hymn of praise to the possibility of cosmopolis. And the film enjoyed considerable success. There was at this time in Canada widespread interest in the challenges of integrating immigrants from cultures different than the still-dominant White Anglo-European one because there were plenty of episodes of racism and xenophobia directed at these newcomers. There was such demand for screenings and discussions of the issues raised in the film that I applied for a research grant to take it 'on the road' across Canada, in partnership with the then executive director, Paula Carr, who had become a dear friend during all this work. We co-created a manual to accompany the film, explaining more about how the CNH had managed to achieve what they had achieved in only 20 years. We took boxes full of the manual with us, and gave them away at the workshops we ran in Halifax, Moncton, Toronto, Winnipeg, and Edmonton.

And then we screened the film at an international conference in Lisbon, and numerous requests for assistance started to roll in from all sorts of organizations across Europe, from housing agencies to police departments, municipalities, and non-profits. As we started to share the inspirational CNH story internationally (and the policies and theories behind it), we always had a lot of explaining to do about Canada's multicultural policy, the City of Vancouver's anti-racism and diversity training policies, and more. And that led to the next decision: to write a book about CNH in this broader context, for an international audience (Sandercock and Attili, with Carr and Cavers 2009). One of the delights of this partnership work with CNH was that we jointly received the BMW Award for Intercultural Learning in 2008, after submitting the film as evidence of CNH's achievement in intercultural work. We were flown to Munich to accept the award: two awards, actually. CNH received the 'practice' award, and Attili and I got the 'theory' award for the film and accompanying essays (as though the making of a film is not a practice in itself). Eventually, CNH was receiving so many international requests for advice and assistance on how to integrate immigrants that they decided to start a social enterprise, which they called 'Where Strangers Become Neighbors', so that they could earn some income for sharing their expertise.

The national and international impact the film had was extremely gratifying, especially for the folks at CNH. In a way, the film had shown them what an amazing place they had collectively created. For me, the delight had been in discovering this microcosm of the cosmopolis dream. The dream had come to earth in Collingwood. I've said that I fell in love with CNH. What I mean is that it activated my heart. *It showed what was possible* when a neighborhood institution adopted not just a milk-toast mantra of creating a welcoming place for everyone, but based the very design (physical and social) of that place on anti-racism, diversity, and equity principles *and* training. The training part was critical to their mission: every new staff member, and every volunteer, had to undergo anti-racism and diversity training. And there was even a program run by youth, for youth, about anti-bullying and anti-racism. But still, the critical part of my brain had to ask: Should I really be generalizing, based on one local institution, about the broader prospects for cosmopolis? Am I becoming an uncritical endorser of Canada's multicultural policy by promoting this story? Have I 'drunk the kool aid'? Was the CNH possible *because* or *in spite of* that policy? What about the White supremacy that was surely as pervasive in Canada as Hage (1998) had argued in the case of Australia?

What brought closure to this cosmopolis project was my return to this big picture/critical thinking on the occasion of an invitation to deliver the annual Laurier Lecture, which was broadcast live for the CBC *Ideas Program* in 2006, and is published here for the first time (Chapter 6). In that essay, 'Mongrel Cities of the 21st Century: Is Multiculturalism the Solution or the Problem?' I raised five critiques of Canada's multicultural policy. I won't rehearse them here. My conclusion was that the policy itself, arguably founded in the 'White nationalist managerial sensibility' of the 1970s that Hage had exposed for Australia, was itself a work in progress, grounded in that sensibility no doubt, but challenged from various quarters to evolve into something better, less culturally hegemonic, more inclusive. I wrote that 'No-one has ever known for sure, (or knows now), whether a country based on such an idea is really possible, but again and again we leap toward the idea, and hope for the best. And maybe making the leap is the whole point. It's a life-affirming, humanity-affirming project'. That's the idealist, the dreamer in me, speaking. A few paragraphs later, I also wrote that 'the leaps that continue to invent and define the intercultural nation continue to be made. When the leap fails, it looks like the destruction of the Africville community in Halifax in the 1960s, or the appalling history of the Indian Residential Schools, which did not end until 1996. When the leap succeeds, it looks like the Collingwood Neighbourhood House'.

What strikes me now, in re-reading those fine words, is that my utopian thinking, which had come to earth on the eastern edge of the City of Vancouver, was still conspicuously lacking any rigorous analysis of the systemic nature of White racism in this country. Although I mentioned Residential Schools, I didn't recognize that Canada is more than a modern nation built on immigration. It is a nation built on the foundational injustice of Indigenous dispossession and subsequent policies of cultural genocide. I didn't mention that those Residential Schools were one part of a tripartite government policy to assimilate Indigenous peoples and destroy their diverse cultures and languages. That failure of recognition on my part came from my settler naïveté, my optimism (faith in my fellow humans), and worst of all, my ignorance of Canada's history of colonization. That was all about to change, as was my life, in this same year 2006, when I was invited to the small town of Burns Lake in north central British Columbia. How and why this was so transformative are the stories I tell in Parts 3 and 4.

From the vantage point of 2022, in the wake of the racism that resurfaced everywhere, not just in the US, during the Trump years; in the wake of Black Lives Matter; in the wake of the police killing of George Floyd in Minneapolis in 2020; and in the wake of the ongoing violence against Indigenous Peoples in Canada (see MMIWG 2019), there was a lesson I'd still not internalized from that classroom crisis in 1992. *Cosmopolis* is an impossible dream while White supremacy is allowed to thrive. The cosmopolis project was a *project of inclusion* that had failed to take into account the systemic nature of White racism in our liberal capitalist democracy. My very way of writing about 'fear of the Other' signaled that I was not 'Other', signaled my Whiteness and its systemic invisibility.

White supremacy is now being challenged in Canada in every arena: in education, in public policy, in the media, in sports, in recruiting, in language. Anti-racism and diversity training, cultural safety and cultural competency workshops, EDI appointments in public and private sectors, are now the order of the day. Surely this is better than where we were as a society 20 years ago when I wrote *Cosmopolis 2*. The depth of change that is needed is profound: it's generational. It involves more than workshops and trainings. But I do not want to abandon the (cosmopolis) dream of more equitable, diverse, and inclusive cities and neighborhoods. Rather, I'm more aware of my own positionality in this discourse, of the nature of the space I'm taking up, and of the need to make new and better arguments.

I think there's a socio-ecological argument that can be made, an argument from a broader sustainability perspective, if you like, in the most holistic sense of sustainability. The parallels between biodiversity and social diversity seem clear and compelling to me. It's now well understood that the more diverse an ecosystem is, the more resilient it also is, and thus more able to withstand disturbance (Gunderson, Allen and Holling 2010; Kimmerer 2013). Couldn't the same be said of social systems? Individuals who 'wall' themselves off from diversity become increasingly paranoid about encountering 'strangers'. Social groups who 'wall' themselves off, often literally in gated communities, define strangers as a menace, a threat. But folks who have regular experiences with/in the presence of strangers are less likely to demonize them and turn them into 'the enemy'. And when we come together in a 'dialogue of differences, the collective becomes smarter than any individual in it' (Palmer 2018, 132).

There's also a socio-psychological argument to be made. I was heartened recently to come upon the work of American-born Sikh feminist, lawyer, and interfaith anti-racism activist Valarie Kaur, who has spent several decades documenting violence against people of color.

Going beyond easy denunciations of White supremacy, she seeks a deeper understanding of the root causes of this violence. People who commit hate crimes, she argues, have themselves been wounded or betrayed by systems they perceive as no longer serving them. That leads her to address system change, but with a recognition that systems are comprised of individuals who reproduce them. So, every person has a role to play in transformative change. In her visionary book *See No Stranger* (2021), she imagines 'the institutions we will build, and the lives we will lead when we affirm that every person *is* a person. Imagine the world we will birth when we see no stranger' (ibid, 299).

Embracing diversity, confronting and overcoming the fear and loathing of those perceived as strangers, gives us a chance to increase not only our personal resilience, but also the resilience of the collective. The bleak alternative, of increasing social polarization and violence, is what we are living through in this present moment: what *New York Times* writer David Brooks has called 'the dark century' (Brooks 2022). I agree with Brooks that the antidote to this trend is the hard work, from the ground up (not from the top down), of cultivating a diverse democracy, in full recognition of the frailties of human nature.[5] And that's where community planning comes in.

I'm done with ideological labels such as 'radical postmodern utopia', the label I gave to my cosmopolis project 25 years ago. But I'm not done with the project itself. These are perilous days on the diversity front. My hope is that we can emerge from these days with our humanity intact. Community planning (community re-building, repair) can play a vital role. We live in a culture of violence. This culture assaults our souls with consumerism, racism, misogyny, homophobia, and more. The remedy (and I hesitate to use that word) lies well beyond the current emphasis on anti-racism and EDI workshops, as important as they are. A larger re-imagining (of the soul of planning, among other things) is needed. I will return to this task in the final Commentary.

Notes

1 A book came out of this weekend event, *Cities for Citizens* (1998), edited by John himself and one of his former PhD students, Mike Douglass, now a prof at the University of Hawaii. I contributed a chapter (Sandercock 1998).

2 Had I done that inner, reflective work then, I might not have volunteered to take on the directorship of the School of Community and Regional Planning at UBC in 2006. I was overwhelmed by the weight of responsibility involved in that role and resigned after a year, not in anger this time but in sad acknowledgement that it was too much for me.

3 Wendy Sarkissian was a highly respected social planner in Australia at this time, renowned for her innovative community engagement techniques. She often worked with artists to break the mold of bureaucratic planning that was then the norm in Australia, and she had received numerous awards for her work from the Royal Australian Planning Institute. She went on to write a useful book about creative community engagement (Sarkissian and Hurford 2010). John Forester uses this same case study of Wendy's work in his most recent book (Forester 2021). Wendy and I had been personal friends since the early 1970s.

4 This is not the place to debate the merits of undergrad versus graduate education for planners. I do now believe there is strong justification for both. But my negativity, then, came specifically out of my experience with *Australian* undergrad programs. I know I shouldn't generalize on that basis, and I'm aware that some of my own students in North America have gone on to teach in undergrad programs there and found it very rewarding.

5 Brooks talks about the 'fragility of democracy' and asks us not only to pay attention to who is trying to tear it down, but also to 'worry about who is building it up' (Brooks 2022).

References

Brooks, D. 2022. The dark century, *New York Times* Available at: https://www.nytimes.com/2022/02/17/opinion/liberalism-democracy-russia-ukraine.html, 17 Feb.

Forester, J. 2021. Ed. How spaces become places. *Place Makers Tell Their Stories* (New York: New Village Press).

Gunderson, L. H., Allen, C. R. and Holling, C. S. 2010. *Foundations of Ecological Resilience* (Washington, DC: Island Press).

Hage, G. 1998. *White Nation: Fantasies of White Supremacy in a Multicultural Society* (Sydney: Pluto press).

hooks, b. 1990. *Yearning: Race, Gender, and Cultural Politics* (Boston: Southend Press).

Kaur, V.. 2020. *See Stranger, N.: A Memoir and Manifesto of Revolutionary Love* (New York: Random House)..

Kimmerer, R. W. 2013. *Braiding Sweetgrass. Indigenous Wisdom, Scientific Knowledge, and the Teachings of Plants* (Minneapolis: Milkweed).

MMIWG 2019. Reclaiming Power and Place: the Final Report of the National Inquiry into Missing and Murdered Indigenous Women and Girls Available at: mmiwg-ffada.ca/final-report/ Ottawa (Government of Canada).

Palmer, P. 2018. *On the Brink of Everything. Grace, Gravity, and Getting Old* (Oakland: Berrett-Koehler Publishers).

Reardon, K. 2003. Ceola's vision, our blessing: the story of an evolving community-university partnership in East St. Louis, Illinois. In: Eckstein, B. and Throgmorton, J. (Eds) *Story and Sustainability: Planning, Practice, and Possibility for American Cities* (Cambridge, MA: MIT Press).

Sandercock, L. 2000. When strangers become neighbours: managing cities of difference, *Planning Theory and Practice* 1(1): 13–20.

Sandercock, l. 1998. The death of modernist planning: radical praxis for a postmodern age. In: Douglass, M. and Friedmann, J. (Eds) *Cities for Citizens* (London: Wiley).

Sandercock, L. 2002. Difference, fear and habitus: A political economy of urban fears'. In: Hillier, J. and Rooksby, E. (Eds) *Habitus: A Sense of Place* (Aldershot: Ashgate).

Sandercock, L. 2003. Dreaming the sustainable city: organizing hope, negotiating fear, mediating memory. In: Eckstein, B. and Throgmorton, J. (Eds) *Story and Sustainability: Planning, Practice, and Possibility for American Cities* (Cambridge, MA: MIT Press).

Sandercock, L. and Kliger, B. 1998a. Multiculturalism and the planning system, Part One, *The Australian Planner* 15(3), 127–132.

Sandercock, L. and Kliger, B. 1998b. Multiculturalism and the planning system, Part Two, *The Australian Planner* 15(4), 223–227.

Sandercock, L. and Attili, G., with Carr, P. and Cavers, V. 2009. *How Strangers Become Neighbours: Integrating Immigrants in Vancouver, Canada* (Heidelberg: Springer).

Sarkissian, W. and Hurford, D., with Wenman, C. 2010. *Creative Community Planning: Transformative Engagement Methods for Working at the Edge* (London: Earthscan Publications).

Wheatley, M. J. 2017. *Who Do We Choose To Be? Facing Reality, Claiming Leadership, Restoring Sanity* (Oakland, CA: Berrett-Koehler).

Young, I. M. *Justice and the Politics of Difference* (Princeton: Princeton University Press).

4

TOWARDS COSMOPOLIS

A Postmodern Utopia (1998)

> Radical postmodernism calls attention to those shared sensibilities which cross the boundaries of class, race, gender, etc, that could be fertile ground for the construction of empathy – ties that would promote recognition of common commitments, and serve as a base for solidarity and coalition.
>
> *bell hooks (1990:27)*

> What scares the opponents of Utopia, like Lyotard for example, is that in the name of such future Utopia the present in its multiple ambiguity, plurality and contradiction will be reduced to a flat grand narrative. … Yet we cannot deal with these political concerns (undemocratic, authoritarian practices) by rejecting the ethical impulse of Utopia but only by articulating the normative principles of democratic action and organization in the present. Will the postmodernists join us in the task or will they be content with singing the swan-song of normative thinking in general?
>
> *Seyla Benhabib (1992:229)*

It is the mission of this book to construct a normative cosmopolis, a postmodern Utopia inspired by the sentiments of hooks and Benhabib. I am not interested in ascribing a built form to this utopia. Rather, I want to imagine a set of organizing principles on which to 'construct' a city/region in which there is genuine connection with, and respect and space for, the cultural 'Other', and the possibility of working together on matters of common destiny, a recognition of intertwined fates. It is a dream informed by critical thinking.

Fear of 'The Other'

In the closing years of the 20th century, it seemed that a new world disorder was emerging, connected with the three dominant socio-cultural forces of the time: the age of migration, the rise of postcolonial and Indigenous Peoples, and the emergence of a range of so-called 'minorities' (women, gays, people of color) hitherto invisible/suppressed as political actors. Linked with the destabilizing effects of global economic restructuring and integration, these

DOI: 10.4324/9781003325451-7

new forces are literally changing the faces of cities and regions, which are becoming much more culturally diverse. This cultural diversity is generating processes of socio-spatial restructuring which are becoming a focal point in the managing of cities and regions, in their urban governance and planning. There would appear to be no way back to an (imagined) homogeneous urban/regional culture and politics, despite the nostalgic appeals of parties of the right with their promises of a return to an earlier time when life was 'simpler' – that is, when women knew their place was in the home and Indigenous, ethnic, immigrant, and other troublesome groups hadn't begun to make claims for their own space. Multi-ethnic, multi-racial, and multi-national populations are becoming a dominant characteristic of cities and regions across much of the globe, and this is causing a profound disturbance to the values, norms, and expectations of many people.

The multicultural city/region is perceived by many to be much more of a threat than an opportunity. The threat is many-layered. It is perceived as economic, as cultural, as religious, as psychological. It is a complicated experiencing of fear of the Other, alongside fear of losing one's job, fear of a whole way of life being eroded, fear of change itself. These fears are producing a new world disorder – rising levels of violence against those who are different (who therefore don't belong), rising levels of racism, and increasingly repressive responses to the claims of insurgent citizenship. From the growing political strength of the right in France, to the United States' and Germany's recent repressive legislation against immigrants, to the bloodshed in West Kalimantan between Indigenous Dayaks and Javanese immigrants, or the attempt by ultra-orthodox Jews to drive secular Jews out of their neighborhoods and out of Jerusalem, these global processes of cultural differentiation and transformation and their spatial expression as conflicts over who can live where, and with what social and economic and cultural rights, is emerging as one of the most profound challenges of the next century.

In the next section of this chapter, I will outline a postmodern utopia which I will call *Cosmopolis*, a construction site of the mind, a city/region in which there is genuine connection with, and respect and space for, the cultural Other and the possibility of working together on matters of common destiny, a recognition of intertwined fates. I will outline the principles of Cosmopolis as an idea/l, rethinking concepts of social justice, citizenship, community, and the public interest, and suggest some necessary principles of urban governance and planning which would help create the space of/for cosmopolis.

Dreaming Cosmopolis: A Construction Site of the Mind

One way of describing my cosmopolis – my postmodern Utopia, which can never be realized but can only ever be in the making – would be by describing what does not yet exist in cities currently experiencing these major socio-cultural transformations. In none of these cities is there a common civic culture which has embraced the social project of tolerance, alterity, and inclusion.

'From Perikles' Athens to David's Paris, the word "civic" has implied an intertwined fate, a crossing of fortunes. It was inconceivable to a Periklean Greek that his or her fortune could be separated from the fortunes of a city' (Sennett 1994:369). In this era of globalization, cities are becoming more important as political-administrative units, in conjunction with their shifting demographic and economic roles as conduits and receptors of global movements of peoples as well as economic transactions. We need to start understanding our cities as bearers of our intertwined fates. We need to formulate within our city a shared notion of a common destiny.

We need to see our city as the locus of citizenship, and to recognize multiple levels of citizenship as well as multiple levels of common destiny, from the city to the nation to transnational citizenship possibilities. We need to see our city and its multiple communities as spaces where we connect with the cultural other who is now our neighbor. If the city is a coming together, a being together of strangers (Young 1990), what holds it together? How does a city with a plurality of cultures, a highly differentiated city, hold itself together?

Social Justice and Postmodern Cities and Regions

There is a prevailing view of justice in the global economy of the late 20th century which holds that just deserts are best arrived at through market behaviors, that a just distribution is whatever the market dictates, and that a just organization of social life, of urban investments, and of resource allocations is best arrived at through the market. While this idea of justice has been around since at least the 18th century, it has enjoyed a revival in the past 20 years of global economic restructuring, particularly in the United States and Britain, and has been boosted by the collapse of the centrally planned economies. But, as has long been observed (most famously by Marx and Engels in the 19th century and Lewis Mumford in the 20th), the urban and environmental consequences of market capitalism have been disastrous, as has the inequality in the distribution of wealth and income. Since the birth of advocacy planning in the 1960s and the rise of an urban political economy critique in the 1970s, the language of distributive justice has infiltrated planning debate, focusing on who is getting what, and where, in terms of access to and distribution of urban goods and services. But these progressive modernist planning concerns, perhaps best expressed in what has come to be called equity planning, have been very much preoccupied with material outcomes, with a concept of territorial justice as the redistribution of urban goods and services under the umbrella of a more general class-based politics (Harvey 1973; Krumholz and Forester 1990).

What is missing from this modernist approach is a broader definition of justice or inequality that includes, but is not limited to, the economic realm. Feminists began to critique the modernist paradigm of social justice and the city in the 1970s, arguing first that the needs of women in the city were different from those of most men, and had been ignored by progressive planners; second, that cities themselves are gendered spaces that embody a patriarchal society which ascribes certain roles and thereby spaces to women; and third, that 'woman' is not a unifying category, that the experiences and challenges of women of color, lesbians, elderly women, poor women, and physically challenged women required a different analysis and a different set of policy solutions. Scholars of color also began to argue that cities and city planning had systematically excluded them from some, and segregated them into other, zones of cities and regions and then ignored them in policy formulations addressing urban problems. These arguments lead to another set of critiques that require us to move beyond debates about justice which are solely focused on resource allocation.

The most eloquent formulation of a postmodern theory of justice is that of political philosopher Iris Marion Young in *Justice and the Politics of Difference* (1990). Young proposes an enabling conception of justice which refers not only to distribution but also to the institutional conditions necessary for the development and exercise of individual capacities and collective communication and cooperation. Injustice then refers to two main forms of disabling constraints: oppression and domination. These constraints include but go beyond distributive patterns to include such matters as decision-making procedures, division of labor, and culture

(Young 1990, 39). For most contemporary emancipatory social movements, Young notes, oppression is a central category of analysis. She suggests that 'oppression is in fact a family of concepts and conditions' which she divides into five categories: exploitation, marginalization, powerlessness, cultural imperialism, and violence. And she insists that oppression is a condition of social groups. Part of what distinguishes Young's definition of injustice is an insistence that many different groups are oppressed in today's world, and that *no single form of oppression can be assigned causal or moral primacy* (my emphasis).

While all of Young's 'faces of oppression' are pertinent to the context of social polarization and spatial differentiation existing in many cities, the categories of cultural imperialism and violence are particularly salient to what I have identified as the 'new world disorder' – rising levels of fear and hostility towards those who are culturally Other. By 'cultural imperialism', Young refers to a social process in which the dominant group's experience and culture is universalized as the norm, and those groups who deviate from it are marked as inferior, deviant, lacking. The resulting oppression is paradoxical in that the culturally dominated groups are both marked by stereotypes of deviance and inferiority and rendered invisible. These same culturally oppressed groups also suffer the related oppression of systemic violence, living in fear of random attacks on their person and/or property because they are deemed not to fit in, not to belong, to have no right to be in certain places at certain times. This group-directed violence is systemic in so far as institutions and social practices encourage, tolerate, or turn a blind eye to its perpetration and perpetrators. Addressing these faces of oppression is clearly part of the constitution of cosmopolis.

A Politics of Difference

If cultural imperialism and systemic violence are features of contemporary urban and regional changes, then a politics of difference is a prerequisite for confronting these oppressions. A politics of difference is a politics based on the identity, needs, and rights of specific groups who are victims of any of the faces of oppression discussed above. The emergence of numerous social movements in the past two decades – feminism, gay liberation, Black power, Indigenous rights – embodies the practice of a politics of difference, an important aspect of which is discursive politics. By that I mean the effort to reclaim and revalorize the meaning of difference by asserting the positive qualities of the particular group and refusing to accept the dominant culture's definition of itself. In asserting gay pride, or Black is beautiful, there is a reversal of the devaluation of difference and an effort to overcome the internal colonization of selves by the dominant culture's definition.

The emergence of this identity politics has not gone unchallenged, by either left or right, and a number of concerns do need to be aired. One is the assumption of group unity and homogeneity that often seems to be implied or stated as essential to identity politics. Clearly none of the social movements which have asserted a positive group identity is in fact a unity. This has been most systematically discussed in the women's movement, where the importance of attending to differences among women has now been well established (Spelman 1988; Anzaldua and Moraga 1981). Another problem with a politics of difference is the alleged impossibility of working on broader agendas of social, economic, and environmental justice so long as oppressed groups insist on fighting only for their group's concerns. But while this may have been a reasonably accurate interpretation of identity politics when it first emerged, it is evident in the late 1990s that many of these social movements are now engaged in broader coalition politics to achieve precisely the broader issues of social justice which are seen to transcend group interest (hooks 1990).

Further difficulties with identity politics may arise from the increasing 'hybridity' of global populations, the 'in-between character of increasing numbers of people' (Bhabha 1994): for example, a Turk who has been living in Australia for 30 years and still identifies with Turkey until she makes a trip 'home', only to discover that she no longer feels 'at home' there; or the third generation 'Asian origin' British person. The point is that a politics of difference is not based in essentialist notions of identity but in situations, historical contexts, in which there are social relations of domination. Hybrid identities are just as vulnerable to stereotyping, vilification, and exclusion. A more serious charge against identity politics is that it, too, is oppressive, both to those inside the group in question (demanding allegiance to specific meanings, norms and behaviors) and to those outside (who may be treated with contempt, or indifference, or exclusion). Anyone who has been involved in social movement politics has experienced these practices, but so too have they experienced the internal struggles against them, as social groups strive to embody a more democratic politics.

How might a politics of difference best be respected in policy formulation and decision making? The answers provided in case studies described elsewhere (Sandercock 1998, chapter 6) are that it's important to develop group-conscious social and environmental policies, and that this is not likely to happen unless oppressed social groups are represented in decision-making forums. Thus, during Harold Washington's tenure as mayor of Chicago, meetings simply did not proceed unless representatives of all those likely to be affected by the discussions were seated at the table (Mier 1993). One of Daniel Cohn-Bendit's achievements as multicultural advisor in Frankfurt was to create a forum through which immigrants could have access to and input into the municipal budget process: likewise in Porto Alegre, with the incorporation of the poorest neighborhoods into the municipal budget process (Abers 1998). Because, in the past, group-conscious policies were used to separate those defined as different and exclude them from access to the rights and privileges enjoyed by dominant groups, a crucial principle of cultural pluralism is that group-specific rights and policies must stand together with general civic and political rights of participation and inclusion. This implies a dual system of rights, which legal scholar Kenneth Karst expresses thus:

> When the promise of equal citizenship is fulfilled, the paths to belonging are opened in two directions for members of cultural minorities. As full members of the larger society, they have the option to participate to whatever degree they choose. They may also look inward, seeking solidarity with their cultural group, without being penalized for that choice.
>
> *(Karst 1986, 337)*

In other words, in order to be different, we must first acknowledge some sense of belonging to a broader polity and society. A politics of difference, in the end, must be able to take on board some (redefined) notion of the public good in that society. This does not necessarily mean the return of the old concept of 'the public interest', but it does demand the creation of a civic culture from among the interactions of multiple publics. But what are the 'paths to belonging' on the road to cosmopolis? We now need to revisit the contested terrains of 'citizenship' and 'community'.

Citizenship: Insurgent, Social, Multiple

Debates about the meaning of citizenship have become important for urban policy in the context of an increasing presence of immigrants in global cities, drawn there by the magnet of economic opportunity or driven from their homeland by the lack thereof, or by various forms of oppression and marginalization. The needs of capital for these forms of labor both create and exacerbate the problem of 'outsiders' dwelling in a territory from which they are conceptually excluded as citizens, and often actively harassed and oppressed, subjected to violent attacks from local citizens. In many parts of the world now, those loosely defined as 'guest workers' have become a permanent presence and survive as blatantly ambiguous groups, excluded from involvement in the places where they live and yet no longer 'at home' in their homelands. This is just one of the reasons for 'the return of the citizen' (Kymlicka and Norman 1994) to the center of political debate today, both in the daily columns of newspapers and in the pages of the books and journals of political philosophers and social theorists (Turner 1993; Karst 1986; Horne 1994; Holston 1995).

> In a world where nationalism is again a threat in Europe, and in an Australia where one of the greatest challenges has been an increase in diversity, we should no longer look for a national identity ... we should seek a civic identity ... We should become an Australia of good citizens.
>
> *(Horne 1994, 32)*

Australia provides an interesting case of an evolving, fluid, and contested notion of citizenship. When its six states joined the federation in 1901, the nation-building exercise rendered Aboriginal people invisible and without citizens' rights, and simultaneously the federal parliament introduced the infamous White Australia Policy, restricting immigration to people of Caucasian ancestry. Neither of these exclusionary founding acts were overthrown until the 1960s. Until then, immigrants were expected to assimilate, and Aboriginal people were subjected to all sorts of formal and informal exclusions (for example, official exclusion from the metropolitan area of Perth, state capital of Western Australia, until the 1950s). Since the 1970s, the White Australia Policy has been eroded in top-down changes of policy, and Aboriginal people have acquired citizenship status, although that has not translated into anything like formal economic and social equality.

While Australia has been evolving as a self-consciously multicultural society since the 1970s (Castles et al., 1992), social and urban policies to give meaning and substance to the official declarations have been slower in coming, and economic rationalism (neo-liberalism), which has severely curtailed public spending, threatens to further inhibit the development of such policies. As the number of immigrants from Asian countries has increased through the 1980s and 1990s there has been, on the one hand, an attempt to open up debate about the meaning of citizenship, and on the other, a backlash, led in 1996 by an independent Member of Parliament, Pauline Hanson, which argues that Australia is being 'swamped by Asians' and that Aboriginal people are being unfairly showered with government handouts. Attempts by the federal Labor Government in 1995 to open up debate on the broader meanings of citizenship, via a Senate Committee inquiry into the possibility of National Citizenship Indicators, has simultaneously reopened the Pandora's Box of always existing racism and xenophobia, indicating that the meanings of citizenship, and attempts to expand the definition, are always contested.

The notion of a contested and *insurgent citizenship* has been elaborated by cultural anthropologist James Holston with specific reference to urban development, migrations, and urban and economic policies. Notions of formal citizenship, he suggests, have become problematic in the context of the massive urban migrations of recent decades. 'As new and more complex kinds of ethnic diversity dominate cities, the very notion of shared community becomes increasingly exhausted' (Holston 1995, 45). He asks, rhetorically, what now constitutes that 'direct sense of community membership based on loyalty to a civilization which is a common possession?' Citizenship changes as new members emerge to advance their claims and as new forms of segregation and violence counter these claims. It is in both grassroots mobilizations and in everyday practices that struggles occur over what it means to be a member of the modern state, and the sites where such struggles occur are what Holston has called 'the spaces of insurgent citizenship':

> These sites vary with time and place: today, in many cities they include the realms of the homeless, networks of migration, Queer nation, auto-constructed peripheries, ganglands, fortified condominiums, employee-owned factories, squatter settlements, suburban migrant labor camps, sweat shops, and the zones of the so-called new racism. They are sites of insurgence because they introduce into the city new identities and practices which disturb established histories.
>
> *(Holston 1995, 44)*

Holston challenges planners and planning to work with these insurgent forms of the social, to seek out emergent sources of citizenship and their repression. 'If modernist planning relies on and builds up the state, then its necessary counter-agent is a mode of planning which addresses the foundations of insurgent citizenship' (Holston 1995, 46). Such a planning would note the contradictions between formal and substantive citizenship and would work on behalf of the expansion of citizenship rights. One possible direction, he suggests, is that of multiple citizenships, based on the local, regional, and transnational affiliations which form the palimpsest of contemporary urban experience.

This insurgence of new citizenship is evident in the new social movements' struggles on behalf of the poor and of women for 'rights to the city', and of gays and ethnic and racial minorities for 'rights to difference'. As these social movements expand citizenship to new social bases, they also create new forms of citizen rights and new forms of collective and social empowerment. This insurgence then is pivotal to any project of rethinking the social in planning, because it reveals the realm of the possible that is rooted in the heterogeneity of lived experience rather than in utopian futures.

Holston's thoughts on insurgent citizenship conclude with a cautionary warning about the dangers of excessive attention to the local at the expense of the (modernist) project of working through the (nation) state. For one thing, the state has and controls resources that may be essential to any locality-based work. And that includes the ability of the state 'to define, and occasionally impose, a more encompassing conception of right than is sometimes possible to find at the local level' (Holston 1995, 49). It is to these paradoxes of 'community' and community action, and to a more encompassing conception of right, that we now turn.

The Idea/l of Community

One of the principal utopias invoked by communitarians and postmodernists is 'community', with little specification of what form it might take. For communitarians, it is usually traditional (families) or territorial (neighborhoods); for postmodernists, it is an identity-based group. Evading such partly normative, partly positive questions inadvertently perpetuates the honorific status of 'community' in popular ideology as a warm and secure alternative to the anomie of modern society. Slides between descriptive and normative uses of 'community' tend to conceal the divisions of the former behind the harmony and consensus of the latter (Storper and Sayer 1997).

Ever since the sociologist Ferdinand Tonnies introduced the concepts of *gemeinschaft* (community: based on mutual aid and trust) and *gesellschaft* (society: based on individual self-interest) into our thinking in 1877, many discussions of the ills of modern society have been dominated by a narrative of loss. The loss of community is perhaps the most common of these narratives, and struggles for the recovery of it, or in defense of 'community', are seen as worthy pursuits. In planning, we write about, teach, and practice community development, community mobilizing, community building, and community-based planning (as well as, or sometimes opposed to, the top-down variety). The very idea of 'working with communities' has acquired a semi-sanctified aura which seems to have the capacity both to inspire and to polarize any group of planners or planning students, with battle lines being drawn between those who want to align themselves with the grassroots struggle and those who want to practice the kind of 'technocratic planning' alleged to reside necessarily in state agencies.

The problem with such narratives and discussions is their uncritical tendency to associate 'community' with some golden age and to define it uncritically as the sole repository of virtue – of meaningful social relations, of human scale, of wholesome (non-market) values – vis-à-vis the (always oppressive) state or the (always rapacious) market. Or, in the case of community as identity group, to seek refuge in the known and familiar, to draw tight boundaries around it, and to pull up the drawbridge. In light of the processes of globalization already discussed – processes which are re-mapping social relations and giving rise to unprecedentedly complex senses of place and belonging (see Chapter 3) – earlier meanings of 'community' begin to seem naïve, if not dangerous. 'Community', the apparently unassailable icon of so much radical and utopian thinking, now needs to be carefully scrutinized.

One starting point could be the observation that 'community', the meaning of which tends to be taken for granted, actually means different things in different contexts, to different people. In contemporary America, there is a tendency to think of it as being bottom-up, anti-corporate, in defense of life space; involving the sharing of a specific heritage or common culture and set of norms; embodying face-to-face relations and a more participatory politics. But community can equally mean 'my region or culture against yours'; my White Australian/British/American culture against your Turkish or Pakistani or Nigerian culture; or my true/authentic/local/working-class neighborhood against your inauthentic/cosmopolitan/gentrified/aestheticized middle-class one (Storper 1998). And there is a long tradition of community- and/or territorially-based movements which have been primarily inspired by exclusionary sentiments, whether racist, or religion- or class-based.

Richard Sennett (1970) has discussed how a 'myth of community' operates perpetually in American society to produce and implicitly legitimize racist and classist behavior and policy. In many cities and neighborhoods, people have (or long for) an image of their locality as one in which people all know one another, have the same values and lifestyle, and relate with feelings of mutuality, trust, and love. In modern American/Australian/British society, such an image is almost always false. There may well be a dominant group with a distinct set of values and lifestyle. But within any territorially based community, there are always those who are regarded as not belonging, as inferior, deviant, or threatening – whether it's the poor who live on the 'other side of the tracks', the Aborigines who dwell on the fringes, the Bengalis or Vietnamese or Algerians who live in the public housing, or the gays who live in the night, in the shadows. The myth, or narrative, of community, the story that describes who was here first, who *really* belongs, operates to produce defensive exclusionary behavior which, subtly or not so subtly, pressures the 'Others' to either leave or make themselves as invisible as possible, to make no claims on public space, to 'know their place', and to know that if they transgress that place, they may be the target of violence.

I am not suggesting that the exclusionary consequences of valuing community are restricted to conservatives/rednecks. Over the past two decades, numerous socialist and feminist organizations, for example, have come apart because they were operating on definitions of 'their community' which were shown, by others also claiming to share socialist and feminist goals, to be exclusionary along class/race/sexual preference lines. And community-based progressive urban politics has frequently come apart because it has not been able to relate to or develop alliances with a broader city or regional context.

This particular critique of 'community' is related to that of Richard Sennett in *The Uses of Disorder*. Sennett identifies 'a myth of a purified community' which is bred out of the way us humans learn at a certain point in our growth how to lie to ourselves in order to avoid new experiences that might force us to endure the pain of experiencing the unexpected, the new, the 'Otherness' around them. From this observation about individual psychology, Sennett makes an intriguing leap to the planning profession as a collectivity, arguing that this past century of city building has proceeded out of this very same craving for ordering, for creating a unified and comprehensible whole from the disparate and messy parts, for eliminating uncertainty and disorder. 'Social history is replaced by the passive product of social planning. Buried in this hunger for planning along machine-like lines is the desire to avoid pain, to create a transcendent order of living that is immune to the variety, and so to the inevitable conflict, between men' (Sennett 1970, 92). What has crippled the very act of planning for large cities is that there has been no provision for the fact of history, for the unintended, for the contradictory, for the unknown, or even for the complexity of the 'ethnographic present' (Holston 1995). The lesson the planning profession needs to learn, Sennett argues, is that it must take responsibility for its acts in a historical, unpredictable society rather than in a dreamworld of harmony and predetermined order. 'Instead of planning for some abstract urban whole, planners are going to have to work for the concrete parts of the city, the different classes, ethnic groups, and races it contains' (Sennett 1970, 102).

This narrative of loss and nostalgia and the urge for fusion and harmony with like-minded others is not confined to the cautious mindset of the Cold War era about which Sennett was writing in 1970. It seems to be the dominant motive and driving force behind the new urbanism movement which blossomed in the United States in the 1990s. The New Urbanism, which has a charter and a paid-up membership and calls itself a movement, is unashamedly neo-traditionalist in its approach to good city form – with its return to Main Street, USA: front porches where people sit and chat with passersby/neighbors; cars separated from pedestrians; and a precise formula for the appropriate (walking distance) proximity of shops, schools, and other facilities. The New Urbanism is an urban design–driven solution to what its advocates see as the two key problems of modern city life: alienation (by which they mean loss of community) and ecological sustainability. It seeks to create instant community through design (the old physical determinism fallacy). But it conspicuously lacks any class, race, or gender analysis of its own assumptions. Its most famous built example, the suburb of Seaside, Florida, is a homogeneous White middle-class community which would typically take a two-income household to make it affordable, leaving little time for anyone to sit on the front porch chewing the fat with neighbors.

There is, however, more than one 'ideal of community', and Young's work is misleading in only presenting and critiquing the well-known White male communitarian philosophers. In the writings of other scholars, (see Chapter 3), an entirely different concept of community emerges, the notion of 'communities of resistance' (hooks and West 1991). Cornell West explains it thus:

> It is important to note the degree to which Black people in particular, and progressive people in general, are alienated and estranged from communities that would sustain and support us ... We confront regularly the question: 'Where can I find a sense of home?' That sense of home can only be found in our construction of those communities of resistance bell (hooks) talks about and the solidarity we can experience within them ... As we go forward as Black progressives, we must remember that community is not about homogeneity. Homogeneity is dogmatic imposition, pushing your way of life, your way of doing things on to somebody else. That is not what we mean by community ... That sense of home that we are talking about and searching for is a place where we can find compassion, recognition of difference, of the importance of diversity.
>
> *(West, in hooks and West 1991, 18)*

What distinguishes these communities of resistance, then, is their refusal of homogeneity, and their quest for a coalition politics.

> We've pretty much come to the end of a time when you can have a space that is 'yours only' – just for people you want to be there ... we've finished with that kind of isolating. There is no hiding place. There is nowhere you can go and only be with people who are just like you. It's over. Give it up.
>
> *(Reagon 1981, 357)*

While there clearly are still such hiding places, as gated communities attest, Reagon's sentiments must be read as part of a utopian postmodern politics, a politics which is emergent, even as it co-exists with its opposite. A rejection of the ideal of the homogeneous community leads us into an investigation of the idea of *multiple publics*, together constituting some form of civic culture, as a basis for the survival of a culturally pluralist form of cities and regions. But before we discuss that transition, there are two further points about 'community', two more compelling reasons for caution when it comes to privileging or romancing 'community'.

Some radical planners have made a point of privileging 'working with communities' as opposed to working for the state, and to bestow on the former a badge of moral superiority and on the latter the badge of 'sell-out'. This polarization among progressive planners is unhelpful and misguided. While it is true historically that most initiatives for social change have originated with grassroots/community mobilization, it is equally clear that those radical challenges cannot be sustained unless they take on the state (local, regional, or national). This is partly because the state is usually the locus of resources, so if the point of radical mobilization is resource redistribution, the mobilization needs to bring about changes in the way the state distributes resources. It is also because laws are still made and changed through state institutions and enduring changes need to be inscribed in the law of the land. And finally, however just and virtuous a community mobilization may be, there is usually a larger picture – a metropolitan or regional scale, a wider society, or polity – which needs to be taken into account.

This is why the Workers Party administration in Porto Alegre insists on 'objective criteria' at city-wide level for budget allocation, alongside the 'subjective' criteria submitted by the mobilized neighborhoods (Abers 1998). Local enactments of democracy often produce anti-democratic results. Sometimes it is only at the level of the city/region, or even nation, that a more encompassing conception of justice can be found and implemented. The lesson of this paradox is that postmodern planners must hold in unresolved, and unresolvable, tension the transformative *and* repressive powers of state-directed planning practices *and* their mirror image, the transformative and also repressive potential of the local, the grassroots, 'the community'.

From 'The Public Interest' to a Civic Culture

If social justice in this postmodern age requires a politics of difference, what happens to the modernist notion of 'the public interest', which has been foundational to planning theory and practice? Is the new cultural politics of difference a self-contained goal, or is it a means to a truly public life, to a revival and redefinition of the public sphere? Or are the pragmatic implications of the rise of civil society, the efflorescence of mobilized groups, likely to be 'a kind of struggle of all against all others, paradoxically facilitated by the relativist celebration of difference as a value in and of itself' (Storper 1998)?

It's a familiar argumentative ploy by anyone defending an existing set of values to suggest that if we let go of these cherished concepts, then the whole fabric of civilization as we know it will unravel. But social critics ever since Marx and Engels have been deconstructing this notion of the public interest, arguing and demonstrating that when this phrase is invoked, it is always some particular group's interest which is being served. In the Marxian exposé, 'the public interest' was always a mask for the interests of the bourgeoisie. In more recent times, feminist political theorists and philosophers have questioned the meanings of 'public' and 'private' and demonstrated that these terms, as employed in public debate, have always been implicitly gendered, with the public sphere represented as a masculine domain from which women and a range of other 'unfit' groups have been excluded (Pateman 1989).

The historic notion of 'the public interest' comes from a frame of reference in liberal political theory in which disinterested experts, working within the institutions of the modern nation state, objectively and rationally analyze a problem and arrive at a solution that is in 'the public interest'. This assumes the ability of a certain chosen, well-educated group to stand outside social processes and decide what is best for everyone else. Economic and social development theories have been based on this unitary notion, which itself is linked to assumptions underpinning the modern nation state and modern nation-building: the myth of unity and sameness, of homogeneity. Class, gender, and race-based critiques have left this particular notion in tatters, as have the lived realities of late-20th-century existence. A defining characteristic of this era is a multiplicity of cultural communities and social groups dwelling in any one city/region, often alongside and yet not connecting with each other, sometimes incapable or barely capable of peaceful co-existence. A precondition for cosmopolis – a city/society which nurtures this cultural diversity – is a politics of inclusion which is grounded in an understanding that there are *multiple publics*, or to use Iris Young's term, a *heterogeneous public*, 'in which persons stand forth with their differences acknowledged and respected, though perhaps not completely understood, by others' (Young 1990, 119).

One of the themes of the new social movements of the late 20th century has been the demand for recognition of cultural differences, and for the space and the right to let such differences flourish. This signifies the end of a modernist age in which such cultural and group-based differences were meant to dissolve as all groups were assimilated into the homogeneous political community of the modern nation state. What the new cultural politics of difference signifies is that the modernist norm of a homogeneous public has become unacceptable and untenable. It requires that people transform their sense of identity in order to assimilate. Self-annihilation, as Young (1990) argues, is an unreasonable and unjustifiable condition of belonging, of citizenship.

But what does this ideal of multiple publics mean, in practice, for politics and planning in cosmopolis? Does group recognition inevitably lead to a flourishing of group-based conflicts, to the disintegration of the 'body politic', as some commentators fear? To argue that this is an inevitable outcome is to believe that conflicts between groups are caused by group differences *per se*, rather than by the relations of domination and oppression which have historically existed between groups. To give meaning to the idea of a heterogeneous public requires a more participatory democracy grounded in the recognition of difference and therefore of formal group representation. A culturally pluralist democracy needs to practice a politics of inclusion by providing mechanisms for the effective recognition and representation of the distinct voices and perspectives of oppressed and disadvantaged groups. Such group representation implies institutional mechanisms and public resources supporting the self-organization of groups so that they achieve collective empowerment and a reflective understanding of their collective experience and interests *in the context of the wider society*. It requires group analysis and group generation of policy proposals and the onus on decision makers to show that their deliberations have taken group perspectives into consideration. And it requires mandatory negotiation over specific policies that affect a group directly, such as land rights for Indigenous Nations or reproductive rights for women (Young 1990, 184).

Historically, policy issues have typically been defined by the assumptions and priorities of the privileged. Specific representation for oppressed groups disrupts this process, because it gives voice to the assumptions and priorities of other groups. Once this disruption of 'other voices' occurs, the dynamics of policy making change, and are potentially enriched in two ways. First, privileged groups do not always understand (or care about) the needs and interests of oppressed groups. Group representation enables those groups to articulate and publicize their own needs. Second, in the process of doing so, in a democratic forum, all participants must express their needs and interests in terms that are persuasive because they appeal to *an overarching principle of justice*. A condition of a heterogeneous public must be that people, as members of groups, call one another to account. This may be painful. And it takes special skills of an interpersonal kind to conduct such discussions, remembering historic conflicts, without becoming embroiled in them all over again.

Above all, explicit in such discussions must be some agreed conceptions of justice, and some sense and recognition of a wider society in which these multiple publics are located. This is not to reinvent a discredited 'public interest', but rather to stress the need for some kind of *civic culture* which embraces a sense of shared/common destiny, an intertwined fate. A precondition for such a civic culture in multicultural cities and regions is the acceptance by all of the right to dignity, however different one might be, which is grounded in a basic humanistic respect for others, for all others, in an enlarged human community. In other words, a heterogeneous public is still a *public*, where participants discuss together the issues before them and come to a decision according to principles of justice which they must also come to some agreement on. In theory, we can imagine the possibility of a 'togetherness in difference'. Different groups might be understood as potentially sharing some attributes, experiences, and goals with other groups. Such an ideal requires, as a basis, that groups participating in a politics of difference understand themselves to be participating in the same society, as part of a single polity, albeit a polity which fosters institutions and procedures for discussing and deciding policies that all can accept as legitimately binding, thereby creating a public in which all groups can and must communicate. This in turn demands the fostering of an everyday politics of dialogue and negotiation as the habit of political participation.

As we approach the end of the final decade of the 20th century, the single undeniable cultural hegemony is that neoliberalism is paramount, and every city, region, and nation has to realize its social ideals as best it can within the constraints of a profit-maximizing world market. In view of this, we have to find ways of collaborating with each other, without retreating to outworn political models, to assert our needs and our social ideals in the face of this global economic rationalization.

In this chapter I have embarked on a journey *towards cosmopolis*, my postmodern utopia. The 'towards' is important. While I believe that utopian thinking is essential if we want to address the economic, environmental, and cultural crisis in which we are now enmeshed, utopia itself can never be realized. It should always and only be in the making – a process. My utopia is not yours. Our utopia is a social project, dialogical and dialectical. History has shown us the rigidities and reinvented oppressions of most actually attempted utopias. I hold cosmopolis in my mind as an always unfinished and contested construction site, one characterized above all by its space for difference.

References

Abers, R. 1998. Learning democratic practice: distributing government resources through popular participation in Porto Alegre, Brazil. In: Friedmann, J. and Douglass, M. (Eds) *Cities for Citizens* (Chichester: Wiley).

Anzaldua, G. and Moraga, C. 1981. (eds) *This Bridge Called My Back* (New York: Kitchen Table, Women of Color Press).

Benhabib, S. 1992. *Situating the Self: Gender, Community and Postmodernism in Contemporary Ethics* (London: Routledge).

Castles, S., Kalantzis, M., Cope, B. and Morrissey, M. 1992. *Mistaken Identity: Multiculturalism and the Demise of Nationalism in Australia* (3rd ed) (Sydney: Pluto Press).

Bhabha, H. 1994. *The Location of Culture* (London: Routledge).

Harvey, D. 1973. *Social Justice and the City* (London: Edward Arnold).

Holston, J. 1995. Spaces of insurgent citizenship, *Planning Theory (Special Issue)* 13, 35–52.

hooks, b. 1990. *Yearning: Race, Gender, and Cultural Politics* (Boston: South End Press).

hooks, b. and C. 1991. Breaking bread: insurgent black intellectual life. In: *West* (Boston: South End Press).

Horne, D. 1994. Ideas for Australia. In: *How to Be Australian* (Clayton: Monash University Centre for Australian Studies).

Karst, K. 1986. Paths to Belonging: the constitution and cultural identity, *Carolina Law Review* 64, 303–377.

Krumholz, N. and Forester, J. 1990. *Making Equity Planning Work* (Philadelphia: Temple University Press).
Kymlicka, W. and Norman, W. 1994. Return of the Citizen: a survey of recent work on citizenship theory, *Ethics* 104(2), 352–381.
Mier, R. 1993. *Social Justice and Local Development Policy* (Newbury Park: Sage).
Pateman, C. 1989. *The Disorder of Women: Democracy, Feminism and Political Theory* (Stanford: Stanford University Press).
Reagon, B. 1981. Coalition politics: turning the century. In: Smith, B. (Ed.) *Home Girls: A Black Feminist Anthology* (New York: Kitchen) Table Women of Color Press).
Sandercock, L. 1998. *Towards Cosmopolis: planning for multicultural cities* (Chichester: John Wiley and Sons).
Sayer, A. and Storper, M. 1997. Guest Editorial Essay, *Environment and Planning D: Society and Space* 15(1), 1–17.
Sennett, R. 1970. *The Uses of Disorder* (New York: Knopf Publishing Group).
Sennett, R. 1994. *Flesh and Stone* (New York: W.W. Norton).
Spelman, E. 1988. *Inessential Woman: Problems of Exclusion in Feminist Thought* (Boston: Beacon Press).
Storper, M. 1998. Civil Society: three ways into a problem. In: Friedmann, J. and Douglass, M. (Eds) *Cities for Citizens* (Chichester: Wiley).
Turner, B. ed. 1993. *Citizenship and Social Theory* (London: Sage).
Young, I. M. 1990. *Justice and the Politics of Difference* (Princeton: Princeton University Press).

5

WHEN STRANGERS BECOME NEIGHBORS

Managing Cities of Difference (2000)

This essay sets out, indirectly, to answer the question posed at the end of Sharon Zukin's *The Cultures of Cities*: namely, whether all of us, children of various diasporas, can find a home in the city (Zukin 1996). I interpret this question as a challenge to planning theory and practice, indeed as one of the most urgent challenges facing the profession. I first ask what kind of a challenge it is, and then what methods we have or might need to invent for dealing with the problem as I understand it.

To my mind the most beguiling formulation of planning yet offered, at least in the English language literature, is Patsy Healey's notion of 'managing our co-existence in shared space' (Healey 1997, 3). Such a formulation speaks with equal clarity about environmental, transport, housing, and other conflicts, reminding us that, whether we like it or not, we do share space on the planet with others who in many ways are not like us, and we need to find ways of co-existing in these spaces, from the next-door neighbor to the street, neighborhood, city, and region. In this essay I explore the particular dilemmas of *co-existing in cities of difference*, and what it might mean to manage difference in ways that could be transformative rather than repressive.

I begin with a brief introduction to the issue of difference in the city and an acknowledgement and discussion of some of the pioneering work of the past decade which has inscribed difference onto the map of planning thought and practice. I then ask in what ways cities of difference pose a challenge to planning systems, policies, and practices. The essay then focusses on one dimension of difference, that of cultural diversity, asking what happens when 'strangers' become neighbors, when the presence of Indigenous Peoples or of immigrants from significantly different cultures begin to make their presence felt in streets and neighborhoods which had hitherto perceived themselves as relatively homogeneous, and how this becomes a problem in and for the planning system. I suggest four possible broad responses to the dilemmas of difference: the courts, the market, social movements, and political dialogue. The latter becomes the focus of the final section, in which I discuss a recent conflict in inner Sydney between Aboriginal and non-Aboriginal residents over land use in the context of existing models of collaborative planning and communicative action. I ask whether those existing models can deal with what I see as the core problem, the dark side of difference, namely fear of the Other, of the stranger, and suggest how they might come to do so.

DOI: 10.4324/9781003325451-8

Difference in the City

The argument that we live in 'cities of difference' is by now a familiar one. Geographers Ruth Fincher and Jane Jacobs published an edited collection with that title in 1998 which covered everything from gender and sexual preference to race and culture, stages in the life cycle, and intersecting differences of gender, class, and immigration in globalizing European cities (Fincher and Jacobs 1998). In the same year, I had argued, in *Towards Cosmopolis*, that three socio-cultural forces – transnational migrations, post-colonialism, and the rise of civil society – had converged to place the concept of difference on the agenda of the planning and design professions. Most presciently, the Royal Town Planning Institute had, in 1992, commissioned two British researchers to explore the sensitivity of the British planning system to the needs of Black and ethnic minorities. Their report, having surveyed over 100 local planning authorities, concluded that there was a great deal of ignorance of the existence and nature of racial/ethnic disadvantage in planning, and they recommended creating an institutional framework which would give greater priority to the issue (Krishnarayan and Thomas 1993). This work perhaps inspired, certainly prefigured, a handful of similar studies in Australia (Watson and McGillivray 1995: Sandercock and Kliger 1998a, 1998b) and Canada (Qadeer 1994, 1997; Wallace and Milroy 1999; Dale 1999; Ameyaw 2000) which have drawn attention to a number of issues: from the overall failure of planning systems to respond to the increasing cultural diversity of the city; to the ways in which the values and norms of the dominant culture are reflected in plans, planning codes and by-laws, legislation, and heritage and urban design practices; to planners' inability to analyze issues from a multicultural perspective or to design participatory processes that bring racial and ethnic groups into the planning process (Ameyaw 2000:105).

Thomas and Krishnarayan have continued to do path-breaking work in this field of cultural diversity and planning, particularly in relation to the ways in which planning is influenced by and implicated in the racialization of social processes (Thomas and Krishnarayan 1994; Thomas 1995a, 1995b, 1997). And work is now emerging from the USA (Burayidi 2000) and Europe (Khakee, Somma and Thomas 1999; Ratcliffe 1999) of both a theoretical and empirical kind, documenting the circumstances and needs of, for example, ethnic minorities, and critiquing urban policies and planning practices which either ignore these needs or actively contribute to the ongoing exclusion of minorities. The work of some Israeli scholars should also be noted as pioneering, both for drawing attention to the plight of minorities like the Bedouin and Ethiopian Jews under the geopolitics of Israeli planning, and for delineating the conflicts between the discourses (and practices) of cultural difference and of human rights (Yiftachel 1994, 1996; Dinero 2000).

Difference, then, has arrived on the agenda of the planning and design professions, and difference, in this context, takes many forms. It acknowledges that population groups, differentiated by criteria of age, gender, class, dis/ability, ethnicity, sexual preference, culture, and religion, have different claims on the city for a full life and, in particular, on the built environment. A politics of difference has emerged, or what philosopher James Tully calls a politics of cultural recognition (Tully 1995), the struggle for the right to difference, as well as the right to a voice in decisions affecting a wide range of groups, including Indigenous Peoples, migrants and refugees, women, and gays and lesbians. And each of these struggles and their associated claims have an impact on, and in, the sphere of the built environment.

From the body of literature referred to above, it is possible to distill four different ways in which culturally diverse cities and regions are a challenge to planning systems, policies, and practices, and why 'difference' comes to be seen as a problem. First, these studies have demonstrated that the values and norms of the dominant culture are usually embedded in legislative frameworks of planning – in planning by-laws and regulations. This is not, on the face of it, a surprising conclusion, since this legal framework evolved at a time when most societies were not yet multicultural and imagined themselves as more homogeneous than now. The planning system thus unreflectively expressed the norms of the culturally dominant majority. It is disturbing, however, to recognize that these values are still driving decisions which in turn are reinforced by the courts. To take a recent case in the United States: a dispute between the Navajo tribe and the US Forest Service over proposed road construction and logging led to claims by the Navajo that this development would violate their rights of religious freedom, which had been established in the American Indian Religious Freedom Act of 1978. The case ended up in the Supreme Court, where Justice Sandra Day O'Connor ruled in favor of the Forest Service. The fundamental issue in dispute was the definition of what constitutes 'religious use' of the land. O'Connor's argument was that the land in question was not used actively for ritual purposes and did not contain a specific religious site. This finding, however, 'begs the central question in planning for a multicultural society: it imposed a Judeo-Christian standard of "exercising one's religion" on Native American cultures' (Meyer and Reaves 2000, 94). What the Justice's ruling did not allow for is that not all religious practices are as bound in time and space as are the church-based belief systems of the dominant US culture.

There is an additional issue relating to legal frameworks and the right to difference. Planning's legal framework in the West has been embedded in a particular conception of democracy as majority rule, and a corresponding belief that the right to difference disappears once the majority has spoken. As one planner whom I interviewed expressed it: 'Local laws and regulations are framed for the majority of the community. If the minority can't fit in, then bad luck' (Sandercock and Kliger 1998a). For the most part, planners have not questioned the modernist paradigm of 'one law for all', in spite of precedents in feminist planning literature, which has challenged the universalism of the legal framework of planning, a framework which has generally been regarded as neutral, or unbiased, with respect to age, gender, religion, and culture. Applying a critical lens to this framework reveals, however, that it is underpinned by all sorts of implicit assumptions – about what constitutes a 'normal household' (the nuclear family); about an 'appropriate urban form' for such households (single family housing); about gender relations and their spatial expression (women occupying domestic space, men public space); about preferred forms of recreation (cricket and football); and so on. These norms conceal not only gender and sexual preference biases but cultural biases as well.

A second way in which the recognition of the right to difference presents a challenge to planning practice is that the norms and values of the dominant culture are not only embedded in the legislative framework of planning, but are also embodied in the attitudes, behavior, and practices of actual flesh-and-blood planners. Burayidi (2000a) has documented six ways in which 'cultural misunderstanding' might occur between planners, who belong for the most part to the dominant culture in the societies of Europe, North America, and Australia, and cultural minorities:

> (i) communication style (cultural differences affect the outcomes of the transactive and social learning processes in planning); (ii) attitude toward disclosure (cultural differences influence the types of information people are willing to share with planners);

> (iii) attitude toward conflict (this has implications for the role that the planner plays as mediator in community conflicts); (iv) approaches to accomplishing a task (this may affect the way in which planners and other professionals undertake teamwork in planning projects); (v) styles of decision making (different cultural groups have different decision-making procedures); and (vi) approaches to knowing (this affects the procedural approach to planning).
>
> *(Burayidi 2000a:5)*

Problems can arise not just from cultural misunderstandings such as these, however, but also from more deep-seated beliefs in the superiority of one's own culture, or the belief that all immigrants should adapt to the mores of the 'host' culture.

A third challenge revealed by the research reviewed earlier concerns about situations in which the xenophobia and/or racism within communities and neighborhoods finds its expression or outlet through the planning system in the form of a planning dispute over, say, the location of a mosque or Hindu temple, the use of a suburban house as a Buddhist community center, the conversion of an abandoned factory into a training facility for Indigenous youth, or the retailing practices of Vietnamese traders (see Sandercock and Kliger 1998a, 1998b). In such conflicts, it is not the planning system per se which is the problem. But the fact that the system becomes an outlet for the deep-seated fears, aversions, or anxieties of some residents means that this does become a problem for the planning system in terms of responding in constructive ways.

A fourth challenge arises when (Western) planners come up against cultural practices which are incommensurable with their own values. One way of interpreting Justice O'Connor's ruling, described above, would be to see it as an incommensurable issue of different perceptions of the value of land between secular mainstream American society and Native American communities.

How can planners and the planning system respond to these four challenges? How is the right to difference accommodated in our cities and neighborhoods, and how might it be? As Wallace and Milroy have noted, neither the planning literature nor experiences in planning practice make it obvious how one *ought* to deal with difference in planning (Wallace and Milroy 1999:55). I want to suggest that there are four broad kinds of response. We might consider overhauling the planning system, either by revising legislation or challenging it in the courts or appeal tribunals, testing whether it is consistent with, say, anti-discrimination legislation or espoused multicultural policies. Wallace and Milroy drew this conclusion in their study of planning issues in Toronto and Southern Ontario, arguing that rather than the 'neutral' framework currently in place, difference needs to be taken as the point of departure. However, the template for this kind of planning, as they note, has not yet been developed (Wallace and Milroy 1999:70). Of course, overhauling any legislation is a daunting task and usually only occurs as a result of powerful lobbying which often takes at least a generation. Think of feminist or civil rights reforms and how long and hard they were fought for. This suggests that unless social movements are actively working on behalf of the rights of cultural minorities in and to the built environment, this kind of change is unlikely to come about.

A second type of broad response to the needs of different cultural groups is through market mechanisms, and here we can see a certain amount of progress in most cities. Most notably, stores open up to provide specialist goods and services (Halal and Kosher butchers,

Asian and African markets for fruit and vegetables, Turkish video stores, Italian bridal wear, Greek pastries, Chinese acupuncturists and herbalists). Shopping malls and whole precincts within a city spring to life providing, say, Vietnamese hairdressers, lawyers, tax accountants, and shoe repairers for predominantly Vietnamese (or Bengali or Iranian) enclaves of the city, whose members are unwilling as yet to venture beyond their own community for these common services.

Other kinds of small businesses emerge, say in housing construction and renovation, to cater for culturally specific kinds of domestic spaces (Jewish builders who know how to do a kosher kitchen; Muslim builders who know about the orientation of bedroom and bathing spaces to Mecca, or about gender-based spatial separations, or the spatial requirements for praying). But what has this market response got to do with planning? Typically, there will be requests for signage, or for a change in regulations concerning retail practices, to allow street vendors or to allow street displays of wares. Planners can ease these neighborhood transitions by facilitating such changes, or they can be obstructive. Alternatively, they can establish mediation processes when local conflicts arise.

This market response, and planning accommodations to it, is all well and good for those who can afford to pay for such goods and services. But we only have to think of the appalling housing situations of certain cultural minorities, or of the poorest members of any immigrant group, to recognize that the market does not, and never will, address all the needs arising from and problems relating to cultural diversity. This was indeed the conclusion drawn from a recent cross-national comparative research project which investigated the socio-spatial exclusion of ethnic minorities in six European countries (Khakee, Somma and Thomas 1999). The specifics of the research inquired into the impact of urban renewal or regeneration policies on immigrant communities, and concluded that whereas 'property-led renewal' has tended to displace ethnic minorities (by demolishing or upgrading inner-city housing), state-led programs have resulted in the ghettoization of immigrants in the worst housing in the worst neighborhoods, and have failed to address associated issues of economic and political integration. The general pattern that emerges from these six national studies shows a strong concentration of immigrant minorities in the worst housing in specific districts of large urban areas, either in inner-city, high-density housing or in peripheral districts. In all six countries (Sweden, Portugal, Italy, the Netherlands, France, and Britain), discussions of immigration are framed by the conception of the 'immigrant as problem'. And, most disturbingly, each national case study notes the intensification of national and cultural forms of racism, an observation which has been reinforced by the political successes of anti-immigrant parties in Austria and Switzerland in 1999 (and in France much earlier), and the accompanying discourses, which both racialize and spatialize the immigration issue, portraying it as an invasion of national space by the Other – those with different everyday practices, which manifest themselves most dramatically in the concrete spaces of streets, shops, schools, housing, and neighborhoods.

In Australian cities in the past two decades, the arrival of increasing numbers of immigrants of Islamic, Buddhist, and Hindu faiths has led to the desire of these immigrants to build their own mosques and temples, and thus to requests for help from local planners in finding a suitable site. It is remarkable how often the sites to which these groups are directed are located on the periphery of the metropolitan area, or in industrial zones, remote from their worshippers, inaccessible by public transport, and often adjacent to incompatible land-uses such as garbage tips, or storage depots, or 'big box' retail outlets.

Thus far I have discussed responses to the right to difference through the courts/legislation and through the market. There are two other (interdependent) possibilities, which will occupy the remainder of this essay. One is the process-based response of establishing a dialogue. The other is an educational response which addresses the different set of skills which might be required if planners are to work in cross-cultural contexts. To introduce this discussion, I'll begin with a small example of conflict drawn from my own research on planning issues in culturally diverse municipalities in Melbourne. This serves as an entrance into the darker side of difference which we must then probe further – namely, fear of 'the Other'.

In one outer suburban municipality with significant numbers of recent arrivals from Vietnam, Cambodia, and China, the Buddhist community sought permission to build a temple and asked the planning staff for help in locating a site. They were directed to industrially zoned land on the periphery, where the temple subsequently built stands empty most of the time because of its inaccessibility in the daily lives of the community, particularly the elderly and transport-disadvantaged. Reacting to this disadvantage, Buddhist monks have purchased suburban houses as residences and have then converted the garage into a small temple and part of the house into a communal kitchen for preparation for feast days. Visiting one such house with a community worker as guide, I encountered a triple-life-sized statue of the Buddha next to a lotus pond in the backyard, and, indoors, 15 or so elderly Vietnamese women making dumplings for a feast day. In the garage was the shrine and place of worship, tell-tale pairs of sandals lined up at the door, and the smell of incense emanating from within. From the street there was no indication that this house was a de facto community center and temple. The local Council, however, had received a complaint from a neighbor about the 'excessive' burning of incense, which was described as air pollution and a health hazard. Having discovered the nonconforming use to which the house was being put, the Council had no choice, legally, but to order the monks to cease using the house for community and religious purposes (Sandercock and Kliger 1998a).

If we think for a moment about the deeper meaning of this story, it would seem to be not at all about a case of air pollution but rather an expression of fear and aversion: fear of change, of the changing face of the neighborhood, of the new neighbors. I want to argue that such fears are increasingly becoming constitutive elements of planning practice in cities of difference. How might planning and planners respond to such fears? In this particular case, there might have been a different outcome if the system had relied more on mediation or negotiation than on complaint mechanisms which immediately become adversarial and through which it is possible to mask darker realities such as xenophobia and racism.

In the next section, I will briefly digress to consider this issue of fear in the city as a prelude to a more detailed case study of a planning conflict generated by cultural differences between Indigenous and non-Indigenous Australians in an inner-city neighborhood of Sydney.

The Dark Side of Difference: Fear of the Other

> The category of the stranger breaks open from the inside the established categories and stereotypes of the local world (the world of the locals). Strangers do not fit into any of the neat containers that they are supposed to fit into, and therein lies an extreme

> irritation. To put it in another way, strangers are those who actually should be excluded according to the stereotypes of social order.
>
> Strangers are … a living refutation of the apparently clear borders and natural foundations through which affiliations and identities are expressed in the nation-state.
>
> The strangeness of strangers appears frightening and enticing.
>
> *(Beck 1998:125–130)*

I've written elsewhere about how the future of planning in cities of difference requires a coming to terms with the existence of fear in the city, specifically fear of 'the Other' – the stranger/outsider/foreigner. (Sandercock 2000b, 2000c). The history of planning could be rewritten as the attempt to manage fear in the city: fear of disorder, fear of disease (and those subjects/citizens thought to cause its spread), fear of women, fear of the working classes, of immigrants, of gays ('polluting the moral order'). The 'solution' has been twofold, involving both exclusion – spatial policing and segregation, keeping certain bodies out of certain areas; and moral reform – the attempt to produce certain kinds of citizens and subjectivities, good Melbournians, good Bostonians, by providing parks and playgrounds, settlement houses, and other 'civilizing' urban facilities. By framing the issue of cultural diversity in the language of '*managing* (cities of) difference', are we still practicing, as Foucauldians would have it, discipline, control, manipulation?

Contemporary planning approaches dealing explicitly with fear in the city revolve around crimes to property and threats to personal safety, and have focused either on urban design (Oscar Newman's 'defensible space') or environmental design (the 'crime prevention through environmental design' literature). Each of these tackles what could be called the hardware of crime prevention rather than the software of fear in the city. Another, more psychoanalytically oriented literature, comes at the issue of fear in a very different way, recognizing, to begin with, that individual identity is often suffused with anxiety, and that these anxieties are projected onto the figure of the stranger, whose very presence seems to challenge and undermine the known social order on which our identity is based. This projection conceals conflicts within the self, mixtures of fear and desire, as Julia Kristeva would have it (Kristeva 1991). In other words, the stranger is also within ourselves. In Kristeva's understanding: 'the foreigner lives within us: he is the hidden face of our identity, the space that wrecks our abode' (Kristeva 1991, 1). This last phrase is a very interesting choice of words: '*the space that wrecks our abode*'. In Bourdieu-ian language, we are talking here about the destruction of the *habitus*, the destruction of all that is familiar and homely, all that we have grown up with and take for granted, including the socio-spatial knowledge of our neighborhoods and, indeed, the nation as a whole.

Armed with these insights about fear, strangers, and ambivalence, I want to turn now to the possibility of a more dialogical, or therapeutic, approach to resolving conflicts arising from or distorted or magnified by this fear of difference, of Otherness. What we might take from the psychoanalytical literature is not that 'difference', and the fear/aversion it generates, is unmanageable, but rather that such fears and anxieties are not being, and cannot be, addressed by the rational bureaucratic regulatory methods that are, or have been, the bread and butter of planning. Should we then declare the whole project of 'managing cultural diversity', or 'managing difference', as fatally flawed? Or do we simply need to think about the problem in a different way?

Managing Intercultural Co-existence in Shared Space

In this final section, I want to argue that *fear in the city is not a problem to be solved*, at least not along the lines of the modernist dreams of control, order, and transparency. Rather, we face the eternal, impossible question of how 'we', all strangers to each other, can live together in the city.

The approach that would seem to make sense, given the discussion in the previous section, is what I'll call a *therapeutic* one: a dialogical approach which brings antagonistic parties together to talk through their concerns. This requires an open and communicative planning process, involving negotiation and mediation in working through a problem with those directly affected. It requires life experience, communicative skills and, in multicultural or multi-ethnic contexts, cross-cultural understanding. While models of collaborative planning and consensus-building have been discussed and practiced and analyzed in the planning literature since the mid-1980s in the USA and UK (see Susskind and Cruikshank 1987; Forester 1989; Innes 1995; Healey 1997), I have two critical differences with this literature. One is that the so-called communicative action or collaborative planning approach is a model that has assumed (following Habermas) that *rational discourse* among stakeholders is both appropriate and achievable. The other is that outcomes from this approach, even when consensus is reached on a way forward, are not necessarily (or even intended to be) *transformative*. To ground this discussion, what follows is an account of a recent cross-cultural conflict over land use in inner Sydney, Australia, and its resolution through such a therapeutic process in which the space is created for speaking the unspeakable, for talk of fear and loathing, as well as of hope and transformation.

The issue concerned the future of a factory site immediately adjacent to the residential area known as The Block in the inner Sydney neighborhood of Redfern. The Block had been a 1970s federal government initiative which had granted urban Aboriginal land rights. This area has received a lot of media and political attention in recent years as housing owned by the Aboriginal Housing Corporation deteriorated, and the area became a center of drug-dealing and drug-taking. Local opinion is now dramatically divided regarding the Aboriginal presence, with some non-Indigenous locals believing, or hoping, that the government will 'clean up' the area before the Olympic Games in 2000, while other non-Indigenous residents are firmly committed to a multi-racial neighborhood as a symbol of a wider reconciliation process in the nation.

In the 1980s, the (local) South Sydney Council had re-zoned the site for community use, which meant that when the factory closed down a decade later, the Council had to acquire the site. It then tried to rush through an approval to demolish the buildings on the site, in sympathy with the conservative White residents' faction who wanted the site to become a park with a prominent police station at its center. This group expressed strong disapproval of any use of the site for Aboriginal purposes. A second group, the Redfern Aboriginal Corporation, wanted the buildings and site used for Aboriginal economic and community purposes, including a training facility. And a third group, White residents calling themselves Redfern Residents for Reconciliation (RRR), supported the Aboriginal group and the larger issue of ongoing Aboriginal presence in the area.

After being embarrassed by resident protests, the Council backtracked and hired a social planning consultant, Wendy Sarkissian, to conduct a consultation process that would result in recommendations for a Master Plan for the 2,200 square meter site[1]. The consultant's initial scoping of the situation suggested to her that there was such hostility between the three identifiable groups of residents that any attempt at a general meeting to start the process would either meet with boycott from one or more groups or end up in violence. Her strategy therefore was to organize a series of meetings. For the first few months, separate meetings were held with each of the three 'camps'. These included small meetings in people's living rooms, larger meetings in more public settings, meetings with children, and meetings with members of the Aboriginal community, using an Indigenous architect as mediator. Eventually, after three months of dialogic preparation, a 'speak-out' was organized in which each group agreed to participate. This latter was the most risky part of the whole process, in that it was the most likely to get out of hand. The point of participation in this event was for people to say what they felt, to speak their feelings, no matter how toxic or painful it might be for others to hear. The hope, implicit in such an event, is that as well as speaking the unspeakable – that is performing a sort of cathartic function for all those carrying anger or fear or betrayal inside themselves – the words will also be heard, in their full emotion, by those whose ears and hearts have previously been closed.

At the time that the speak-out was held, three months into the process, there had deliberately been no discussion, and certainly no design drawings, of alternative uses of the site. The consultant's intention was to encourage the 'real issues' at stake to be aired, prior to any site-specific discussions. The 'real issues' ranged from resentment on the part of conservative Whites at the Aboriginal presence in 'their' neighborhood, to concerns with personal safety and children's well-being related to the presence of drug-dealing and drug-taking, to, on the part of Indigenous people, anger and sadness at 200 years of domination by 'White fellas' who even now had little understanding of their history and culture. At one point in the speak-out, the consultant herself was verbally attacked by a tearful Aboriginal woman storyteller who demanded to know how the consultant thought she could shift 200 years of racist history in a few months, with a few meetings. There is no satisfactory answer to such a profound question, only the honest answer in this case, which was the attempt to create the space, in one place, at one point in time, where perceptions might shift, where public learning might occur and some larger transformation take place.

Such a shift did in fact take place. Interestingly, before the speak-out, the consultant had been criticized by White residents sympathetic to the Indigenous desires for the site for her overly therapeutic approach, for too much talk about feelings. But clearly what had been happening during this initial period of meetings and listening was the creation of a safe space in which parties could meet and speak without fear of being dismissed, attacked, or humiliated. The speak-out would not have been possible without this preparatory work, which simultaneously involved the building of trust in the consultant and her team. The speak-out itself also had to be designed as a safe space, and this was achieved in part by ceremonializing the activities.

It was only after the speak-out (and the painstaking communicative preparation for it) that Sarkissian was able to move the process on, to enable joint group discussions and negotiations, to forge a set of principles for deciding the future use/s of the site, and finally to a set of meetings to draw up guidelines to present to the Council. One of the operating principles guiding the consultant was the determination not to force closure before there was the possibility of a genuine agreement, rather than a mere 'deal', an unsatisfactory compromise. That agreement finally came, not without pain, after nine months of talk and a budget of about A$50,000. The outcome, ten guidelines for a master plan, was a compromise, but it was also a breakthrough of sorts, in that the White conservatives backed off from their opposition to *any* Aboriginal use of the site and agreed to some training facilities.

This kind of planning work, involving dialogue and negotiation across the gulf of cultural difference, requires its practitioners to be fluent in a range of ways of knowing and communicating: from storytelling to listening to interpreting visual and body language. It would seem to be a model which is very relevant to the new complexities of nation-building and community development in multicultural societies. It is perhaps the *best* model in situations where direct face-to-face meetings are unthinkable or unmanageable due to prior histories of conflict and/or marginalization. In such cases, the use of narrative, of people telling their own stories about how they perceive the situation, becomes a potential consensus-building tool for unearthing issues unapproachable in a solely rational manner.

For most complex and highly charged public policy issues, sound expert analysis plays an important role in shaping the possibilities for agreement. So too do well-planned, well-executed, face-to-face negotiations. But when the parties involved have been at odds for generations or come from disparate cultural traditions, or where there is a history of marginalization, something more than the usual tool-kit of negotiation and mediation is needed, some 'method' which complements but also transcends the highly rational processes typical of the communicative action model. In the case just discussed, that 'something more' was the speak-out, which provided an occasion for dealing with history in highly personal, narrative, and emotional ways. There are other possible methods, using drama, for example, or other more symbolic or non-verbal means of storytelling and communicating deeply felt emotions. Indigenous people are often, with good reason, preoccupied with the unacknowledged and therefore unfinished business of the past. It is particularly important for them to be able to tell their stories. But all parties involved in planning disputes have a story, and there is growing recognition of the importance of the telling and hearing of stories in the process of conflict resolution. Narratives about the past can be vital in navigating long-standing, cross-cultural disputes.

A more democratic and culturally inclusive planning not only draws on many different ways of knowing and acting, but also has to develop a sensibility able to discern which ways are most useful in what circumstances. What has been missing from most of the collaborative planning/communicative action literature is this recognition of the need for a language and a process of emotional involvement, of embodiment, of allowing the whole person to be present in negotiations and deliberations. There are, however, two notable exceptions to this absence, two scholars to whom we can turn for guidance: Howell Baum (1997, 1999, 2000) and John Forester (1999, 2000).

Baum argues that when such language and behavior is disallowed or discouraged by planners insisting that participants be rational, or that discussions follow a logical order, they will elicit only superficial participation. 'Told to be rational, people assume they have been told not to be themselves. They may feel relieved. Planning will not require them to reveal or risk what matters' (Baum 1999, 12). Baum also suggests that it is important for planners working in emotionally charged situations not to try to suppress conflict, for to do so is to sabotage the work of grieving and healing which needs to be done as part of a process of change. Helping people to discuss their fears, he argues, is a way of seeing past them toward the future. Baum emphasizes that the planning process must create a transitional space between past and future, 'where participants can share the illusion of being apart from time. They need to imagine stepping away from past memories without feeling they have lost their identity or betrayed the objects of memory … They must be able to imagine alternative futures without feeling obliged to enact any of them' (Baum 1999, 11). This is what he calls the 'serious play' of a good deliberative planning process.

What also interests me about the philosophy underlying this therapeutic approach is the possibility of transformation, that is of something beyond a merely workable trade-off or band-aid solution. Much of the negotiation and mediation literature, argues Forester (1999:90), remains economistic, more concerned with trading and exchange than with learning, more concerned with interest-based bargaining and 'getting to yes' than with the broader public welfare. But, just as in successful therapy there is breakthrough and individual growth becomes possible, so too with a successful therapeutically oriented approach to managing our co-existence in the shared spaces of neighborhoods, cities, and regions there is the capacity for collective growth. Or, to move from the language of therapy to that of politics, there is the possibility of social transformation, of a process of public learning which results in permanent shifts in values and institutions.

John Forester's work since the mid-1990s, culminating in *The Deliberative Practitioner* (1999), is arguably the most important work to date in advancing the idea of a democratic deliberative process which can produce transformative results. He is concerned with a transformative theory of social learning in which not only our arguments change in dialogues and negotiations, but we change as well. The importance of explicitly addressing traumas and grievances, feelings of loss and betrayal, of 'not leaving the emotions at the door' when one enters a negotiation is also emphasized by Forester. While he is explicit that 'we cannot turn public deliberations into therapy sessions' (Forester 1999, 219), he is equally clear that:

> in the shadow of traumatic histories, public deliberations will have to enable ... acknowledgement and mourning of loss as well as the articulation and reinterpretation of ends. When we learn about the significant historical experiences of others and articulate our own in public settings, we may change ourselves as well as our strategies and senses of priorities. ... We need to learn about strategies, about norms, and about selves, transforming all three as we go.
>
> *(Forester 1999:218)*

It hardly needs to be said that the success of this kind of planning work depends very much on the skills and wisdom of the practitioners involved. It would take another essay to deal adequately with the kinds of preparation appropriate to this kind of work on deep-seated conflicts in cross-cultural contexts. At the very least, it involves training in negotiation and mediation, facilitation and consensus-building, organizing and working with groups of different sizes and different kinds of internal conflict. It involves some understanding of individual, group, and community psychology, as well as group and community dynamics; and some experience doing research in and about communities, with community members. Again, the works of Baum and of Forester have taken these matters furthest. Baum suggest that we might learn from anthropologists' methods. First, they tell us that getting to know another (group, or culture) takes more than a few meetings and/or a needs assessment survey. Understanding, and building trust, depends on spending time in a community, and it calls for in-depth talk and not just discussions with formal leaders. Second, they tell us that every group we encounter has a culture (as does every observer/planner), which may be thick or thin, thoroughly or only partially defining and directing actions. Learning about any culture is an empirical and time-consuming task which requires a special attitude (Baum 2000:133). Planners need to learn about culture: what it is and what shapes and maintains it, how and why it changes, and how one's own culture affects one's ability to understand that of others. Baum also recommends courses in cultural anthropology, community psychology, group process, and organizational behavior (Baum 2000, ibid.).

What, then, might we conclude about the wider significance of this Sydney case study, and of the therapeutic approach described here? One point is the cost and time involved. These are significant commitments of resources, and cannot be expected to be available for every dispute that arises. My argument is that this approach is the best model in cases where prior histories of conflict have made more traditional negotiation techniques irrelevant. A second point relates to scale and significance. The conflict in this case was ostensibly over the fate of a 2,200-square-meter site within one local municipality. The symbolic stakes, however, were considerably greater, especially for Indigenous people, whose ongoing presence in the inner city seemed to be the hidden agenda and longer-term threat. So this small, localized conflict was, in the minds of most of those involved, about core issues of history, injustice, and national identity. At whatever scale such conflicts arise – neighborhood, city, or region – there is a strong case for using this therapeutic approach.

The final point to be teased out of this Sydney case study relates to the perennial question of how change comes about. I do not want to leave the impression that planners, if only they become 'therapists' in their practice, will rid the world of racism, sexism, and other plagues. Remember that the politically conservative South Sydney Council initially attempted to rush through its preferred solution, and would have succeeded had not some of the White residents who supported the Indigenous community mobilized, confronted, and embarrassed the Council into conceding a more consultative approach. *A political space had to be created for this approach through political action.* Occasionally, planners themselves can create such a space, but in highly charged, highly politicized conflicts, it is more likely that the planner's role will be to 'design' the space which has been created through political action. In the case of Redfern, once the space was created, the skills, courage, and commitment of the consultant were what made it possible to work through to a solution which dealt with the kinds of fears and anxieties previously discussed, rather than ignoring, marginalizing, or overriding them. Herein lies my concluding observation: that the work of planners in 'managing difference' is the work of negotiating fears and anxieties, mediating memories and hopes, and facilitating change and transformation.

Note

1 The following account is based primarily on discussions with the consultant, Dr. Wendy Sarkissian, in October 1999.

References

Ameyaw, S. (2000). Appreciative planning: an approach to planning with diverse ethnic and cultural groups. In: Burayidi, M. (Ed.) *Urban Planning in a Multicultural Society* (Westport: Praeger).

Baum, H. (1997). *The Organization of Hope* (Albany: State University of New York Press).

Baum, H. S. (1999). Forgetting to plan, *Journal of Planning Education and Research* 19(1) (Fall), 2–14.

Baum, H. (2000). Culture matters - but it shouldn't matter too much. In: Burayidi, M. (Ed.) *Urban Planning in a Multicultural Society* (Westport: Praeger).

Beck, U. (1998). *Democracy Without Enemies* (Cambridge: Polity Press).

Burayidi, M. (Eds.) (2000). *Urban Planning in a Multicultural Society* (Westport: Praeger).

Burayidi, M. (2000). Urban planning as a multicultural canon. In: Burayidi, M. (Ed.) *Urban Planning in a Multicultural Society* (Westport: Praeger).

Dale, N. 1999. Cross-cultural community based planning: negotiating the future of Haida Gwaii, in L. Susskin, S. McKearnan and J. Thomas-Larmer (eds) *The Consensus Building Handbook* (Thousand Oaks, CA: Sage).

Dinero, S. (2000). Planning for and ethnic minority: the Bedouin Arabs of Israel. In: Burayidi, M. (Ed.) *Urban Planning in a Multicultural Society* (Westport: Praeger).

Fincher, R. and Jacobs, J. (Ed.) (1998). *Cities of Difference* (London: Guilford Press).

Forester, J. (1989). *Planning in the Face of Power* (Berkeley: University of California Press).

Forester, J. (1999). *The Deliberative Practitioner* (Cambridge, MA: MIT Press).

Forester, J. (2000). Multicultural planning in deed: lessons from the mediation practice of Shirley Solomon and Larry Sherman. In: Burayidi, M. (Ed.) *Urban Planning in a Multicultural Society* (Westport: Praeger).

Healey, P. (1997). *Collaborative Planning* (London: Macmillan).

Innes, J. E. (1995). Planning theory's emerging paradigm: communicative action and interactive practice, *Journal of Planning Education and Research* 14(3), 183–189.

Krishnarayan, V. and Thomas, H. (1993). *Ethnic Minorities and the Planning System* (London: Royal Town Planning Institute).

Khakee, A., Somma, P. and Thomas, H. (Eds.) (1999). *Urban Renewal, Ethnicity and Social Exclusion in Europe* (Aldershot: Ashgate).

Kristeva, J. (1991). *Strangers to Ourselves* Roudiez, L. S. (Translated by) (New York: Columbia University Press).

Meyer, P. B. and Reaves, C. R. 2000. Objectives and Values: planning for multicultural groups rather than multiple constituencies, in M. Burayidi (ed) *Urban Planning in a Multicultural Society* (Westport, CT: Prager), 83–99.

Qadeer, M. (1994). Urban planning and multiculturalism in Ontario, Canada. In: Thomas, H. and Krishnarayan, V. (Eds) *Urban Planning in a Multicultural Society* (Westport: Praeger).

Qadeer, M. A. (1997). Pluralistic planning for multicultural cities: The Canadian Practice, *Journal of the American Planning Association* 63(4), 481–494.

Ratcliffe, P. (1999). Ethnicity, socio-cultural change, and housing needs, *Journal of Planning Education and Research* 19(2), 135–143.

Sandercock, L. (2000a). Negotiating fear and desire: the future of planning in multicultural societies, Keynote Paper, Urban Futures Conference, Proceedings, July (Johannesburg: University of Wittwatersrand).

Sandercock, L. (2000b). Difference, fear, and habitus: a reflection on cities, cultures, and fear of change, Keynote Paper, Habitus Conference, Perth, September.

Sandercock, L. and Kliger, B. (1998a). Multiculturalism and the planning system, part one, *The Australian Planner* 15(3), 127–132.

Sandercock, L. and Kliger, B. (1998b). Multiculturalism and the planning system, part two, *The Australian Planner* 15(4), 223–227.

Susskind, L. and Cruikshank, J. (1987). *Breaking the Impasse* (New York: Basic Books).

Thomas, H. (1995a). Ethnic minorities and planning, *The Built and Natural Environment* 6, 4–8.

Thomas, H. (1995b). "Race", public policy and planning, *Planning Perspectives* 10(2), 123–148.

Thomas, H. (1997). Ethnic minorities and the planning system. A study revisited, *Town Planning Review* 68, 2.

Thomas, H. and Krishnarayan, V. (Eds.) (1994). *Race Equality and Planning: Policies and Procedures* (Aldershot: Avebury).

Tully, J. (1995). *Strange Multiplicity. Constitutionalism in an Age of Diversity* (Cambridge: Cambridge University Press).

Wallace, M. and Milroy, B. (1999). Intersecting claims: planning in Canada's cities. In: Fenster, T. (Ed.) *Gender, Planning and Human Rights* (London: Routledge).

Watson, S. and McGillivray, A. (1995). Planning in a multicultural environment: A challenge for the nineties. In: Troy, P. (Ed.) *Australian Cities. Issues, Strategies and Policies for Urban Australia in the 1990's* (Cambridge: Cambridge University Press).

Yiftachel, O. (1994). The dark side of modernism: planning as control of an ethnic minority. In: Watson, S. and Gibson, K. (Eds) *Postmodern Cities and Spaces* (London: Blackwell Publishing).

Yiftachel, O. (1996). The internal frontier: Territorial Control and Ethnic Relations in Israel 1, *Regional Studies* 30(5), 493–508.

6

MONGREL CITIES OF THE 21ST CENTURY

Is Multiculturalism the Solution, or the Problem? (2006)[1]

I want to talk this evening about the power, and the limitations, of an idea. That idea is multiculturalism, although I'm going to suggest another name – interculturalism - to clarify and expand on its evolving meaning. There have been three great animating ideas in the West in the past three centuries. The American revolution gave us 'life, liberty, and the pursuit of happiness' and an evolving form of liberal democracy. The French revolution gave us 'liberty, equality and fraternity' and a republican form of democracy. And the Russian revolution gave us socialism: equality, but without democracy. I want to make a very controversial argument tonight – namely that what I think of as Canada's 'quiet revolution' since the 1970s, and still evolving, has given us a fourth great animating idea for organizing society: multiculturalism. This idea is the one that is the most appropriate to the new world social order of the 21st century, a century in which the whole world will, finally, become urban, and in which most of the world's cities, certainly in the West, are what I call 'mongrel': that is, multi-ethnic, multi-racial, hybrid.

Multiculturalism is of course much more than an idea. It's a political philosophy and a social project, a way of imagining how we might manage to co-exist peaceably in the increasingly culturally diverse shared spaces of streets and neighborhoods, cities, and regions. But the concept is not only controversial, it is also not well understood, and has different meanings in different countries. So I'll clarify what the idea means at its best, how it has evolved in Canada, and why I think it is *the* idea that is most likely to get us successfully through the challenges of the 21st century and this unprecedented Age of Migration which we are experiencing.

The multicultural city/region is seen by some as more of a threat than an opportunity. The threats are perceived as multiple: psychological, economic, religious, and cultural. It's a complicated experiencing of fear of 'the other' alongside the fears of losing one's job, fear of a whole way of life being eroded, and fear of change itself. In Europe, the emergence of multi-ethnic cities shatters citizens' self-understanding of national and collective identity – what it means to be Dutch, Danish, German, even British. These fears have produced rising levels of anxiety about and violence against those who are different, who are seen as 'not one of us', as 'not my people'.

Looking back at the last two decades of the 20th century from the vantage point of the 21st, we can see the confluence of several elements: the collapse of communist states; a post-Marxist

DOI: 10.4324/9781003325451-9

appreciation of the latent energies in civil society exceeding the unity and power of command economies; an acceleration of population flows accompanying the globalization of economic life, as affluent members of a new managerial class enjoy transnational mobility while impoverished postcolonials migrate to the center of former empires; the eruption of new claims to positive identity among groups whose previous identifications along lines of race, gender, sexuality, religion, or disability were experienced as marginalizing and degrading; and the increasing porosity and uncertainty of territorial boundaries and national identities.

These gathering drives to civil society, multiculturalism, the return of the colonial repressed, transnational mobility, and pluralist tolerance in turn encounter their dialectical opposite, a series of aggressive responses that demand a return to a unified faith, race, reason, nation, gender duality or 'normal' sexuality (Connolly 1995). With a little help from the events of September 11, 2001, real and metaphorical border patrols appear to be tightening in the early years of the 21st century as this battle between the impulse to fundamentalism and the impulse to pluralization and tolerance plays itself out in our cities in complex ways. While fundamentalist terrorists attack what they see as the manifold decadence of the West, there are powerful fundamentalist forces within Western nations seeking to purify their own societies of all that is seen as Other or 'unnatural' – feminists and homosexuals, atheists and Muslims, immigrants, and multiculturalists.

But there's something else happening in the West, as well. Declining birth rates, an aging population, and depleted social security funds mean that we need the influx of able-bodied men and women to maintain a growing labor force and economy. This demographic and economic reality, combined with the culturally diverse source countries of emigration, is what ensures that the 21st century is, indisputably, the century of multicultural cities and societies, empirically speaking. It will also be the century of struggle for multiculturalism and against fundamentalisms of all kinds – that is, the belief in cultural and/or religious purity. Nobody has expressed this more eloquently than Salman Rushdie, writing in defense (and explanation) of his novel *The Satanic Verses*.

> If *The Satanic Verses* is anything, it is a migrant's eye view of the world. It is written from the very experience of uprooting, disjuncture and metamorphosis (slow or rapid, painful or pleasurable) that is the migrant condition, and from which, I believe, can be derived a metaphor for all humanity. Standing at the center of the novel is a group of characters most of whom are British Muslims ... struggling with just the sort of great problems that have arisen to surround the book, problems of hybridization and ghettoization, of reconciling the old and the new. Those who oppose the novel most vociferously today are of the opinion that intermingling with a different culture will inevitably weaken and ruin their own. I am of the opposite opinion. The Satanic Verses celebrates hybridity, impurity, intermingling, the transformation that comes of new and unexpected combinations of human beings, cultures, ideas, politics, movies, songs. It rejoices in mongrelization and fears the absolutism of the Pure. Melange, hotchpotch, a bit of this and a bit of that is how newness enters the world. It is the great possibility that mass migration gives the world ... The Satanic Verses is for change-by-fusion, change-by-conjoining. It is a love song to our mongrel selves.
>
> *(Rushdie 1992, 394)*

This Laurier Lecture is intended as a love song to our 'mongrel cities' and to the political philosophy that undergirds their thriving: namely, multiculturalism. So first I want to delve further into the multiple and contested meanings of this philosophy and dwell a little on its evolution in Canada – on its opponents as well as its supporters. Then I want to take us to one of the most culturally diverse neighborhoods in Vancouver and talk about how this neighborhood has succeeded in building an intercultural community over the past 20 years of rapid growth and immigration. Here, I think, is a model, at the micro-scale, of the lived experience of this ambitious political philosophy, a model full of relevance for addressing the anxieties that are unsettling European cities at this moment in history. Finally, I return to this animating idea and talk about it as a leap of the imagination, and about the pay-off that comes from making the leap.

An Evolving Idea and Its Paradoxical Meanings

At the beginning of the last century, Wilfred Laurier's Liberal government decided that immigration was Canada's destiny, and addressed the question of immigration and identity by telling new arrivals, 'Let them look to the past, but let them also look to the future: let them look to the land of their ancestors, but let them also look to the land of their children'. These words are as inspiring today as they were remarkable then, and they also succinctly capture the tight rope that Canadian multicultural policy has teetered along as it has sought to adjust to new waves of immigration from non-Anglo-European source countries.

Today, worldwide, there is a distinct sense of crisis in the air surrounding multiculturalism, with detractors in the UK, the Netherlands, and elsewhere arguing that it has resulted in a ghettoizing of societies into ethnic silos. While this may have been precisely the intent of the policy in the Netherlands, this was not so in Canada, although there are those who argue that this has been an unintended effect of multicultural policies. In recent years there has been a questioning not only of specific policies but also of the very idea itself as a guiding philosophy for the nation. I want to briefly outline these debates and suggest why they necessarily agitate us.

Canadian multiculturalism has served as a guideline for government policy since 1971, and also as a framework for national discourse on the construction of Canadian society: that is, what kind of country we are, and aspire to be. Initially conceived as a way of accommodating separatist impulses in Quebec, the policy has had to evolve to take on board, literally, the arrival of increasing numbers of immigrants from 'non-traditional' (non-Anglo-European) countries. This sense of an evolving multiculturalism indicates that it is a *work in progress* rather than a finalized or permanently settled philosophy. I will argue that this is a strength of Canadian multiculturalism rather than a weakness, and that the unsettled nature of the philosophy reflects a productive unsettling of the nation in the face of three decades of increasingly culturally diverse immigration.

Canadian multiculturalism has encouraged individuals to voluntarily associate with the culture and tradition of their choice, essentially declaring as a human right the right to *retain* one's cultural heritage, not only in private but also in public. There has been significant spending, through multicultural grants, to support the maintenance of various cultures and languages and to encourage diverse cultural festivals in public places, as well as the symbolic gesture of public artworks that recognize and celebrate the multiple peoples who make up the nation. This is surely a remarkable change from conventional strategies of nation-building, and many immigrants who have arrived in Canada after periods of stay in European countries comment on how much less racism there is here, how much more welcome they feel.

As remarkably harmonious as Canadian society undoubtedly is, and in no small part thanks to multicultural philosophy and policy, there are at least *five significant criticisms* of this philosophy and its associated policies. One comes from Indigenous communities who argue that their claims, which go beyond calls for cultural recognition, to demands for land and sovereignty, cannot be addressed within a multicultural political framework. A second critique is aimed at the apparent 'culturalist' assumptions that define immigrants (but rarely the host society) by ethno-cultural markers; focus on ethnicity as a primary identification; ignore differences within, not to mention the evolving nature of any culture; and allegedly encourage ethnic separatism and competition. A third criticism has been that the apparent tolerance expressed in multiculturalism has masked an ongoing and institutionalized racism in Canadian society directed at visible minorities; that the posture of equality upheld by multiculturalism is actually a veil that conceals significant economic and political inequalities; and that the gaiety of multicultural festivals is a sedative that dulls the senses to the continuing marginalization of immigrants from the economic and political mainstream of national life.

I accept all of these as valid and powerful critiques of *actually existing* multiculturalism. But they do not tell the full story of what has been achieved, nor do they exhaust what David Ley has called 'the potential reach of multiculturalism', to which I'll return. There's a fourth and even more fundamental critique that I need to mention, because it helps us to understand the necessarily paradoxical nature of multiculturalism. In both Canada and Australia, multiculturalism was introduced by the state as a way of *managing* increasingly diverse streams of immigration, immigration that was understood as essential for the continued economic development of the nation. When a nation state adopts multiculturalism as its guiding philosophy, there is an underlying concern that there's a new problem for the state, a problem that needs managing. If the age of global migrations unsettles the established order of things (notions of belonging, of identity, ways of life taken for granted, thought of as 'normal'), then the state responds either by attempting to restore the old order through repressive and exclusionary policies, or by writing new rules for shaping and managing the new order of things. The new rules re-write some old definitions of belonging and citizenship and create new, different boundaries. Multicultural legislation is an attempt to define, and perhaps to limit, the extent to which the nation will change as a result of immigration. It reflects and addresses a profound unsettling of norms, and fear of change, on the part of the host society, *at the same time as* it appears to celebrate, and perhaps genuinely desires, this change and seeks to move cautiously towards a new national identity.

So, multiculturalism and the associated legislation is at once very pragmatic in its attempt to manage a new situation, and simultaneously very idealistic in seeking to create new ways of defining the nation. Further, it is likely to be, and should always be, contested, at one extreme by those who wish to see no change to the nation as they understand it, and at the other extreme by newcomers as they come to experience exclusion in various ways. In other words, multiculturalism is not an entirely altruistic project, and the language of a virtuous tolerance in which it is couched needs to be constantly challenged by scrutinizing its actual effects in every policy field – which leads to a fifth and final criticism of actually existing multiculturalism. There is an extraordinary discrepancy between discourses of immigration and multiculturalism on the one hand and labor market practices on the other. Multicultural rhetoric celebrates the pluralist immigrant society. But the labor market valorizes all things Canadian – Canadian education credentials, Canadian work experience, and Canadian English accents – while devaluing all things and people defined as 'foreign'. The recent 2003 report of the Standing Committee on Citizenship and Immigration, *Settlement and Integration: A Sense of*

Belonging, Feeling at Home, for example, highlights the common immigrant experience of de-skilling and under-employment, concluding that the stated goals of Canada's immigration policy are being derailed by employment practices.

These five critiques reveal serious shortcomings as well as a dark side to the Canadian experience with implementing multiculturalism. Since the philosophy has been inscribed in legislation and in the Charter of Rights and Freedoms, policies across the board need to respond to new mandates. That inevitably takes time, since it requires institutional change, changes in the training of policy makers as well as service providers, and changes in everyday life: how the host society treats newcomers, in stores, on public transport, as neighbors, in a multitude of everyday mundane ways.

I want to switch now from critique to hope, exploring the idea of multiculturalism as an unfinished and evolving social and political project. What's striking to me since arriving here 5 years ago as an immigrant myself is the subtle evolution in the interpretation of the philosophy itself, clearly in response to some of the criticisms noted above. It seems to me that by stealth rather than by any well-publicized debate, Canada's approach to the integration of immigrants in the first few years of the 21st century has shifted *from a multicultural to an intercultural foundation*, and that the emphasis now is less on funding to preserve immigrants' cultural heritage and more on creating an intercultural civic culture. This does not mean that funding for the preservation of cultural heritage has ceased, but rather that there is a recognition of the longer-term importance of bridging rather than reinforcing cultural differences.

This is by far the most promising approach to managing our co-existence in the mongrel cities of the 21st century, and one that could be a model for other countries. An intercultural perspective, as a normative project, has at least two dimensions: a social imaginary of inventing and learning new ways of living together in all of our differences, and a political project capable of overcoming the weaknesses of 20th-century multiculturalism by constructing a different sense of citizenship and of belonging. I'll briefly outline each of these before moving on to some concluding thoughts.

Interculturalism as a political project recognizes *the right to difference*, expressed as the legitimacy and specific needs of minorities and marginalized groups, but also recognizes the need to contest this right against other rights – human rights, women's rights, and gay rights, for example. Where the host culture has already established such rights, accepting them must surely be a required condition of immigrants wishing to join this new political community.

Given that the lived complexities of immigrant integration occur in the city, and usually the larger cities of any nation, some scholars have argued for a notion of urban or local citizenship, understood as the practices through which individuals or groups formulate and claim new rights or struggle to expand or maintain existing rights to the city.

The right to local citizenship is the right of all residents to inhabit and appropriate public space and the right to participate as an equal in public affairs, to be engaged in debating the future of cities and neighborhoods and creating new intercultural spaces and built forms and new ways of being together. This involves nothing less than expanding the spaces of democracy through participation at the local level, and a model of democracy that implies no closure – that is, no permanent state of integration and harmony towards which we are working but an always contested engagement with, and continually redefined notion of, the common good and shared destiny of the citizens of the city. This in turn involves an understanding of politics as, (in the words of Chantal Mouffe), 'the construction of the political community in which we desire to live, and this is a permanently unfinished project'.

The cultural embeddedness of human beings seems to be inescapable, and some form of cultural identity and belonging is unavoidable. And yet we are capable of critically evaluating our own culture's beliefs and practices and of understanding and appreciating, as well criticizing, those of other cultures. We are capable of imagining and desiring cultural change (for example, anti-racism campaigns, the civil rights movement, the feminist movement). At the heart of an intercultural perspective is the belief that no culture is perfect or can be perfected, and that all cultures have something to learn from and contribute to others. So *intercultural dialogue* is a necessary component of cultural and societal growth and development. Cultures are always evolving, dynamic, and ultimately hybrid. And since diversity exists within as well as between cultures, no pure form of cultural identity is capable of being the foundation of membership in political community. This implies *the end of mainstream*, in politico-cultural terms, and the birth of plurality as the basis of political community.

A sense of belonging in an intercultural society cannot be based on race, religion, ethnicity, or any other such marker of identity/difference. Rather, that sense of belonging must be based on a *shared commitment to a political community*, and specifically to a political community that remains perpetually open to re-definition as its membership changes. But there must be agreed-on procedures for debate and for resolving conflicts, and there must be legal and institutional protections against discrimination. A shared commitment to a political community also requires an *empowered* citizenry, which in turn means addressing prevailing inequalities of political and economic power as well as developing new stories about and symbols of national and local identity and belonging.

That is a brief outline of the political components of an intercultural project. The social imaginary that must accompany this is expressed in four words, 'how strangers become neighbors': that is, discovering new ways of living together that are not merely tolerant of difference, at a distance, but involve genuine respect for and connection with other cultures, a *being in relation with*, which is only possible at the scale of the neighborhood.

To explain this, I want to talk very briefly about an exceptional local institution in Vancouver, the Collingwood Neighbourhood House (CNH).

Building a Community Where None Existed

The CNH was established 21 years ago in a neighborhood undergoing rapid population growth due to immigration from predominantly non-Anglo/European countries. By 2004, native English speakers comprised a mere 27% of the population. The initial neighborhood response to this rapid socio-cultural change was a fairly typical one: people started locking their doors and expressing sentiments such as 'what are they doing here' and 'why don't they go back where they came from'. The major catalyst in turning around these fears and anxieties was the CNH, which was founded by local community members with the mission of building a community where none had existed, creating a place for everyone, irrespective of cultural origins or other differences. The resident-run board that governs the place and shapes its programs and services has always been representative of the diversity of the community. And the programs have always been based on an intercultural philosophy. In other words, all programs and services must be open to everyone, rather than catering to specific groups defined by age, ethnicity, religion, and so on.

The CNH now offers hundreds of programs and services, with a staff of 90, 400 volunteers, and a clientele of just over 100,000 users in 2005, in a neighborhood whose population is 48,000. (In other words, residents make multiple use of the place.) What is so special about this place, to its many users as well as staff and volunteers, is that it creates a sense of welcome and of belonging, it helps people to connect, to move out of their own cultural group or demographic and get to know others. It is referred to by many users and staff as a 'blessed place'. Its model is one of co-operation (rather than competition) with other local agencies and facilities. And its philosophy is that of empowering local residents to shape their own destiny through working together and building relationships. What strikes me as so compelling about this model is that it works in apparently very prosaic ways. People get to know each other and move out of their cultural isolation by sharing/taking part in everyday activities such as language classes, child care, recreational activities, and skills-training programs. During these activities, they begin to share stories, of their immigration experience, for example, which create bonds across cultures and ages and religions. Intercultural exchanges through children, food, and the arts seem to provide a non-intimidating initiation into deeper, more long-term involvement in the community and the organization. The CNH also provides regular anti-racism education programs and teaches, through its consistent policies and actions, that community is built through inclusion rather than through drawing boundaries. All staff and volunteers undergo annual anti-racism and diversity training. The CNH story is instructive in what it takes to work towards living with diversity, beyond the American model of 'indifference to difference', towards actually building an intercultural community. It is a living, breathing example of the daily and ongoing negotiation of difference through coming together on common projects and in everyday activities of survival and the reproduction of life.

In Conclusion: Leaving the Shore – a Leap of Faith

Multiculturalism is a leap of the imagination. From its beginnings, in Canada at least, people had an idea of what a good country could be. The idea involved welcoming rather than shunning strangers and not requiring them to abandon their own cultures and lifeways in order to become a member of Canadian society. Most people couldn't describe this idea even now in more than fuzzy terms, or by associating it with food and festivals, the ethnic bazaar. There is, yes, a huge element of expediency and opportunism in the concept, going back to the 1970s. And it's always been a risk. No one has ever known for sure (or knows now) whether a country based on such an idea is really possible, but again and again we leap toward the idea, and hope for the best. And maybe making the leap is the whole point. It's a life-affirming, humanity-affirming project.

But the idea doesn't truly live on the philosopher's page. It lives when it's expressed in acts. And the country doesn't live unless we make the leap from our tribe or focus group or gated community or demographic, and land on the shaky platform of that idea of a good country which all kinds of different people share. The Hopi Elders sent out a message in 2000, as follows:

> There is a river flowing now very fast.
> It is so great and swift
> That there are those who will be afraid.
> They will try to hold on to the shore.
> They will feel they are being torn apart
> And will suffer greatly.

Know that the river has its destination.
The Elders say we must let go of the shore,
Push off into the middle of the river,
Our eyes open and our heads above the water,
And I say, see who is there with you,
And celebrate.

At this time in our history
We are to take nothing personally,
Least of all ourselves,
For the moment we do,
Our spiritual growth and journey come to a halt.

The way of the lone wolf is over.
Gather yourselves.
Banish the word struggle from your attitude and vocabulary.
All that we do now
Must be done in a sacred manner and in celebration.
We are the ones we have been waiting for.[2]

This leap of faith that I'm talking about, or the pushing away from the shore that the Hopi Elders envisage and encourage, is a leap that's made in public, and it's made for free. It's not, in the end, a product or a service. It's a way of being in the world, a way of relating. You do it for reasons unexplainable by *economics or patriotism* – the two driving ideas of our time. It's done in public spaces, face to face, where anyone is free to go. It's not done on TV or over the Internet or by phone.

The places you'll see it are on city sidewalks, in school playgrounds, on subways, at bus stops, in public parks, in parking lots, at sports clubs, in queues in stores, at sports events, in high-school gyms, in child care centers. In those places, and others like them, the leaps that continue to invent and define the intercultural nation continue to be made. When the leap fails, it looks like the destruction of the Africville community in Halifax in the 1960s. When the leap succeeds, it looks like the Collingwood Neighbourhood House.

The leap requires physical presence and physical risk: to speak to, smile at, or just even meet the gaze of the stranger, to begin a process of welcoming. To see no stranger. The risk, psychologically, is that we might all be, *will be*, changed in some way by this encounter. But the pay-off, in terms of dreams realized, of understanding, of people connecting instead of colliding, can be so glorious as to make the risks seem miniscule.

There's a magic in this idea and the acts that embody it, acts like the creation of the Collingwood Neighbourhood House. There's also a public act of courage. And those acts are out there each day along with their opposites. It's a public service to be brave.

Notes

1 This paper was given as the invited Laurier Lecture 2006, in partnership with the World Urban Forum in Vancouver that year, and was broadcast on CBC Radio's *Ideas Program*.
2 Hopi Elders Prophecy, June 8, 2000, https://www.awakin.org/v2/read/view.php?tid=702.

PART III

Expanding the Language of Planning

FIGURE 7.1 *Joyful Thoughts #36*, sculpture by Dorothy Gillespie. Gift from John on my 50th birthday. Photo, Leonie Sandercock.

Commentary: The Storytelling Project: 1986–2022

> A rhyme's
> a barrel of dynamite.
> A line is a fuse
> that's lit.
> The line smolders,
> the rhyme explodes –
> And by a stanza
> a city
> is blown to bits.
>
> (Mayakovsky 1925 in Blake 1960: 195)

I had an epiphany on the road to Wollongong in 1984. I was doing research on the social impacts of economic restructuring in the coastal steel and coal mining town of Wollongong (90 miles south of Sydney, Australia), when I realized, with a power of epistemological detonation akin to Mayakovsky's poem, that the research as formulated wasn't going anywhere. My political economy framework appeared to me as a ghostly ballet of bloodless categories (class, labor, capital). Listening, in homes and pubs and union offices, to the stories of the men who had lost their jobs, I realized that the research could only be animated through the telling of their stories. I changed the research plan, hired a research assistant to help with in-depth interviews, read a book of poems by a Wollongong lad who told obliquely of the ordeals of some of the retrenched miners and steelworkers. But still, two years into the research grant, I was unable to write the expected academic book. I had a macro-political economic framework that carried one narrative, and a micro-sociological and psychological set of field data that carried a myriad of individual stories, and I didn't know how to put the two together. I didn't know how to make a good academic story out of these two seemingly incompatible sources. I didn't even have a clue where to look for illumination. I gave up on the project, and my sense of failure and guilt was such that I started to ask that awkward but potentially transformative question, 'why am I doing this?'

Within a year, I had resigned from my professorial position in Urban Studies in Sydney and moved to Los Angeles, where I enrolled in a Master of Fine Arts in Screenwriting at the fabled Film School at UCLA. My experience of the social sciences in general and of planning education in particular, up to that point – the intolerance of diverse ways of knowing and writing – convinced me that my epistemological crisis could not be resolved from within academia. Film making, on the other hand, spoke directly to the emotions yet also, in the hands of directors like John Sayles, involved '*thinking* in pictures' (Sayles 1987). Perhaps, I thought, film could bridge the unbridgeable: intellect and emotions. Sayles, the writer/director/editor of such politically saturated feature films as *Matewan*, *The Brother from Another Planet*, *City of Hope*, and *Lone Star*, would be my role model.

So here I was in my mid-thirties, a student again, in a field about which I was clueless. I had studied literature as an undergraduate, but screenplays certainly weren't considered 'literature' in the 1960s in Australia. I had no idea what was involved in writing a screenplay, but was eager to learn. Ten people each year were admitted into this program. My classmates were all in their early twenties and had already written pilots for TV shows and had a 'portfolio'. I got in because one place each year was reserved for someone who had had some other life/

work experience and had interesting reasons for wanting to switch to scriptwriting. I felt exactly like I had felt starting out as an undergraduate at 17, half a lifetime ago – an impostor, no longer for reasons of social class, but from sheer inexperience. I hadn't ever watched the TV shows my classmates aspired to write for. They in turn were not enthralled by my hero, John Sayles.

As terrifying as it was to find myself in this new learning environment, it was also exhilarating. For three academic terms a year, for two years, we had to write an entire two-hour feature film script, while simultaneously studying Aristotle's *Poetics* alongside the contemporary authorities on the art and craft of screenwriting. This really was *learning by doing.* I learned fast, and I loved it. It was liberating and thrilling, and it felt meaningful because I was choosing to write stories about issues that mattered to me. One of my scripts was based on the true story of the 'Bread and Roses' strike in 1905 in a textile mill in Lawrence, Massachusetts; another on an (also true) attempt to set up a utopian community (Octagon City) in Kansas in the 1850s based on the principles of anti-slavery and vegetarianism. I also wrote stories of love and loss, based on my own life experience. Although I had no idea at this point whether I could ever earn a living as a screenwriter, I had managed to acquire an agent in Hollywood by the time I graduated and was willing to give it my best shot.

For a half dozen years after I'd graduated, I led two very different lives: one as a part-time screenwriter in Hollywood, the other as a part-time academic at the University of California's Graduate School of Architecture and Urban Planning (GSAUP). Despite an early success in getting my first post–film school script optioned and subsequently produced as an ABC TV Movie of the Week, by 1994 I felt I was losing my moral compass, second guessing what producers might want to buy instead of writing what I wanted to write (Commentary, Part I). After attending a planning theory conference in Norway in 1994, I resolved to return to academia, somehow, somewhere (Commentary, Part II).

Along with this new resolve, I made myself a promise. I would apply what I learned in film school to my academic teaching and writing, concentrating on the power of story as a way of learning, a form of inquiry, and as a communicative device. At this time, I *believed in* the power of stories, but in a completely fuzzy and stubbornly un-analytical way. I was afraid I might spoil the magic if I *thought too much* about why story is important, how it works, in what circumstances it works, and what kinds of work stories do. It took another decade (Sandercock 2003) for me to actually analyze why story is so important and useful in planning (Chapter 7 here). But *thinking in pictures* was already influential in my approach to the writing of *Towards Cosmopolis* (1998) and *Cosmopolis 2: Mongrel Cities* (2003) in that I sought the collaboration of the visual artist Peter Lysiottis, who created a brilliant set of provocative, jarring images to complement (as opposed to illustrating) those books.

After a less than happy stint as a department head of Landscape, Environment and Planning at RMIT University in Melbourne (Commentary, Part II), I returned to North America in 2001, to my new home in the graduate School of Community and Regional Planning (SCARP) at the University of British Columbia in Vancouver. As a newcomer to Canada, I was eligible for funding from the federal government's Canadian Foundation for Innovation (CFI), which was typically used to fund electron microscopes, cyclotrons, and other expensive laboratory equipment, as well as to build laboratories. The grants were generous. My head of department, Tony Dorcey, encouraged me to apply. But I don't need all that money, I said, let alone a laboratory. I just need time, peace, to do my research and write my academic stories.

Enter Serendipity.

I was invited to give a keynote address at the International Network of Urban Research and Action (INURA) conference in 2001. For the first few days, the conference took place in a derelict factory (these are true urban activists) in Florence. Then we moved to a villa in the Tuscan countryside, recently taken over by socialist organic farmers trying to start an *agriturismo* business. At the end of the week, PhD students were given a brief chance to display their wares. Resisting the temptation to take a much-needed nap in the shade of an olive tree, I attended their session. And that day I saw a presentation by Giovanni Attili, then a doctoral student (now a professor) in Planning at the University of Rome, La Sapienza, that rocked my world. As part of his doctorate, Attili was constructing a two-hour interactive hypermedia CD rom depicting refugee landscapes in a neighborhood in inner Rome.

'And by a stanza, a city is blown to bits' (Mayakovsky 1925/1960: 195). Here was my second epistemological detonation. While I was still poking around the campfire of storytelling as the written and spoken word, Attili's feet were firmly planted in 21st-century storytelling modes. With the impact of his handful of dynamite thrown into the campfire, I was being shown the cascading potentialities of multimedia in depicting a complex and unsettled urban reality composed of many stories, images, sounds, music, a multi-sensory experience, a complex tapestry, artfully woven together. I was impressed. We talked. I suggested keeping in touch. And three years later we were working on our first collaborative film project (Chapter 8).

I applied for that CFI grant. My 'laboratory' would be a multimedia/film studio and my equipment would be digital video cameras, microphones, servers, laptops, and editing software. I hoped these 21st-century tools would enable us to depict the city of spirit, the city of memory and the city of desire (Sandercock 1998); to explore the complexities of the mongrel (multi-racial, multi-ethnic) cities now emerging globally; and to think through the challenges that such cities pose to planners and planning. My educational mission was to nurture a planning imagination for the 21st century, which I defined (in *Cosmopolis 2*, 2003) as creative, audacious, political, and therapeutic – a long way from the techno-rational model that had dominated 20th-century planning praxis and pedagogy.

My research partnership with Giovanni Attili (Vanni) was one answer to this personal as well as educational yearning for the 'juice', the emotional language, that could help animate the work of planners. Vanni came to Vancouver as a post-doctoral fellow, and we set about teaching a class on 'digital ethnography' to 15 Masters students. Vanni taught the videography skills in my 'laboratory', which had been grandiosely named the Vancouver Cosmopolis Laboratory in the successful grant application, while I orchestrated the research, which took place in the low-income Collingwood neighborhood on the eastern edge of the City of Vancouver, with 48,000 residents, only 27% of whom spoke English as their first language. Here was our first 'experiment' with multimedia as a means of social research, as a form of meaning making, as a tool for community engagement and community development, and as a catalyst for policy dialogue (Chapter 8).

It had been a long road, from Wollongong to the UCLA film school and beyond, but now I felt I had come full circle, back to film making, but now applying it as a planning intervention. As it turned out, a new path was just beginning with my new creative partner. And just as *learning by doing* had been the approach in film school, now Vanni and I set out to learn by doing with our digital ethnography projects. That learning is the subject of Chapters 8 through 11. Each new project built on the experience gained in the previous one, as we developed our collaborative approach to film making in and with communities: film making as a form of community-based action research. Those chapters articulate the ideas and schools of thought we were drawing on, as is expected in refereed academic essays. What is rarely, if ever, dwelled on though, in that kind of writing, is the crucial role of the relationships that underpinned these projects and without which none of them would have come to fruition. So that's what I'd like to emphasize here.

Relationships: Chance and/or Serendipity?

There were two non-academic relationships out of which the first two film projects evolved. I learned about the extraordinary community-building work of the Collingwood Neighbourhood House (CNH) from Steven Dang, a Masters student whom I was supervising shortly after joining UBC in Vancouver. I asked Steven to take me there and introduce me, and thus began a series of conversations with the executive director, Paula Carr. Paula had dedicated almost 20 years to building that local institution into the magical place it is, thanks to her relational skills and political savvy. From the moment I walked through that Welcome Arch entrance, carved by local resident artists from Indigenous and other cultures, it struck me as a microcosm of the 'cosmopolis' I had been imagining (Commentary, Part II). A couple of years into our friendship, sealed with many wonderful dinners cooked by Paula, CNH faced a budget crisis due to a change in Provincial politics. Paula wondered whether UBC might be able help in some way. This coincided with my invitation to Vanni to come to UBC as a Postdoctoral Fellow to work on a film project with me. Given that he had just made a 30-minute documentary about the increasingly immigrant neighborhood where he lived in inner Rome, the Pigneto, it seemed to make sense that we might make a film about the Collingwood neighborhood and the role of the CNH in helping strangers to become neighbors. Paula was interested but non-committal until she'd had a chance to meet Vanni and to see his film about the Pigneto. We screened the film for the staff at CNH, and even though Vanni's English was pretty minimal in early 1985, he passed the personal and professional test with flying colors. The folks there warmed to him, and we were given the go-ahead. We spent a year hanging out in Collingwood, getting to know many people, eating with them, interviewing them, growing connected and invested in supporting what CNH stood for.

After the local success of the film, and Paula's and my 'road trip' to five Canadian cities where we used the film as a catalyst for workshops about the policy issues it raised, came the next serendipitous personal connection that led to the next film project. Not long after arriving in Vancouver, I'd been introduced to Nathan Edelson, a senior planner with the City of Vancouver, whose professional 'turf' was the most marginalized neighborhood in the city, known as the 'downtown eastside'. We shared a social justice orientation to planning work, and Nathan invited John and I to dinner. By the end of the evening, Nathan's wife Norma Jean McLaren and I felt like we were old friends. She had been doing anti-racism and diversity training work for the Province of British Columbia for the previous decade, and had been a consultant with CNH at some point. Norma Jean had a way of making whomever she was talking with feel they were the most important person on the planet. Nobody embodied the skill of listening as well as she did. And she was a gifted storyteller. The 'work stories' that she shared with me that night prompted me to invite her as a guest to the course I had just created at SCARP, 'Cross-cultural Planning: educating the heart'. What began as one guest spot grew over the next several years into us co-teaching the class, and it is no exaggeration to say that students *loved* her, deeply. Her level of care for them as embodied individuals, each with gifts they were bringing to planning, was not something most of them had ever experienced in post-secondary education, and it set a new bar for me.

Norma Jean was enthusiastic about the film Vanni and I had made about CNH. Thus, on one of her work trips to north central BC, where she was doing anti-racism and diversity training with the Royal Canadian Mounted Police (RCMP) and a few community members in the small town of Burns Lake, she shared a copy of that film with Pauline Goertzen, a young woman known locally as 'the anti-racism lady' for her activism in an openly racist town. When Pauline watched the film, its story (about how the CNH had worked to combat racism and xenophobia in Collingwood) ignited something for her. She invited me to Burns Lake, hoping I would be interested in what was happening (past and present) in that town – hoping, in fact, that I might be interested enough in those stories to want to make a film. I spent a week there, listening to a wide range of people (Indigenous and non-Indigenous) whom Pauline introduced me to, and at the end of that week I couldn't *not* commit to that as my next project.

Serendipity at work again.

Loving Attachment

Relationships and serendipity. Some people come into one's life for a reason. Up until that life-changing trip to Burns Lake in 2006, my social justice lens had been focused exclusively on the city, on the urban. It was time for my eyes to be opened, as well as my heart, to Canada's foundational injustice, the colonization of its First Peoples. I had read about, but never experienced first-hand, the same story with respect to Aboriginal people and the 'discovery' of Australia in 1788, yet had somehow fallen for the 'official story' of Canada as a country with a deep respect for human rights and no history of colonization. I was ashamed when I realized my ignorance and am embarrassed even now to look back on it. But, as I learned once we had finished *Finding Our Way*[1] and were doing community screenings in Burns Lake and other small towns like it on Highway 16 and beyond, my ignorance was shared by the vast majority of the 'settler' population. This was true even after the 2008 apology, by the then prime minister, to Indigenous peoples for the genocidal policy of Indian Residential Schools.

Finding Our Way could not have been made without Pauline's volunteer support over the next few years. And the making of the film, in turn, involved me spending a year getting to know people in and around Burns Lake, building trust slowly, across the 'two solitudes' – Indigenous and non-Indigenous. By the time we were in conversation with a potential distributor, Sylvia Lisitza (head of the non-profit company Moving Images), she suggested that we add an epilogue, interviewing key people in the film and getting them to talk about why they agreed to have us tell their story in an age when it is increasingly controversial for non-Indigenous people to be telling the stories of Indigenous people. We followed her advice, interviewing the chiefs of the two First Nations and the senior policy advisor for the Cheslatta Carrier Nation, asking why they wanted us to tell their stories. Their answer was two-fold. They wanted their story told in hopes of educating youth – their own, as well as non-Indigenous youth throughout BC if possible – about the true history of their treatment at the hands of the Canadian state and of the local settlers. They didn't expect their situation to change, but their hopes were for change in the hearts and minds of the next generations. They had no-one with technical film expertise in their communities, so they were grateful for our interest. And they trusted us to tell their stories because, as Chief Rob Charlie says in the epilogue, 'you took the time to get to know us', and because we had promised to work in partnership with them.

Vanni and I had become deeply involved in the stories we were hearing. As we were filming, we could not hold back our tears as we listened to what our Indigenous interviewees were telling us about their and their ancestors' and their children's lived experience under an ongoing regime of colonization, administered by the Department of Indian Affairs but based on the White supremacy that (still) pervades the country. In the same way as had happened while working on the CNH film, we had become involved in this story of colonization and were never impartial, although we did have to walk a careful line in our relationships with some of the settlers in the town because we wanted their voices in the film too.

What I want to emphasize here – and I have written about this elsewhere (Sandercock and Attili 2012) – is that we were becoming not only politically invested in our stories but also emotionally attached to the people who were sharing these stories. At a planning conference in 2011, in response to a question from a young woman of color about the 'danger' of working with communities we care about (her expressed fear of being unable to be impartial), I came up with the term 'loving attachment'. I believe that to do this kind of community-based action research well, it is *not only permissible but also necessary* to develop these bonds of loving attachment. This concept of loving attachment so resonated with that audience that a few of us subsequently co-curated a set of essays on the topic for *Planning Theory & Practice* (Porter, Sandercock and Umemoto 2012).

★

Creative Serendipity

My intent in this Commentary has been to shift attention away from the intellectual ideas, concepts and theories that inform the chapters in Part III and re-direct attention to what is usually invisible in academic writing: the relationships that made all of this work possible. One thing that stands out for me as I look back on this evolving approach to community-based action research, to the power of story, and to the use of film as a planning intervention is the role of what I've come to call *creative serendipity*, which I've sought to highlight in this Commentary.

I think of serendipity as something more than mere chance. Timing is everything in life: in love, as in work. People come into our orbit. Do we connect with them, or not? Some magic of timing might be at work (i.e. chance), but so too is intentionality, recognition, readiness. When I'm on a path of inquiry, I am curious about everything, receptive, attuned, looking for connections. Creative serendipity is *recognition*, enabled by an open mind and an open heart, followed by an act of *connection*. Sometimes a leap of faith is involved, otherwise known as intuition: a job opportunity, a lover, a creative partner. The building of the relationships that I've described in this Commentary began with this creative serendipity. This is a way of being in the world. It is arguably a crucial ingredient of social innovation. It's also a source of joy, wonder, and delight.

Teaming up with another person, whether in love or in work, is always both a risk and a mystery.

In Vanni, I had found the perfect creative partner and kindred spirit for my project of expanding the language of planning. We set out to explore the possibilities of multimedia in urban policy and community planning, beyond the flatlands of conventional cartography, a search for more expressive languages, languages more attuned to the embodied nature of urban social experience, ways of capturing the multiple stories and polyphonies of urban life. We aimed to give attention to the hidden life of places and to invisible inhabitants; to direct attention to those things which cannot be said, to listen for those things that are muted; to develop a sensitivity to, a window on, and a respect for the uncelebrated; to search for the remarkable things that are not remarked on; to listen to the city with a third ear; to observe with a fresh gaze that can make our capacity for astonishment come alive again; to think with all five senses; to make the familiar strange; to investigate the self-evident; to offer an alternative story to the glossy official accounts; to show the implications of the past in the present; and to link individual biographies with larger social and historical forces.

We also started from a theoretical position critical of the dominant forms of representation of the city (the cartographic anxiety) and the kinds of knowledge and experience regarded as legitimate, and thus to questions of exclusion and silencing. It is in the silences and absences that inequitable relations and gross political complicities are hidden. We each felt an urgency to invent new descriptive and analytical tools which could give centrality to people, focusing not only on individual practices but also on the collective practices (such as those of the Collingwood Neighbourhood House) through which citizens create their own meaningful life spaces in the face of larger social historical forces. From this perspective, we decided to explore tools which not only capture such everyday experiences of meaning making but also give citizens more opportunities to participate in conversations about their cities and communities and more influence in shaping, improving, and protecting them.

Our own collaborative investigations and evolving experiments with multimedia in turn led us in search of a network of like-minded folks. That search came together in our edited collection *Multimedia Explorations in Urban Policy and Planning: Beyond the Flatlands* (2010), a book which unabashedly reflects our own and others' excitement about the ways in which multimedia can be used by activists, non-governmental organizations (NGOs), immigrant and Indigenous communities, planning scholars and educators, wherever urban policies and community planning strategies are being debated and communities are struggling to shape, improve, and protect their life spaces. Our capacity for astonishment comes alive again through the contributors to that book, and in seeing our students take up these tools and use them in their Masters and doctoral theses as well as in their community engagement and activism (see Hallenbeck 2019; Ortiz Escalante 2019).

My hope is that this work has begun to create *spaces of possibility* for the next generations, as well as achieving some tangible, material impacts in the short to medium term.

Note

1 The three-part film depicts the lived experience of colonization of two small First Nations in and near Burns Lake, as well as providing the historical policy context that amounted to cultural genocide.

References

Hallenbeck, J. 2019. 'The water we call home: five generations of indigenous women's persistence Along the Salish Sea', Unpublished PhD thesis, Department of Geography, UBC.

Mayakovsky, V. 1925. Conversation with a tax collector about poetry. In: Blake, P. (Ed.). 1960. *The Bedbug and Selected Poetry of Vladimir Mayakovsky* (Bloomington, IN: Indiana University Press).

Ortiz Escalante, S. 2019. 'Planning the everyday/everynight: a feminist participatory action research with women nightshift workers', Unpublished PhD thesis, School of Community & Regional Planning, UBC.

Porter, L., Sandercock, L., Umemoto, K., Umemoto, K., Bates, L. K., Zapata, M. A., Kondo, M. C., Zitcer, A., Lake, R. W., Fonza, A., Sletto, B., Erfan, A. and Sandercock, L. 2012. What's Love got to do with it? Illuminations on Loving Attachment in Planning, *Planning Theory and Practice* 13(4), 593–627.

Sandercock, L. 1998. *Towards Cosmopolis: Planning for Multicultural Cities* (Chichester: John Wiley and Sons).

Sandercock, L. 2003. Out of the Closet: the importance of stories and storytelling in planning practice, *Planning Theory and Practice* 4(1), 11–28.

Sandercock, L. and Attili, G. 2010. (Eds) *Multimedia Explorations in Urban Policy and Planning: Beyond the Flatlands* (Heidelberg: Springer).

Sandercock, L. and Attili, G. 2012. Unsettling a settler society: film, phronesis and collaborative planning in small town Canada. In: Flyvbjerg, B., Landman, T. and Schram, S. (Eds) *Real Social Science: Applied Phronesis* (Cambridge: Cambridge University Press).

Sayles, J. 1987. *Thinking in Pictures* (Boston: Houghton Mifflin).

7

OUT OF THE CLOSET

The Importance of Stories and Storytelling in Planning Practice (2003)

Introduction

Allow me to introduce you to Martha Quest, the protagonist of Doris Lessing's novel *The Four-Gated City* (1969). Martha is a Marxist intellectual recently arrived in 1950s London from Rhodesia. She sees London as socially deprived and ugly. For Iris, however, a local woman with whom Martha is staying, the neighborhood is a living archive, textured and animated by layers of history and memory (Donald 1999: 122).

> Iris … knew everything about this area, half a dozen streets for about half a mile or a mile of their length; and she knew it all in such detail that when with her, Martha walked in a double vision, as if she were two people, herself and Iris, one eye stating, denying, warding off the total hideousness of the whole area, the other, with Iris, knowing it in love. With Iris, one moved here in a state of love, if love is the delicate but total acknowledgement of what is … Iris … had lived in this street since she was born. Put her brain, together with the other million brains, women's brains, that recorded in such tiny loving anxious detail the histories of windowsills, skins of paint, replaced curtains and salvaged baulks of timber, there would be a recording instrument, a sort of six-dimensional map which included the histories and lives and loves of people, London – a section map in depth. This is where London exists.
>
> *(quoted in Donald 1999, 122)*

Martha, we might say, 'knows' London through theory. Iris knows it through immersion, through all of her senses, through empathy and love. In the paragraph quoted there is an implicit epistemology of the city, a distinctly feminist epistemology, which is at first sight appealing. It outlines a way of knowing the city through the senses and emotions rather than through (Marxist) theory. In the declaration of the last sentence, 'This is where London exists', there is a truth claim, a claim that to really know the city, you need to know the stories in the heads of its entire female population. If we take this literally (which of course we're not meant to), then it's obviously ridiculous, since we could never know the stories of all those

DOI: 10.4324/9781003325451-11

'other million brains'. But even if we did, we would still have to interpret them, make our own sense of them. Is it even true, as a generalization, that inside women's heads there exists a particular kind of knowledge about the city? And why is that knowledge – Iris's perspective in the novel – inherently more valuable than Martha's more analytical perspective? My own answer to this last question is that it's not, that both are valuable and that it's not helpful to think of these different ways of knowing the city as mutually exclusive, or as superior/inferior. But that is the way 'story' is usually thought of, as 'soft', as a woman's way of knowing, as inferior, lacking in rigor.

In response to this kind of marginalizing of story in the social sciences, feminists and others have reasserted its importance, both as epistemology and as methodology. What I want to argue in this essay is that story has a special importance in planning that has neither been fully understood nor sufficiently valued. In order to imagine the ultimately unrepresentable space, life and languages of the city, to make them legible, we translate them into narratives. The way we narrate the city becomes constitutive of urban reality, affecting the choices we make, the ways we then might act. My argument will be deceptively simple. Stories are central to planning practice: to the knowledge it draws on from the social sciences and humanities; to the knowledge it produces about the city; and to ways of acting in the city. Planning is *performed* through story, in a myriad of ways. I want to unpack the many ways in which we use stories: in process, as a catalyst for change, as a foundation, in policy, in pedagogy, in explanation and critique as well as a justification of the status quo and as moral exemplars.

My approach is not uncritical. Despite increasing attention to and use of story in some of the newer academic fields (feminist and cultural studies, for example), I don't see it as the new religion. We still need to question the truth of our own and others' stories. We need to be attentive to how power shapes which stories get told, get heard, carry weight. We need to understand the work that stories do, or rather that we ask them to do, in deploying them, and to recognize the moral ordering involved in the conscious and unconscious use of certain plots and character types. I believe that a better understanding of the role of stories can make us more effective as planning practitioners, irrespective of the substantive field of planning. Story and storytelling are at work in conflict resolution, in community development, in participatory action research, in resource management, in policy and data analysis, in transportation planning, and so on. A better understanding of the role of stories can also be an aid to critical thinking, to deconstructing the arguments of others. Stories can often provide a far richer understanding of the human condition, and thus of the urban condition, than traditional social science, and for that reason alone, they deserve more attention.

In short, I want to make two bold arguments in this paper. One is about the importance of story in planning practice, research, and teaching. The other is about the crucial importance of story in multicultural planning. Much of what planners do, I will argue, can be understood as performed story. Yet the importance of story has rarely been understood, let alone validated, in planning. Story is an all-pervasive, yet largely unrecognized force in planning practice. We don't talk about it, and we don't teach it. Let's get this out of the closet. Let's liberate and celebrate and think about the power of story. Let's appreciate its importance to the 21st-century multicultural planning project, as a way of bringing people together to learn about each other through the telling of stories.

Very few scholars within the planning field have investigated the work of story in planning, and even then, they have considered only aspects of it (Forester 1989, 1993, 1999; Mandelbaum 1991; Throgmorton 1996; Marris 1997; Eckstein and Throgmorton 2003). In coming out of the closet about the importance of story, I want to be systematic about the ways, implicit and explicit, in which we use story, and to demonstrate what I mean when I say that planning is performed through story: in process, in foundational stories, in stories as catalysts for change, in policy, and, finally, in academic stories, as method, as explanation, and as critique.

Planning as Performed Story

Story and Process

For many planning practitioners, the role of story is central, although not always consciously so. Those who do consciously make use of story do so in diverse, and often imaginative and inspiring, ways. The best way to demonstrate this is by using some examples – of story as process, and of story being used to facilitate process. But these examples are so varied that I'll use sub-headings as guides.

Community Participation Processes

In community participation processes, planners orchestrate an event in such a way as to allow everybody, or as many people as possible, to tell their story about their community, neighborhood, school, or street. We tend to refer to this as drawing on local knowledge, and there are various techniques for eliciting people's stories, such as small group work with a facilitator for each group. What is not always clear is how these collected stories will be used in the subsequent process, but the belief operating here is that it is important for everybody to have a chance to speak and to have their stories heard. This is linked with an argument about the political and practical benefits of democratizing planning.

If a participatory event is a way of *starting* a planning process, its purpose is most often about getting views and opinions, so the story-gathering is likely to be followed by an attempt to find common threads that will help to identify priorities. If, on the other hand, the participatory event is a response to a pre-existing conflict that needs to be addressed before planning can move ahead, then the gathering of rival stories takes on more import. In such a situation, practitioners will usually meet separately with each involved person or group and listen to their stories of what the problem is before making a judgment about when and how to bring the conflicting parties together to hear each other's stories. In extreme cases, where the conflict is long-standing, relating to generations or even centuries of oppression or marginalization, this is very difficult work, but when done well can be therapeutic, cathartic, even healing.[1]

Mediation, Negotiation, and Conflict Resolution

In one growing branch of planning practice – mediation, negotiation, and conflict resolution – there is a raft of techniques and procedures for facilitating storytelling, and the hearing of stories, in conflict situations[2]. In this kind of work, the ability of a practitioner to make the space for stories to be heard is more important than the ability to tell stories. And it is here that the importance of listening to others' stories, and the skills of listening in cross-cultural contexts, is at a premium.

Forester describes a case in Washington State where the mediator, Shirley Solomon, brought together Native Americans and non-Native county officials to settle land disputes. A critical stage in that mediation was the creating of a safe space in which people could come together and 'just talk about things without it being product-driven' (Solomon, quoted in Forester 2000: 152). Solomon ceremonialized this safe space by creating a talking circle and asking people to talk about what this place meant to them. Everyone was encouraged to tell their story, of the meaning of the land, the place, to them and their families, past, present, and future – the land whose multiple and conflicting uses they were ultimately to resolve. It was this story-ing that got people past 'my needs versus your needs' and on to some 'higher ground', moving toward some common purpose. Solomon describes this stepping aside to discuss personal histories as both simple and powerful, as a way of opening surprising connections between conflicting parties. Or, as Forester has it, storytelling is essential in situations where deep histories of identity and domination are the context through which a present dispute is viewed. Stories have to be told for reconciliation to happen (Forester 2000:157). In terms of process, too, the design of spaces for telling stories makes participants from different cultures and class backgrounds more comfortable about speaking, and more confident about the relevance of the whole procedure. A tribal elder who was present at Solomon's mediation said to her: 'In those meetings where it's Roberts Rules of Order, I know that I either have nothing to say, or what I have to say counts for nothing' (quoted in Forester 2000:154).

Core Story

Another interesting development of the use of story in practice is what Dunstan and Sarkissian (1994) call 'core story'. The idea of core story as methodology draws on work in psychology which suggests that each of us has a core story: that we do not merely tell stories but are active in creating them with our lives. We become our stories. When we tell stories about ourselves, we draw on past behavior and on others' comments about us in characterizing ourselves as, say, adventurous, or victims, or afraid of change, or selfish, or courageous. But in telling and re-telling the story, we are also reproducing ourselves and our behaviors. Social psychologists argue that communities, and possibly nations, have such core stories that give meaning to collective life (see Houston 1982, 1987). Culture is the creation and expression and sharing of stories that bond us with common language, imagery, metaphors, all of which create shared meaning. Such stories might be victim stories, warrior stories, fatal flaw stories, stories of peace-making, of generosity, of abandonment, of expectations betrayed.

In their work in evaluating the success of community development on a new outer suburban estate developed by a public agency in an Australian city, Dunstan and Sarkissian used an array of research tools: attitude and satisfaction surveys, interviews, and focus groups, as well as census and other 'hard' data. When they came to analyze this material, they found contradictions that were not likely to be resolved by collecting more details. In order to go beyond the details and the quantitative scores on 'satisfaction', they explored the notion of core story, drawing on heroic, mythic, and meta-poetic language. They scripted such a story of heroic settlers, of expectation and betrayal, of abandonment, and took the story back to the community, saying 'this is what we've heard'. The response was overwhelming and cathartic. 'Yes, you've understood. That's our story'. The task then, as the social planners defined it, was to help the community to turn this doomed and pessimistic story around. They asked the community how they thought their story might/

could/should be changed. Underlying this was a belief that core stories can be guides to how communities will respond to crisis or to public intervention. As with individuals, some tragic core stories need to be transformed by an explicit healing process or else the core story will be enacted again and again. Renewal and redemption are possible, Dunstan and Sarkissian argue. New chapters can be written if there is the collective will to do so. They suggest four steps towards renewal. The first is a public telling of the story in a way that accepts its truth and acknowledges its power and pain. The second is some kind of atonement, in which there is an exchange that settles the differences. The third is a ceremony or ritual emerging out of local involvement and commitment by government (in this case municipal and provincial) that publicly acknowledges the new beginning. The fourth is an ongoing commitment and trust that a new approach is possible and will be acted on (Dunstan and Sarkissian 1994: 75–91).

This fascinating case study offers some illumination to a more general puzzle in participatory planning: how to turn a raft of community stories into a trustworthy plan, one that is faithful to community desires. To turn the light on inside the black box of that conversion surely requires planners to take their plan back to the community storytellers and say, 'this is how we converted your stories into a plan. Did we understand you correctly?' In a community or constituency where there is only one core story, this is a more straightforward process than in a situation where what the planners have heard is two or more conflicting stories. In the latter situation, there is far more working through to do, in order to prioritize and to reach some consensus about priorities.

Story as Foundation, Origin, Identity

There's a related notion of foundational story, a mytho-poetic story of origins, a story that cities and nations tell about themselves. This is particularly relevant to planning in multi-ethnic, multi-racial contexts in which conflicting notions of identity are at play.

In the winter of 2002, I was working with Charles Landry and Phil Wood in Birmingham, England, at the invitation of the City Council. Partly in response to race riots in other northern British cities in the preceding summer, Birmingham's politicians were concerned about 'getting it right' in relation to 'managing' ethnic diversity. As I met with various groups in the city, from the city planning staff to workers in a variety of community development programs, to young Black men and Muslim women, I began to hear very different versions of Birmingham's identity. There was a fairly widely accepted founding story on the part of some Anglo residents (who referred to themselves as the 'indigenous' population) that Birmingham was an *English* city (not a multicultural city), and that those who were there first had greater rights to the city than the relative newcomers from the Indian subcontinent, the Caribbean, and so on. This profoundly political question of the city's changing identity clearly needed the widest possible public debate. I suggested that at some point the city was going to have to re-write its foundational story, to make it more inclusive and open to change. The planning staff were very much implicated in this debate. At the community coalface, and especially in non-Anglo neighborhoods, these predominantly Anglo-Celtic planners were either reproducing the founding story of 'British Birmingham', or helping to change that story by making their policies and programs reflect and respect the diversity of the 'new city'.

This is not an isolated example anymore, but a situation increasingly common across Europe in this age of migrations. The need to collectively change these old foundational stories, and represent those changes in the built environment itself, is one of the contemporary challenges facing planners.

Story as Catalyst for Change

Stories and storytelling can be powerful agents or aids in the service of change, as shapers of a new imagination of alternatives. Stories of success, or of exemplary actions, serve as inspirations when they are re-told. I've lost count of the number of times I have told 'the Rosa Parks story',[3] either in class or in a community or activist meeting, when the mood gets pessimistic and people feel that the odds are too great, the structures of power too oppressive and all-encompassing. When Ken Reardon (1998, 2003) tells or writes his East St. Louis story, he is amongst other things conveying a message of hope in the face of incredible odds. This 'organizing of hope' is one of our fundamental tasks as planners, and one of our weapons in that battle is the use of success stories, and the ability to tell those stories well, meaningfully, in a way that does indeed inspire others to act.

In multicultural contexts, there is usually a dominant culture whose version of events, of behavior, and practices represents the implicit norm. It is also usually the case that those engaged in planning – as a state-directed activity – are members of the dominant culture, and are therefore less likely to recognize, let alone question, dominant cultural norms and practices. For a society to be functionally as well as formally multicultural, those norms occasionally have to be held up to the light and examined and challenged. The most effective way to do that is through story. Canadian planner Norman Dale has written of the critical importance of hearing the stories of the Haida people (an Indigenous community on an island archipelago off the Northwest coast of Canada) in what was meant to be a cross-cultural community economic development project in the Queen Charlotte Islands (now known by their pre-colonial name, Haida Gwaii), sponsored by the provincial government. After a series of formal meetings with local residents, Dale was struggling to create a space in which the lone Haida representative (whose name was Gitsga) would feel empowered to say anything. Gitsga seemed to have taken a vow of silence, and was on the verge of pulling out of the consultation process when Dale sought him out and encouraged him to return. At the next consultation meeting, there was some informal chat among the White folks, before the real meeting began, about the artistic and environmental merits of a sculpture that had been erected on a rock offshore. It hadn't occurred to anybody to ask the Haida people what they thought. When Gitsga broke his silence to volunteer the information that the rock was sacred to the Haida, there was genuine shock and consternation, leading to an opening up of the whole community economic development planning process to the involvement of the Haida (Dale 1999).

I have one more example of the use of story in planning practices – in the process of policy analysis, formulation, and implementation – before I turn to academic storytelling about planning.

Story and Policy

Here I am aided by my colleagues and friends, James Throgmorton and Peter Marris, each of whom has done a lot of thinking about the connections between story and policy. In *Witnesses, Engineers and Storytellers: Using Research for Social Policy and Community Action* (1997), Peter Marris argues that the relationship between knowledge and action is not straightforward, and that knowledge itself cannot, has not ever, determined policy. In analyzing various types of and approaches to social policy research, Marris asks why so little of the research produced on poverty, for example, has affected policy. His answers are several. One is that academics are powerful critics but weak storytellers. That is, they fail to communicate their findings in a form that is not only plausible but persuasive. By contrast, he notes that community actors have great stories to tell, but no means of telling them, except to each other. So the wrong stories win the debates. Storytelling, he says, is the natural language of persuasion, because any story has to involve both a sequence of events and the interpretation of their meaning. A story integrates knowledge of what happened with an understanding of why it happened and a sense of what it means to us. Stories organize knowledge around our need to act and our moral concerns. The stories don't have to be original, but they must be authoritative – that is, provide reliable evidence marshaled into a convincing argument. The best stories are both original and authoritative.[4]

To be persuasive, the stories we tell must fit the need as well as the situation. Policy researchers compete with everyone else who has a story to tell, and their special claim on public attention lies in the quality of their observation as well as the sophistication of the accumulated understanding through which they interpret their data. But this truthfulness is not, in itself, necessarily persuasive. Good stories have qualities such as dramatic timing, humor, irony, evocativeness, and suspense, in which social researchers are untrained. 'Worse', says Marris, 'they have taught themselves that to be entertaining compromises the integrity of scientific work' (Marris 1997: 58). Writing up policy research is hard work: it's hard to tell a good story while simultaneously displaying conscientiously the evidence on which it is based. But, Marris insists, the more social researchers attend to the storyteller's craft and honor it in the work of colleagues and students, the more influential they can be. We have to be able to tell our stories skillfully enough to capture the imagination of a broader and more political audience than our colleagues alone.

There are two notions of story at work here. One is functional/instrumental: bringing the findings of social research to life through weaving them into a good story. The other is more profound: storytelling, in the fullest sense, is not merely recounting events but also endowing them with meaning by commentary, interpretation, and dramatic structure.

While Marris seems to confine his advocacy of storytelling to the publishing of research results, James Throgmorton's work addresses the next step, the arts of rhetoric in the public domain of speech and debate. The lesson he wants to impart is that if we want to be effective policy advocates, then we need to become good story creators and good storytellers, in the more performative sense. In *Planning as Persuasive Storytelling* (1996), Throgmorton suggests that we can think of planning as an enacted and future-oriented narrative in which participants are both characters and joint authors. A crucial part of Throgmorton's argument is that this future-oriented storytelling is never simply persuasive. It is also constitutive. The ways in which planners write and talk shape community, character, and culture. So, a critical question for planners is what ethical principles should guide and constrain their efforts to persuade their audiences.

Marris's and Throgmorton's work has very important implications for policy research and recommendations. If planners want to be more effective in translating knowledge into action, they argue, then we had better pay more attention to the craft of storytelling in both its written and oral forms. That means literally expanding the language of planning to become more expressive, evocative, and engaging, and to include the language of the emotions. 'Academic story telling', writes Finnegan, 'is ugly in its stark, clichéd monotone manner. We tell the dullest stories in the most dreary ways, and usually deliberately, for this is the mantle of scientific story telling: it is supposed to be dull' (Finnegan 1998, 21). What Finnegan alleges of academic storytelling is equally true of bureaucratic storytelling. Policy reports produced by government planning agencies, and also by consultants for those agencies, are cut from the same clichéd cloth. They are dry as dust. Life's juices have been squeezed from them. Emotion has been rigorously purged, as if there were no such things as joy, tranquility, anger, resentment, fear, hope, memory, and forgetting at stake in these analyses. What purposes, whose purposes, do these bloodless stories serve? For one thing, they serve to perpetuate a myth of the objectivity and technical expertise of planners. And in doing so, these documents are nothing short of misleading at best (dishonest at worst) about the kinds of problems and choices we face in cities.

To influence policy, planners need to learn story, or rather, an array of storytelling modes. But where to learn it? What is the academy teaching?

Story as Critique and/or Explanation

There is a false binary in our heads that separates planning documents, social scientific research, and theorizing from storytelling, rather than allowing us to appreciate the ways in which each of these employs story. Planning documents – from maps, to models, to GIS, to plans themselves – do in fact all tell a story. Sometimes the story is descriptive, or poses as descriptive: 'this is how things are', 'these are the facts'. But there is no such thing as mere description, or pure facts. There is always an author, the planner as policy analyst, who is choosing which facts are relevant, what to describe, what to count, and in the assembling of these facts, a story is shaped; an interpretation, either consciously or unconsciously, emerges. Facts are usually marshaled to explain something and to draw some conclusions for action.

Along with the explaining usually comes a valuing. Things were better before, or after, such and such, which then suggests we should go backwards or forwards. Evocative plots of rural superseded by urban, community by alienation, tradition by modernity, or community triumphing over capital, residents over bureaucracies, squatters over the forces of law and order are moving stories with which individual readers can identify, positioning themselves in a larger historical narrative. There are stories of times of transition, of new eras, of an old order passing, of lost Golden Ages. And there are some stories, but not many, which foretell a happy ending, if only ... if only 'we' would do such and such, then we could live happily ever after.

In other words, academic urban stories and theories evoke basic narrative plots that are familiar to us from other contexts (from fairy tales to movies) and which resonate with us morally as well as intellectually, satisfying or disturbing or challenging us. My point here is not to say that these stories are therefore worthless. On the contrary, they are illuminating and instructive precisely because of these underlying plots, which are all exercises in valuing human activities as part of a moral ordering of life and social organization. As with planning documents, the more alert we can be to the underlying story or stories, the better we are able to evaluate them. We need to understand the mechanisms of story, both in order to tell good stories ourselves, and to be more critical of the stories we have to listen to.

To imagine the future differently, we need to start with history, with a reconsideration of the stories we tell ourselves about the role of planning in the modern and postmodern city. In telling new stories about our past, our intention is to reshape our future. If we can uncouple planning history from its obsession with the celebratory story of the rise of the planning profession, then we may be able to link it to a new set of public issues – those connected with a dawning appreciation of a multicultural heritage and the challenge of planning for a future of multicultural cities and regions.

The multicultural city cannot be imagined without a belief in inclusive democracy and the diversity of social justice claims of the disempowered communities in existing cities – communities of migrants, Indigenous Peoples, poor people. If we want to work towards a policy of inclusion, then we need to start from a sound understanding of the exclusionary effects of planning's past practices and ideologies. And if we want to plan in the future for and with a heterogeneous public, acknowledging and nurturing the full diversity of all the social groups in the multicultural city, then we need to develop a new kind of multicultural literacy. An essential part of that literacy is familiarity with the multiple histories of urban communities and the multiple histories of newcomers, especially as those histories intersect with struggles over space and place claiming, with planning policies and resistances to them, with traditions of Indigenous planning, and with questions of belonging and identity and difference.

Story and Pedagogy

There are a variety of ways of using story in pedagogy – that is, specifically in the training of planners. I have been inspired by new teaching ideas that emerged from feminist studies in the 1980s. During my decade teaching graduate students in the Urban Planning Program at UCLA, I used the device of 'life stories' as a way to explore difficult issues of identity and difference. My students, in their diversity, mirrored the social and cultural diversity of that city, and this occasionally led to tensions in classes. I began each semester asking students to write short stories about the ways that race, or gender, or ethnicity, or disability had shaped their lives. We then shared those stories in class, and drew on them during the semester, as a way of connecting the personal with the political. I have also used the idea of a 'housing autobiography' when teaching undergraduates about housing issues, asking students to craft a story about the houses they've lived in and how that might have shaped their ideas of the ideal house and neighborhood. When I did this at the University of Melbourne, with students from Hong Kong, Singapore, and Malaysia in class as well as Anglo-Australians, it worked very well in bringing out cultural and class stereotypes of the 'normal' house and neighborhood. In general, I have found that the more creative I can be in the classroom – using music, images, and so on – the more creative is the response of my students in their own papers and in their thinking. They start connecting unusual things, thinking laterally, which is definitely what is needed from planners in the 21st century. It was partly feminism, partly my film school experience, and partly the need to find as many ways as possible to connect with my multicultural students that led me to experiment with story. The results, in terms of wonderful papers and presentations from my classes, have taught me a lot about the creativity that so often lies dormant, undernourished, or even discouraged through our academic straightjackets of 'appropriate', 'objective', 'scientific' papers.

For two decades, the outstanding planning educator John Forester has been a tireless story-gatherer, collecting the details of the working days and lives of a wide range of practitioners in North America and a handful from Israel and Europe, using interviews to get them to describe what it is that they do, always in terms of action rather than theorizing. With minimal editing, these 'work stories' have been published both as transcripts for pedagogical purposes and also, with detailed commentary and interpretation, incorporated into Forester's books as the foundation of his understanding of and theorizing about planning (Forester 1989, 1999). His underlying concern, as I interpret his recent work, has been puzzling over what it means to 'respect difference'. He sees the danger of respect conceived as the mere acceptance or appreciation of difference; in that form, respect can stymie dialogue and mutual learning. He is acutely aware that planning conflicts are often about more than resources – such as land, money, facilities. They are also about relationships, and this involves not only personality and politics but also race, ethnicity, and culture. To learn about how to work successfully in such cross-cultural or multicultural situations, he has sought out practitioners with good stories to tell. One such story, that can do double duty for me in writing here about pedagogy and story in multicultural contexts, is the work of Marie Kennedy, who teaches community development planning at the University of Massachusetts in Boston.

Kennedy's undergraduate students are primarily of a working-class background, urban, and older (average age 39). The class works with grassroots community organizations in the Boston area around issues defined by those organizations, and in the process, students learn planning skills. The project that Kennedy describes in her interview (Forester, Pitt and Welsh 1993, 110–122) was in the city of Sommerville, adjacent to Cambridge and Boston, population around 100,000. Sommerville was in transition from a predominantly White ethnic working-class district to a city with a significant new immigrant population, as well as a new liberal/radical, more educated White group who were moving in from Cambridge. In the previous ten years, the population had changed from 95% White ethnic working class to 25% foreign born (Haitian, Vietnamese, Central American) and this was accompanied by increasing racial tension and incidences of racial violence. Kennedy was approached by the Mystic Welcome Project, an organization of newcomers in the Mystic housing project, the largest public housing project in Sommerville. The question was how to build, or rebuild, a sense of community cooperation and support in this neighborhood. There was also the challenge of how to bring together several neighborhood-based organizations in the same area who had nothing to do with each other. For Kennedy, the agenda was clear. 'We are explicitly going in with an agenda to build a healthy multi-racial, multi-ethnic community. So, the goal is out front. We will have many discussions, and some of them will be heated' (ibid, 118).

What I want to draw from this story is how Kennedy prepared her students to work in this situation. The students themselves were diverse in terms of age, gender, race, and ethnicity, and they were mostly working class. For the first month of a one-year course, she met with the students and concentrated on their own attitudes towards immigrant communities and newcomers, and their own attitudes towards Sommerville as a place.

> We spend the initial time getting their biases and preconceptions on the table ... I feel strongly that no matter who we are, we bring our previous experience, our baggage, our preconceptions with us into any planning situation. The first step is to get real clear about what you are bringing ... You can either set your baggage aside in order to clearly hear and listen to other people's experience, or you can check it out against other people's opinions and against facts to see whether your preconception is born out or not.
>
> *(ibid, 113)*

The students all individually took walking tours through the neighborhood and had to figure out how the neighborhood affected them, whom they saw there, what racial and ethnic and socio-economic groups they noticed, what they saw in the physical environment, what their assumptions about it were, whether they thought it would or would not be a 'nice' place to live. Kennedy got all the students to write about Sommerville, and to write about their attitude to newcomer groups and individuals. The writing was done anonymously, and discussed collectively. There was a lot of disagreement among students about their impressions. What came out in the discussions was how different students' backgrounds – growing up, or not, in a public housing project; living, or not, in a neighborhood of newcomers; and so on – had shaped their reflections on Sommerville. Students were asked to think about the experience of becoming a minority newcomer, maybe the only family that is different from the now-majority community. Some could draw on their backgrounds as minority members to talk about this, and educate their fellow White students. Gradually this led into discussions on housing policy, immigration policy, a needs analysis of the area, and so on. But a whole semester was spent in this kind of preparation before the students started to work with the community group, in the community.

This is a deeply informative account of what it takes to work as an agent of social change in a changing neighborhood, and how important it is to examine one's own preconceptions. It gives us some idea of the detailed personal work that needs to be done in preparation for working in multicultural environments. In this learning stage, the stories that students tell about themselves, and hear from each other, are crucial in peeling back layers of preconceptions and assumptions about 'others', and about physical, residential environments different from whatever one is accustomed to. In turn, when I read this account of Kennedy's work, I learn new ways of approaching the training of community development planners. Her 'work story' helps my work. Forester's gathering of such stories helps us all. Stories teach. But what do they teach?

Forester's work, in spite of its empirical base, is both ethically and normatively saturated. He is not merely describing what planners do, in their own words. He wants planners to do good and make a difference, and he searches for stories from practitioners which demonstrate these possibilities. His purposes shape his collection of stories. His pedagogical aims in passing these stories on to his students are not simply to convey the skills of these practitioners but also to inspire his students with how those skills are used: that is, for what moral purposes. In Kennedy's case, the purpose is building healthy multi-racial, multi-ethnic communities. And herein is perhaps the oldest and most traditional use of stories, as moral exemplars.

Conclusions

There are of course limits to the power and reach of stories and storytelling in planning. Two need to be mentioned in closing. One concerns scale; the other, power itself. I am not claiming that storytelling works in situations of extreme conflict that divide nations. My examples are drawn from local and regional contexts and from scenarios in which planners have a role and some leverage. Nor am I claiming that storytelling is so powerful that it can or should replace other planning tools. Persuasive storytelling is one form of power at the disposal of planners, but it takes its place in a force field in which there are other powers at work, including the powers of misinformation, deception, and lying, which are deployed by fellow planners as well as outside forces opposing planning interventions.[5]

Nevertheless, this essay has argued that stories and storytelling are central to planning practice, that in fact we can think about planning as performed story. I have shown stories working as, and in, *planning processes*, where the ability to tell, to listen to, and to invent stories is being nurtured, alongside the equally important ability to create and design the spaces for stories to be heard. When stories work as *catalysts for change*, it is partly by inspirational example, and partly by shaping a new imagination of alternatives. I've explored the notion of *foundational stories* that need to be rewritten, whether at the level of the nation, the city, or the neighborhood. I've explained how story could be critical in *policy research and analysis*, as well as how the mantle of scientific storytelling may actually handicap our policy causes. I've shown how academics use story, *as explanation and as critique* of planning practices, and how these stories too can make a difference – can uphold as well as question the status quo. I've explored various ways that stories are used in the *training of planners*: personal stories, practical stories, moving and inspiring stories. I have also suggested the crucial importance of story in multicultural planning and demonstrated how particular applications of story contribute to the multicultural planning project.

I am advocating both a creative and a critical approach to stories and storytelling. I've talked about the need to deconstruct the 'official story' in relation to planning history. Using stories in planning practice must be open to the same process of critique, including an alertness to the ways in which power shapes which stories get told, get heard, and carry weight. Critical judgment will always be necessary in deciding what weight to give to different stories, as well as what stories are appropriate in what circumstances. The telling of stories is nothing less than a profoundly political and potentially transformative act.

> The storyteller, besides being a great mother, a teacher, a poetess, a warrior, a musician, a historian, a fairy, and a witch, is a healer and a protectress. Her chanting or telling of stories … has the power of bringing us together.
>
> *(Minh-ha 1989, 140)*

Notes

1 Forester (2000), Dale (1999), and Sandercock (2000) all detail such cases.
2 See Susskind, McKearnan and Thomas-Larmer (1999)
3 Rosa Parks was the African American woman who, in Alabama in 1955, refused to move to the back of the bus when white folks boarded. This act of civil disobedience catalyzed a year-long boycott of the bus service by Blacks, and gave birth to the Civil Rights Movement.
4 Marris cites Herbert Gans' *The Urban Villagers* and Michael Young and Peter Willmott's *Family and Kinship in East London* as good examples.
5 For some hair-raising examples of how planners use their data dishonestly and deceptively, see Flyvbjerg, Holm, and Buhl (2002).

References

Dale, N. 1999. Cross-cultural community based planning. Negotiating the future of Haida Gwaii. In: Susskind, L., McKearnan, S. and Thomas-Larmer, J. (Eds) *The Consensus Building Handbook* (Thousand Oaks: Sage).

Donald, J. 1999. *Imagining the Modern City* (Minneapolis: University of Minnesota Press).

Dunstan, G. and Sarkissian, W. 1994. Goonawarra: core story as methodology in interpreting a community study. In: Sarkissian, W. and Walsh, K. (Eds) *Community Participation in Practice. Casebook* (Perth: Institute of Sustainability Policy) pp. 75–91.

Eckstein, B. and Throgmorton, J. (Eds.) 2003. Story and sustainability: planning, practice, and possibility for American cities (Cambridge, MA: MIT Press).

Finnegan, R. 1998. *Tales of the City. A Study of Narrative and Urban Life* (Cambridge: Cambridge University Press).
Flyvbjerg, B., Holm, M. S. and Buhl, S. 2002. Underestimating costs in public works projects: error or lie?, *Journal of the American Planning Association* 68(3), 279–295.
Forester, J. 1989. *Planning in the Face of Power* (Berkeley: University of California Press).
Forester, J. 1993. Learning from practice stories: the priority of practical judgment. In: Fischer, F. and Forester, J. (Eds) *The Argumentative Turn in Policy Analysis and Planning* (Durham: Duke University Press).
Forester, J. 1999. *The Deliberative Practitioner* (Cambridge, MA: MIT Press).
Forester, J. 2000. Multicultural planning in deed: lessons from the mediation practice of Shirley Solomon and Larry Sherman. In: Burayidi, M. (Ed.) *Urban Planning in a Multicultural Society* (London: Praeger).
Forester, J., Pitt, J. and Welsh, J. (Eds.) 1993. *Profiles of Participatory Action Researchers* (Ithaca: Department of City and Regional Planning, Cornell University).
Houston, J. 1982. *The Possible Human* (Los Angeles: Tarcher).
Houston, J. 1987. *The Search for the Beloved: Journeys in Sacred Psychology* (Los Angeles: Tarcher).
Mandelbaum, S. 1991. Telling stories, *Journal of Planning Education and Research* 10(1): 209–214.
Marris, P. 1997. *Witnesses, Engineers, and Storytellers: Using Research for Social Policy and Action* (Baltimore: University of Maryland, Urban Studies and Planning Program).
Minh-ha, T. 1989. *Woman Native Other* (Bloomington: Indiana University Press).
Reardon, K. M. 1998. Enhancing the capacity of community-based organizations in East St. Louis, *Journal of Planning Education and Research* 17(4): 323–333.
Reardon, K. 2003. Ceola's vision, our blessing: the story of an evolving community/university partnership in East St. Louis, Illinois. In: Eckstein, B. and Throgmorton, J. (Eds) *Stories and Sustainability: Planning, Practice and Possibility for American Cities* (Cambridge, MA: MIT Press).
Sandercock, L. 2000. When strangers become neighbors: managing cities of difference, *Planning Theory and Practice* 1(1): 13–30.
Susskind, L., McKearnan, S. and Thomas-Larmer, J. (Eds.) 1999. *The Consensus Building Handbook* (Thousand Oaks: Sage).
Throgmorton, J. 1996. *Planning as Persuasive Storytelling* (Chicago: University of Chicago Press).

8

DIGITAL ETHNOGRAPHY AS PLANNING PRAXIS

An Experiment with Film as Social Research, Community Engagement, and Policy Dialogue (2010)

Introduction: From Film as 'Propaganda' to Film as Collaborative Praxis

The relationship between film and modern town planning can be traced to the 1920s when what Ciacci (1997, 2010) calls 'Town Planners' Cinema' emerged. Ciacci cites examples from Germany, France, Italy, and the United States in which the tools of cinema were in the service of a thoroughly modernist, expert-driven approach to place-making and managing urban change. The very composition of the discourse of these films was intended to convince 'the public' of the social duties and potential of town planning. While the ostensible purpose of the films was to contribute to debate within the field, usually on the occasion of a public exhibition for the circulation of ideas for resolving issues of urban change, their actual mission was to persuade the public of a particular planning scheme which had already been decided and was being proposed as 'the solution' by 'the experts'. For example, in the 1939 film *The City*, made by the American Institute of Planners with a commentary written by Lewis Mumford, the representation of the squalid living conditions in the industrial suburbs justified the decentralization of families of American workers to small green cities. In the 1930 film *Die Stadt von Morgen (The City of Today)*, made in Berlin, the detailed list of urban problems justified a social town plan for inhabitants who would thus learn to be its citizens (Ciacci 2010).

While theoretically the tools of cinema could extend participation in the project for urban change to the largest possible number of people, Ciacci's deconstruction of their actual use delivers a sobering message. Town planning films have been primarily made for 'propaganda value': to explain to and persuade an audience of choices already made by the expert urbanists. This raises profound questions about the purposes of and possible audiences for documentaries and other video interventions in the urban arena.

Unsurprisingly, a counter trend developed in reaction to such 'official' or top-down uses. 'Participatory video' emerged in Latin America, inspired partly by Paulo Freire's thinking in *The Pedagogy of the Oppressed* (Freire 1970). In Canada, the 'Fogo Process' was part of a 1970s initiative called Challenge for Change, in which the National Film Board of Canada sponsored a series of films 'intended to raise public consciousness about the rights and needs of

DOI: 10.4324/9781003325451-12

disenfranchised and disadvantaged groups' (Weisner 1992: 68). Challenge for Change did not set out to make progressive films *about* social issues but rather to use the film-making process as a *form* of social change. The intent was to use film production and distribution as a means of empowering politically and socially disenfranchised people.

Fogo Island, off the coast of Newfoundland, Canada, was a community of 5,000 people who, by the late 1960s, were facing the collapse of their local fishing economy, deterioration of public infrastructure, widespread functional illiteracy, and differences in religion and tradition that divided the community. To address the problems Fogo Island was facing, the government had proposed to relocate the entire community to a nearby 'development town'. Colin Low, a film maker, and Don Snowdon, an academic, came up with the idea of using video as a way to facilitate dialogue between residents and government officials. In doing so, they invented a revolutionary film-making process centered on some key community-based concepts. They required every subject to have full editorial rights over their appearance, and every effort was made to ensure that the government would respond to the community-produced video (Frantz 2007, 104).

While there was never any formal evaluation of the Fogo project, shortly after the films were produced, a fishing cooperative and a school were established on Fogo Island and the community was not relocated. The question then arose as to whether projects like Fogo could work in urban areas if properly designed, and, if so, to what ends (Weisner 1992, 71). Participatory video is still an evolving action research methodology, immensely helped by recent technological advances in cameras and editing software, and related cost reductions. Within the last decade there has been a 'democratizing' of film/video making, along with many other important tools of the digital revolution, all of which amount to technological developments that offer enormous communicative and activist possibilities for urban planning and policy, if only we can grasp the centrality of storytelling to our mission (Chapter 7) and envision equally democratic ways of harnessing this potential.

This essay explores some of the potential applications of multimedia[1] in the urban policy and planning fields in the early 21st century, almost 100 years since the emergence of 'Town Planners' Cinema'. This is an epistemological as well as a pragmatic exploration, probing the capacities of multimedia as a mode of inquiry, as a form of meaning making, as a tool of community engagement, and as a catalyst for public policy dialogues. This is not an entirely new epistemological excavation. What is new are the tools that we are exploring and the approach that we are developing, which we describe as 'digital ethnography' (Attili 2007, 2010a; Sandercock and Attili 2010), a qualitative inquiry using film and multimedia languages and informed by collaborative planning theory and situational ethics.

After outlining this approach, we then explain the research question we chose to work on: 'how do strangers become neighbors?' and our research 'site', one of the most culturally diverse neighborhoods in the City of Vancouver, and a specific local institution, the Collingwood Neighbourhood House (CNH). The next section describes and discusses the three-stage research process, beginning with the making of the film *Where strangers become neighbors: The story of the Collingwood Neighbourhood House and the integration of immigrants in Vancouver* (Attili and Sandercock 2007). The final section offers reflections on the successes and limitations of this project.

Epistemological Shifts

The beginning of an epistemological shift in the field of planning was foreshadowed in the early 1970s. John Friedmann (1973) outlined a 'crisis of knowing', hammering the limitations of 'expert knowledge' and advocating a new approach which he called 'mutual learning' or 'transactive planning', an approach which could appreciate and draw on local and experiential knowledge in dialogue with expert knowledge. Accompanying a broader post-positivist movement in the social sciences (Stretton 1969; Rabinow and Sullivan 1987; Bourdieu 1990), pushed further along by feminist and postcolonial critiques (Said 1979; hooks 1984; Kelly 1984; Trinh 1989), planning scholars began to see the need both for an expanded language for planning and for ways of expanding the creative capacities of planners (Landry 2000, 2006; Sandercock 2003, 2005a, 2005b; Sarkissian and Hurford 2010) by acknowledging and using the many other ways of knowing that exist: experiential, intuitive, and somatic knowledges; local knowledges; knowledges based on the practices of contemplation and dialogue; and knowledges expressed in visual, symbolic, ritual, and other artistic ways. An 'epistemology of multiplicity' (Chapter 2) would nurture these other ways of knowing, without discarding or dismissing more traditional forms of scientific or technical reasoning.

The 'story turn' in planning has been one response to this epistemological crisis. In the past two decades, a growing number of planning scholars have been investigating the relationship between story and planning (Chapter 7). These investigations highlight how planning is performed through stories, how the communicative dimension is central to planning practices, and how story can awaken energies and imaginations, becoming a catalyst for involving urban conversations, for deep community dialogues. There are diverse ways in which multimedia can nurture community engagement and community development as well as oppositional forms of planning (Sandercock and Attili 2010). Multimedia tools also create the opportunity for urban researchers to discover new realities, to expand the horizons of qualitative and quantitative research, and to represent the city in multidimensional and polyphonic ways (ibid). And multimedia products can offer transformative learning experiences, 'educating the heart' through mobilizing 'a democracy of the senses' (Back 2007). In this paper, we outline our own approach and reflect on our first attempt at applying it.

Digital Ethnography as Planning Praxis

Digital ethnographies are self-reflexive analytical practices aimed at portraying lives and stories and creatively expressing meaningful narratives. The polyphonic ethnographic narrative is the result of a series of in-depth interviews acquired after a process of building rapport and trust, an activity based on a deep interaction with people who entrust their stories to us. It is an interpretive practice, the aim of which is to make sense of phenomena in terms of the meanings people bring to them (Denzin and Lincoln 2005; Denzin 1997). Learning from our predecessors in visual anthropology (see Asch 1992; Banks 2001; Ruby 1982, 1995, 2000), we are very aware, in our practice of digital ethnography, of ethical issues and power relations.

This praxis is a very involving process in which the researchers may develop close relationships with interviewees, and may remain close during the phases of data analysis, feedback, and editing of the film and beyond, to the arenas of dissemination and subsequent action. As researchers, we consciously expose ourselves to reactions from the research environment – both positive and negative – and we derive benefit from the learning effect which is built into this strategy. In this way, the researcher becomes part of the phenomenon being studied, perhaps experiencing the uncomfortable sense of being both inside and outside the project. This existential reality invokes a series of important ethical issues and demands self-reflexivity around relationships of power, including the power to narrate.

Situational Ethics

Reflexivity is at the heart of an ethical process. The question of ethics has been central to ethnographic film making for many decades (see Asch 1992; Banks 2001; Geertz 1979; Ruby 1982, 1995, 2000) and is equally pertinent to multimedia projects in planning. Typical ethics reviews in university research settings cover such formal and universalized issues as inclusion and exclusion, potential risks to interviewees, and protocols for feedback with interviewees. We challenge this universalizing approach, which has been based on an assumption that social research is value-free and that its ethical dilemmas could be addressed through positivistic and standardized procedures. We argue that research is a contextualized activity in which values play a significant role. We agree with Denzin (2003) that:

> participants have a co-equal say in how research should be conducted, what should be studied, which methods should be used, which findings are valid and acceptable, how the findings have to be implemented and how the consequences of such actions are to be assessed.
>
> *(Denzin 2003, 257)*

This reflects a different commitment on the part of the researcher, who is no longer answerable to a 'remote discipline or institution but to those he or she studies' (Denzin 2003, 258). This approach implies a constant redefinition of what needs to be done to address ethical challenges in social research. We are dealing with embodied relationships that embrace the research participants. From this perspective, the research cannot be seen as a pre-established or pre-approved set of procedures. Rather, it needs to be built in collaborative ways through a mutual learning process. Researchers and communities have a co-responsibility in defining the proper strategies to deal with ethically sensitive issues.

There are two ethical dimensions of special significance to multimedia producers: the politics of voice, and the issue of authorship/ownership. One aspect of the ethical complexity of documentary or ethnographic film making resides in the very act of advocating for someone else. Planning has been historically imbued with this sense of assuming the needs of others, of claiming to understand others. Advocacy planners exalted this mission, traditionally working for marginalized communities and expressing their needs 'in a language understandable to his client and to the decision makers he seeks to convince' (Davidoff 1996, 307). This assumption has been widely challenged, especially by feminist and postcolonial critics (Trinh 1989; hooks 1990). Who is speaking, who is spoken of, and who listens is a result as well as an act of political struggle (Alcoff 1995, 105).

Multimedia offers ways of addressing this dilemma through critical reflection on the politics of voice, a reflection which asks at the start as well as throughout a project, who is speaking, who is spoken of, and who is listening. These considerations have important implications for planning theory. The advocacy planning approach signified a historically disruptive shift which was able to make visible and integrate the inevitably political dimension of the work that planners do. Nevertheless, it appeared shackled by a certain naivety in failing to address the core of the (power) relationship between planners and those being planned for. Assuming a heightened ethical awareness connected with the concept of asymmetrical reciprocity, advocacy planning can now redefine itself: as a contextualized and situational practice in which planners and communities co-develop strategies to address shared values and needs. In other words, advocacy planning matures into collaborative planning. In the next section we discuss how these critical reflections applied to our film project.

A second ethical dilemma, familiar in classical anthropology and ethnography, is the oft-repeated charge that its practitioners are the a-moral plunderers of the stories of 'Others', typically those with little power, for personal as well as scientific aggrandizement and without any accompanying sense of reciprocity. Eventually, an auto-critique developed within anthropology, and a lot more emphasis is now placed on thinking through what constitutes an ethical relationship in terms of research design and methods, authorship and ownership, trust and reciprocity. All of these issues are germane to multimedia production in general, and to our project. Our approach, informed by a situational ethics, is that all of these questions must be talked through before a project begins, and protocols should be established collaboratively. If an individual, organization, or community is willing to have their story told by an outsider/researcher, a collaborative partnership of understanding needs to be developed. Extended dialogue needs to occur around the following questions: What's in it for those whose story is being told? Will the project be designed in such a way as to ensure their needs are likely to be realized? What control will they have over how their interviews are used? How will they be consulted or involved throughout the production and post-production process? Will authorship be shared? Will ownership of the multimedia product be shared, including any potential profits? What's in it for the researchers/producers? Will there be an ongoing relationship after the production is finished? What are the action components of the project? Is the researcher prepared to acknowledge the gift of this story, and what can s/he offer in return?

While it is unlikely that such a collaborative approach can ever be a perfectly symmetrical power relationship, researchers can make an effort in this direction by openly discussing the questions identified above and co-creating both a process and an outcome that addresses a shared sense of purpose and meaning making. This was our intent throughout the three-year research project with the Collingwood Neighbourhood House, as the following narrative explains.

The Research Question and Collaborative Research Protocols

Not long after moving to Vancouver, I (Leonie) was introduced to the work of the Collingwood Neighbourhood House (CNH) by a Masters student, Steven Dang, who then introduced me personally to the executive director, Paula Carr. I was drawn to CNH because of its mission of social inclusion and its reputation as having created a welcoming place and having done a lot of work to overcome the anti-immigrant sentiment that had existed in the neighborhood 20 years earlier. As a result of a change in provincial government, CNH was now faced with budget cuts to many of their programs and services. Through a series of conversations between myself and Paula Carr, the idea emerged of a university–community collaboration in which I would use the resources of my new multimedia laboratory to make a film telling the CNH story. Our thinking was that a well-crafted story could appeal to funders in a more emotionally powerful way than was available by reading annual reports, which tend to be dry, statistics-based documents whose primary purpose is to account for how funds are spent.

I was about to start a research collaboration with Giovanni Attili from the University of Rome, who was interested in similar research themes and had been experimenting with multimedia as part of his doctoral thesis (Attili 2008, 2010b). From the very beginning, there was both an advocacy and collaborative dimension of our film/research project. Our research agenda was twofold. We wanted to understand how this neighborhood of 48,000 people had overcome the anti-immigrant sentiment that existed there (and in the city at large) in the 1970s and 1980s, and succeeded in becoming a remarkably integrated intercultural community – how a once predominantly Anglo-European population had adapted itself to the arrival of large numbers of newcomers from East and Southeast Asia, the Middle East, Latin America, and Africa. How did this process of immigrant integration actually work, on the ground? How did immigrants begin to establish a new life and develop a sense of belonging to a new place? How did strangers become neighbors? How important was CNH's role in this? And we also wanted to 'experiment' with the use of film, not only as a mode of inquiry but also as a way of communicating research results and of catalyzing community dialogue around these issues in other places. There was always an action component of our research agenda, in that we believed that the CNH story was a potentially inspirational one which deserved to be told across Canada. (We didn't yet imagine how relevant it might be in European countries experiencing similar unsettlings as a result of immigration from culturally diverse source countries.)

CNH folks were comfortable with our research objectives, which seemed compatible with their own desire to document their achievements – in building community, and in using a community development approach as a way of integrating immigrants – for fund-raising purposes and for the sharing of institutional learning. We reached informal rather than contractual agreements on a timeline and research protocols. The research work would involve a group of University of British Columbia Masters students, supervised (in a course called 'Digital Ethnographies in Urban Planning') by Giovanni and myself. Groups of three to four students (15 in total) would work together, researching issues jointly identified by CNH and the researchers. Each group of students would make a five-minute video as their 'product', which would be screened at the annual general meeting of CNH.

After the class ended, Giovanni and I continued with the research, aiming for a 'premiere' screening at the 20th anniversary of CNH in November 2005. We would then donate as many copies of the video (in DVD format) as CNH requested, to use however they wanted to use them. At this stage it did not occur to us that our research product might have commercial value or potential, so that aspect was not discussed. We did, however, discuss how the research would proceed in collaboration with CNH, and the executive director designated the coordinator of Settlement Services, Val Cavers, as our liaison. Val would assist us in accessing both CNH archives and initial local contacts whom we might interview. In exchange for this assistance, I offered a 'consultant's honorarium' from my research grant. The final piece of our research protocol involved protecting interviewees. We wrote into the Consent Forms that everyone whom we interviewed would have a chance to see how we used their interviews at an early rough-cut stage, and if they were not happy, they could insist on being taken out of the film.

Research, Filming, and Post-Production

While I worked with students on the research questions, the storytelling approach, and the neighborhood context, Giovanni was introducing them to the philosophical elements of the digital ethnographic approach. He was also teaching students practical skills: how to use a video camera, do sound recording, conduct interviews, and learn editing software. We did walking tours of the neighborhood, and then brainstormed research themes for each group, themes that would each contribute to an understanding of the larger research question.

Over the next three months, students spent a lot of time interviewing people in the neighborhood, oldtimers and newcomers, as well as staff, volunteers, members of the resident-run board that governs CNH, and residents, city planners and politicians who had been involved in the initial community planning process that had resulted in the residents' desire to create the CNH as both a gathering place and a facility that could provide services for families and children. Once the work of the students was finished, we started to work on the next phase, identifying further research, expanding the horizon of the interviews, and reflecting on the shape and content of the story we wanted to tell: the story of a neighborhood that, just 20 years earlier, had been locking its doors, afraid of change, and telling immigrants to go back where they came from, but is now a welcoming place. We spent three more months talking with many people in an ever-widening web of relations. We were developing a sense of identification with and affection for the neighborhood and were, on a number of occasions, moved to tears by the stories we were being told, the passionate intensity of the lived experiences of upheaval, arrival, struggle, and achieving a new sense of belonging. At the same time, we were also trying to think critically and analytically about what we were absorbing and how to interpret it, always aware that our own voices would ultimately mingle and dialogue with those of our interviewees, both implicitly, in the way that we chose to edit; and explicitly, in the narration that I was beginning to write.

The decision to use voice-over narration came about in the post-production stage. When we began the project, our reflections on the politics of voice suggested to us that we wanted this community and this organization to tell their own story, in their own words, rather than us speaking for them. But once we began editing, we realized that a narrative voice was necessary both to thread these pieces together, and to connect some of the micro-stories with the larger socio-cultural and political economic forces at play in Canada and globally. We were, always consciously and reflexively, trying to be in and out at the same moment, walking a tightrope between empathic compassion, admiration, and critical distance. The voice-over narration also serves to make transparent that our own interpretive voice is the frame of the film.

Research Findings

We have written at length about the answers we found to our substantive research question: how do strangers become neighbors? (Sandercock and Attili 2009). Here we focus on the second and third parts of our research agenda: the experimental use of film as a research tool and as a communicative device.

Our approach to editing was a two-stage process, in the first of which we asked the following questions: What is the story we want to tell? What is the answer to our initial question (how do strangers become neighbors?) and how can we best tell that story? The first challenge was how to tell the story in a powerful way, as narrative, using as much as possible the voices of the actual community. We then essentially had a 'talking heads' story. The second stage was the visual storytelling, the layering in of all the digital material we and the students had been collecting (photographs, old video footage, newspaper clippings, posters, location footage, statistics), and the creative invention of visually stimulating ways of placing all this on a screen, often telling multiple stories in one frame through the use of multiple images alongside and around a 'talking head' or voice-over narration.

The theme that emerged most powerfully from the multiple interviews was how this neighborhood had been transformed over a 20-year period, from initial fear, anxiety, and even hostility towards new immigrants to eventual acceptance and working together to build a community where none had previously existed. It was clear that the central actor in this story was the Collingwood Neighbourhood House itself, but the story of the CNH was in turn a story about the people who founded it, their vision and mission, and their ongoing efforts at outreach and social inclusion. We also realized that it was important to explain how and why this particular local institution was established in 1985, as that story reflected something about the planning culture in the City of Vancouver (a larger contextual element of explanation). The short answer to the question of how strangers had become neighbors in Collingwood was the community development approach of the CNH (as opposed to being simply a service and program delivery agency) and the vision of social inclusion. Our challenge was how to explain this community development process and the gradual shifting in attitudes which it produced.

We decided to construct the narrative both as a 'then and now' story and as a sequence of stories around themes, programs, and events which illustrated the community development approach. Beginning with the 'origins story' of how the CNH came into being, we then proceeded to construct sequences focusing on the growth of the CNH, both physical and programmatic; the development of a Community Leadership Institute that helped local residents to acquire leadership skills in order to take on important roles in the community; the development of the Buddy Program, which involved matching local youth with newcomers and then developing programs taught by youth themselves about anti-racism and bullying; the unearthing of local artistic skills through the Arts Pow Wow, a community cultural development process that began with an inventory of residents' skills and then involved mobilizing those skills in an array of community development activities led by these local artists (from gardening to music and dance to lantern making, puppetry, and storytelling, all of which are used to enhance local celebrations); the story of the reclamation of a local park from drug dealers and gangs, and its transformation through a participatory design process led by local residents; and the story of the annual celebration of cultural diversity in 'Multiweek', one week in February during which the many cultures that co-exist in this neighborhood come together to share traditions and customs.

As we reviewed our digital interviews, we realized that there were five women whose personal stories of leaving, arrival, struggle to find their place, and achieving a sense of belonging through the CNH could drive the narrative forward and provide a very involving emotional dimension. We used their voices to convey, through their life stories, what the CNH had been able to achieve. We also realized that in order to weave these sequences together, we needed a voice-over narration, which could also convey some analytical and thematic information such as 'the age of migration' and the global forces operating on a neighborhood, as well as the national political philosophy of multiculturalism that shaped federal and provincial funding and enabled the CNH to access resources for the kinds of programs it was inventing. So, our own interpretive voice was added to provide these additional contextual dimensions, as well as to make our own part in the collaboration more visible.

As we worked on the film, it became more and more evident to us that we were making a kind of advocacy documentary, one that might be used in the policy and funding arenas to raise awareness of the importance of a local, place-based, community development approach to foster the social integration of immigrants, as well as specifically helping the CNH with its funding crisis. In other words, we understood that we were acting politically in making this film, in assuming the responsibility of narrating an inspiring story, showing how a local institution managed to help a neighborhood to overcome its fear of strangers and, in the process, to do remarkable community building work.

Every single choice we made during the editing process was connected with these emerging objectives, which shaped not only the construction of the story itself but also the way it was told visually. Once we had a rough-cut (65 minutes in length, although we had set out to make a 30-minute film), we invited members of the resident-run board of the CNH, the executive director, and the coordinator of Settlement Services to watch it with us and give us their feedback. Had we adequately captured their story? Had we misrepresented anything? Had we left out important elements? What could be cut? Did the film work at an emotional as well as an informational level?

The response from this first 'audience' was both exuberantly supportive and critical. Overall, the verdict was that the film was emotionally very powerful, because of the use of the five personal stories. The most significant critical feedback we received was that we had not included any Indigenous voices or indicated how much outreach work the CNH had done with Indigenous residents (almost 1,000 of the 48,000 residents). We had been trying all summer to reach the Elder whom we had been told was the appropriate person to speak to, but we could not make contact. We tried again, with no luck. We knew the importance of the protocol of speaking first with Elders, and having Elders speak for that community. Our compromise was to construct a sequence which described the proposal to build a Native Housing Coop in the neighborhood, the initial opposition to that, and how that was resolved; and to show the Coop as it exists today, an important and beautiful part of the neighborhood. Part of this story, too, was how in the process of working through this conflict, non-Indigenous residents came to appreciate that their Indigenous neighbors were bringing assets to this community as well as needs.

Once we made these changes, we invited the CNH folks back for a second screening. We now had a 70-minute film, which we felt was too long for prospective audiences. We had to think about who our target audience was, what length would work for different audiences, and how much 'information' we needed to convey, about programs and services for example, as opposed to the more involving personal stories. We knew that some viewers would want to know how the CNH was financed, and what kinds of programs were offered. But conveying that information is hard to do in an interesting way, and always tends to detract from emotional involvement in the narrative. Here we had to acknowledge and accept community priorities, as opposed to our artistic instincts about storytelling. We also saw a creative challenge in how to present more quantitative types of information in visually compelling and understandable ways.

Dissemination of the Film/Research

The single most important audience, initially, was the local community, and they were the first to see the film, at the 20th birthday celebration of the CNH in November 2005. On that occasion, their gym was converted into a big movie theater holding 400 people. CNH rented a huge screen and professional sound equipment. It was the first official projection of the documentary, and the audience's reaction was exuberant. Many people were in tears after the screening, deeply moved by the inspiring story. There was incredible pride as people realized what they were part of, seeing the bigger picture of a complex story of which they only knew the small segments in which they had personally been involved. The consciousness of being part of a bigger story was both empowering and an important step in a renewed sense of belonging. For many, it was akin to a ritual passage that enabled them to make sense of the previous 20 years of history of this institution. It was both energizing and an opportunity, subsequently, for further reflection by CNH staff and board about the future.

The screening generated a lot of interest in the film, and we began to receive a lot of requests for further screenings. We now had to think about the distribution and dissemination issue: how could we get the film out there, without ourselves traveling with it everywhere? Simultaneously we were discussing with CNH how they would like to use the film. We all wanted it shown as widely as possible, because of its educational value and policy relevance. CNH agreed that the ideal distributor would be the National Film Board of Canada (NFB).

While we waited for the NFB to review the film, I worked with CNH's executive director, Paula Carr, on a grant application with the idea of producing a manual to accompany the film, and then using both film and manual in a series of workshops in different Canadian cities. We got the grant, from the Metropolis research network,[2] and hired Val Cavers, who had just retired as coordinator of Settlement Services at CNH, to research and write the manual. We agreed that the manual would try to convey more details of how the CNH had done what it had done, and that it would be written as much as possible through stories.

Our Metropolis grant funding enabled Paula and I to visit four cities (Edmonton, Halifax, Moncton, and Toronto) to run half-day workshops using the film as a catalyst for community dialogue about immigration and integration, and leaving participants with copies of the manual to take back to their organizations.[3] The collaboration with the Metropolis research network was an important part of this stage of the project. Each of the Centers of Excellence advertised our workshops through their local networks, so we were able to reach exactly the audiences we hoped to reach: local organizations involved with immigrants and refugees, non-profits, and faith-based organizations as well as municipalities that were becoming increasingly concerned about the challenges of the arrival of significant numbers of newcomers. Provincial and federal policy makers also attended in each city, as well as folks from the arts sector and university researchers. We used the same format for each workshop: a brief introduction about why we'd made the film, followed by a screening, followed by 90 minutes or so of facilitated discussion, tailored to the interests and concerns of the specific audience.

Discussions were extraordinarily animated, and the response was extremely positive. In all post-workshop evaluations, 90% of attendees rated the workshops in the top category (on a scale of one to five) and rated the value of the film at the same level. We received many invitations to screen the film and facilitate discussion, from government agencies to community meetings to keynotes at academic conferences,[4] as well as to show it at several international film festivals.[5] Perhaps most gratifying of all was the use of the film as the 'evidence' in support of CNH's self-nomination for the BMW Award for Intercultural Learning in the 'Practice' category in 2007. CNH was awarded First Prize, and the executive director and the chairman of the board of CNH were flown to Munich to receive the award, the prize money, and a lot of attendant publicity. It was the first time in the ten years of this award that an organization outside of Europe had received the prize.

Finally, we decided that in order for the film to have any significant impact in Europe, we would need to write a book to accompany and contextualize the film for European audiences, setting it in its metropolitan planning context as well as the national political, legal, and philosophical contexts (Sandercock and Attili 2009). The book, which includes the DVD, was again a collaborative enterprise between ourselves and Paula and Val, the executive director and the former coordinator of Settlement Services.

Evaluation

How can we gauge the successes and weaknesses of this project? Given the intent expressed earlier, the project could be assessed by the community dialogue it produced; by the range of organizations currently using the film for education and training; by the number of invited screenings and awards; and in the longer term, by community and government responses. Outcomes are hard to identify and even harder to measure in this kind of research, but here is what we have observed, evaluated, and heard, three years after the film was finished and two years after its official release by the NFB.

Impact on the Local Community

One more or less immediate impact was the sense of psychological empowerment experienced by local residents and the CNH. Even those interviewees who had been involved with the CNH from its inception, or over many years, had a kind of epiphany about the scope and achievement of the organization and the way it had transformed so many lives. Among those interviewees who were less frequent users of programs, or had been volunteers at a certain stage of their lives in the neighborhood, there was an eye-opening sense of the larger history of which they had been a part. This realization, and the enormous pride it engendered, has been energizing and renewing for staff, volunteers, and the board, as well as local residents.

It was a surprise to us, as film makers, that the film encouraged the local community to reflect about themselves, and helped to further motivate and empower people. In a sense, we were performing what anthropologist Barbara Myerhoff (1982, 1988) has described as the role of witnessing, creating a space and a medium through which people could show to outsiders as well as to each other who they are, why they matter, what the nature of their past and present lives is and has been. Myerhoff, who made a celebrated film about Elders in a Jewish community center in Los Angeles, describes this as a search for 'arenas of visibility' for people who feel that their lives have been increasingly marginalized (Myerhoff 1988, 271). This witness role is particularly important when people have been rendered invisible by historical processes beyond their control, such as the effects of migration. And that was exactly what our interviewees described in the film: the migrant experience on arriving in a new country, of feeling 'invisible', of being 'nobody'. We learned about the necessity and power of witnessing as a part of film making, and how film can be 'a service as well as a record' (Myerhoff 1988, 282).

The subsequent publicity and attention focused on CNH both nationally and internationally, especially as a result of the BMW Award for Intercultural Learning, has brought an enormous sense of recognition and generated new opportunities. In late 2008, Paula Carr put forward a proposal to the board to start a social enterprise within CNH, offering their acquired expertise to other non-profits and government agencies as consultants in 'creating welcoming communities'. In this work, the film is being used as a catalyst and training tool.

Wider Impact

As word has spread about the film through screenings, conference presentations, and keynotes, a number of organizations have purchased the film and are using it in training with their staff. These organizations include the municipalities of Halifax, Nova Scotia; Richmond and Vancouver, British Columbia; the planning department of the City of Auckland, New Zealand; the City of Oslo and the Police Department of Oslo; the Norwegian National Housing Authority; the Race Relations Office of the Northern Ireland Housing Executive; and the Salem-Keizer Community Development Corporation in Oregon, USA. This is a random sample, known to us only because these organizations approached us for copies rather than purchasing from the NFB.

There has certainly been great interest and lively debate at all of the screenings of the film that we have attended, whether in universities, conferences, or communities. Anecdotal evidence from these occasions is that the film works well as a provoker of dialogue, as an inspiration, and a learning tool that demonstrates the effectiveness of community development approaches to particular issues. But the only systematic evidence that we have is from the aforementioned four workshops organized using our Metropolis research grant. The 90% approval rating for both workshop and film after each of these workshops seems to confirm both the power of the story itself and the power of film as the storytelling medium.

Critical Reflections

In developing our approach to digital ethnography as planning praxis, our intent is both to evoke and to provoke: to evoke the richness and diversity of urban life through a polyphonic multimedia approach, and to provoke an urban conversation or a community dialogue about the subject matter. There is always an action research agenda, which is context specific and affects how we 'frame' the project, ethically and politically. What follows is a self-evaluation, bearing in mind these intentions.

On reflection, there were a number of failures and omissions in the storytelling, some of which we only realized through dialogues with various audiences. Perhaps the most frequently asked question at screenings was the relative absence of conflict in the story. Our answer to this expressed our own frustration at not succeeding in getting interviewees to tell their stories of their experiences with racism or xenophobia *on camera*. We heard a lot more, in informal conversation with residents, about levels and forms of racism and xenophobia, but understandably interviewees were reluctant to open up about this, presumably because to explain details would likely implicate people with whom they may have ongoing interactions. But the unfortunate effect of this reluctance – along with the fact that, ten minutes into the film, one resident tells how a few hundred of the oldtimers, who were racist and didn't like the changes happening in the neighborhood, just 'up and left and moved out to Surrey'[6] – is to create the misleading impression that a harmonious immigrant neighborhood was then possible because of White flight. This was not in fact the case. The numbers of Anglo-European residents remained almost constant over the past two decades, while the overall population increased due to immigration. But the resulting absence of discussion of race-based conflict is a problem in filmic terms, resulting in less tension and less of a sense of what was at stake here in the past two decades. It is even more of a problem in terms of social learning. Other communities could learn more from the film if they had a better understanding of how great the challenges were back in the 1980s.

One solution would have been to insert a voice-over narration outlining the different forms and manifestations of race-based conflicts that had existed, and that we had learned about from interviewees. It did not occur to us to do this at the time: a major failure of critical perspective on our part. In trying to understand this omission, from a distance now of several years, we realize that we were too single-mindedly focused on the inspirational story of the CNH, while ignoring the inevitable dark side of the (of any) community. Even so, we should have given much more emphasis, in the film, to the extensive and ongoing anti-racism training conducted by CNH for board members, staff, and volunteers, and their many proactive efforts involving anti-racism work – for example, in the immediate aftermath of 9/11.

There was a second dimension of conflict implicit in our story, involving resistance to some of the activities that CNH sought to initiate. There is one such case study in the film, concerning initial opposition to the building of the Native Housing Coop. That is a story with a happy ending. But there were other instances – such as the introduction of a breakfast and shower program for homeless people – which we did not describe. As we began screenings of the film, and became more aware of this shortcoming, we sought to address it in the manual and the book, foregrounding stories of specific conflicts and providing a detailed description of how these conflicts were handled by CNH. We are more aware now than we were during the making of the film of how attached we became to CNH and how much we wanted to create an inspiring narrative. What we overlooked at the time, in spite of having always been aware of this in theory, was that stories of struggle are even more inspiring than stories of achievement with no apparent struggle.

The second most frequently asked question concerned the apparent dominance of women in CNH activities. This was the result of obtaining more lively interviews from the women than from most of the men we interviewed, resulting in an editing/selection process that gives an inaccurate portrayal of the levels of involvement of men in CNH programs and governance. Reflecting on what we learned from this, in future projects we would either spend more time preparing with interviewees who are intimidated by the camera, or we would record them on audio tape and create audio montages of voices, in order not to exclude people on the grounds of their shyness or stiffness on camera.[7] Interestingly, though, we found that people's level of articulation and expressiveness was not closely related to class and education. This is a low to moderate income neighborhood in which 73% of the population does not speak English as their first language, yet most people were wonderfully engaging interviewees, irrespective of their English language fluency.

A third omission is the absence of voices from the Indigenous residents of the neighborhood, as already explained. A related shortcoming was that our voice-over narration goes straight into the story of late-20th-century immigration to Canada, without acknowledging the original inhabitants of this land. Because we were worried about the length of the film, and felt that a couple of sentences on this topic could easily be regarded as 'tokenism', we chose not to mention it at all: a regrettable decision. In Canada, Indigenous Peoples are still locked in a struggle with provincial and federal governments over land and sovereignty. The history of colonization and the attempted cultural genocide of Indigenous Peoples is not well known within Canada, let alone beyond, and remains the dark side of Canadian history (see Chapter 9).

The three shortcomings just identified relate specifically to the content of the film. Looking at the process, though, and the approach we are developing for digital ethnography as planning praxis, there are four important lessons that we would like to reiterate. The first is the embracing of a *situational ethics* which enables us to develop a collaborative approach with those whose story we are telling, at every stage in the process. The second is close attention to the *politics of voice*, to ensure that we are not speaking for others, along with a parallel awareness that our own interpretive voice is the frame of the film, and should be transparent. A third lesson is the importance of identifying, at the outset of a project, one's desired *audience/s* and developing a plan for getting the film to that audience. And finally, when contemplating distribution, be wary of entering into contracts with commercial distributors who will insist on owning the copyright. We learned the hard way. After we had signed our contract with the National Film Board of Canada, the CNH asked us if we could make a 15-minute version of the film for their own use in short meetings with funders. The NFB refused us permission to use our own material in this way (seeing this as competition). Lesson four, then, is to *keep control/copyright* of one's material.

This essay inevitably reflects our own excitement about the ways in which multimedia can be used by activists, NGOs, immigrant and Indigenous communities, planning scholars, and educators wherever urban policies and planning strategies are being debated and communities are struggling to shape, improve, or protect their life spaces. This is an exploration of a new frontier in the urban planning and policy fields, a frontier 'beyond the flatlands' of imperial cartography and narrative.

Notes

1 *Multimedia* refers to the combination of multiple contents (both traditional and digital: texts, still images, animations, audio and video productions) and interactive platforms (offline interactive CD-ROMs, online websites and forums, digital environments) in the urban policy and planning fields.
2 The Metropolis Research Network is funded jointly by the Canadian government (Department of Citizenship and Immigration) and the Social Sciences and Humanities Research Council, and is now in its third 5-year funding cycle. There are five Centres of Excellence in five cities, each involving a collaboration between universities, government agencies and NGOs.
3 The Manual tells how the CNH developed its unique approach of community development through relationship-building; what kinds of programs and services it offers; where it gets its funding; how it trains staff and volunteers; how the resident-run Board operates; how they develop community leadership; why its approach is intercultural rather than multicultural; and the values that have shaped this very special place.
4 We personally screened the film at thirty-two venues between Nov 2005 and July 2009.
5 The film was awarded an Honourable Mention (International Federation of Housing and Planning, International Film and Video Competition, Geneva, 2006) and a Special Mention (Berkeley Video and Film Festival, 2006).
6 Surrey is an outer suburban municipality which, thirty years ago, had very few immigrants but now is the fastest growing area in the region and just as culturally diverse as Collingwood.
7 Thanks to Jessica Hallenbeck for this insight.

References

Alcoff, L. M. 1995. The problem of speaking for others'. In: Roof, J. and Wiegman, R. (Eds) *Who Can Speak? Authority and Critical Identity* (Urbana: University of Illinois Press).

Asch, T. 1992. The ethics of ethnographic film making. In: Crawford, P. I. and Turton, D. (Eds) *Film as Ethnography* (Manchester: Manchester University Press).

Attili, G. 2007. Digital ethnographies in the planning field, *Planning Theory and Practice* 8(1), 90–98.

Attili, G. 2008. *Rappresentare la Citta dei Migranti* (Milan: Jaca Book).

Attili, G. 2010a. Beyond the flatlands: digital ethnographies in the planning field. In: Sandercock, L. and Attili, G. (Eds) *Multimedia Explorations in Urban Policy and Planning: Beyond the Flatlands* (Dordrecht: Springer).

Attili, G. 2010b. Representations of an *unsettled* city: hypermedial landscapes in Rome. In: Sandercock, L. and Attili, G. (Eds) *Multimedia Explorations in Urban Policy and Planning: Beyond the Flatlands* (Dordrecht: Springer).

Attili, G. and Sandercock, L. 2007. *Where Strangers Become Neighbors: the Story of the Collingwood Neighborhood House and the Integration of Immigrants in Vancouver* (Montreal: National Film Board of Canada (50-minute documentary).

Back, L. 2007. *The Art of Listening* (Oxford and New York: Berg).

Banks, M. 2001. *Visual Methods in Social Research* (London: Sage).

Bourdieu, P. 1990. *Other Words: Essays Towards a Reflexive Sociology* (Cambridge: Polity Press).

Ciacci, L. (Eds.) 1997. *Il Cinema degli Urbanisti* (vols. 1, 2) (Modena: Comune di Modena).

Ciacci, L. 2010. Film works wonders: analysis, history and town plan united in a single representation. In: Sandercock, L. and Attili, G. (Eds) *Multimedia Explorations in Urban Policy and Planning: Beyond the Flatlands* (Dordrecht: Springer).

Davidoff, P. 1996. [1965]. Advocacy and pluralism in planning. In: Campbell, S. and Fainstein, S. (Eds) *Readings in Planning Theory* (Malden: Blackwell Publishing).

Denzin, N. K. 1997. *Interpretive Ethnography: Ethnographic Practices for the 21st Century* (Thousand Oaks: Sage).

Denzin, N. K. 2003. *Performance Ethnography: Critical Pedagogy and the Politics of Culture* (vol. 2) (Thousand Oaks: Sage).

Denzin, N. K. and Lincoln, Y. S. (Eds.) 2005. *The Sage Handbook of Qualitative Research* (Thousand Oaks: Sage).

Frantz, J. 2007. Using participatory video to enrich planning process, *Planning Theory and Practice* 8(1), 103–107.

Freire, P. 1970. *Pedagogy of the Oppressed* (New York: Herder and Herder).

Friedmann, J. 1973. *Retracking America* (New York: Doubleday Anchor).

Geertz, C. 1979. From the Native's point of view: on the nature of anthropological understanding. In: Rabinow, P. and Sullivan, W. M. (Eds) *Interpretive Social Science: A Reader* (Berkeley: University of California Press).

hooks, b. 1984. *Feminist Theory: From Margin to Center* (Boston: South End Press).

hooks, b. 1990. Marginality as site of resistance. In: Ferguson, R., et al. (Eds) *Out There: Marginalization and Contemporary Cultures* (Cambridge, MA: MIT Press).

Kelly, J. G. 1984. *Women, History, and Theory* (Chicago: University of Chicago Press).

Landry, C. 2000. *The Creative City* (London: Earthscan Publications).

Landry, C. 2006. *The Art of City Making* (London: Earthscan Publications).

Myerhoff, B. 1982. Life history Among the elderly: performance, visibility, and re-membering. In: Ruby, J. (Ed.) *A Crack in the Mirror. Reflexive Perspectives in Anthropology* (Philadelphia: University of Pennsylvania Press).

Myerhoff, B. 1988. Surviving stories: reflections on number our days'. In: Kugelmass, J. (Ed.) *Between Two Worlds* (Ithaca: Cornell University Press).

Rabinow, P. and Sullivan, W. M. (Eds.) 1987. *Interpretive Social Science: a Second Look* (Berkeley: University of California Press).

Ruby, J. 1982. The Mirror. In: Crack, A. (Ed.). *Reflexive Perspectives in Anthropology* (Philadelphia: University of Pennsylvania Press).

Ruby, J. 1995. The moral burden of authorship in ethnographic film, *Visual Anthropology Review* 11(2), 77–82.

Ruby, J. 2000. *Picturing Culture: Explorations of Film and Anthropology* (Chicago: University of Chicago Press).

Said, E. 1979. *Orientalism* (New York: Vantage Books).

Sandercock, L. 2003. *Cosmopolis 2: Mongrel Cities of the 21st Century* (London: Continuum).

Sandercock, L. 2005a. A planning imagination for the 21st century, *Journal of the American Planning Association* 70(2), 133–141.

Sandercock, L. 2005b. A new spin on the creative city: artist/planner collaborations', *Planning Theory and Practice* 6(1), 101–103.

Sandercock, L. and Attili, G. 2009. *Where Strangers Become Neighbors: the Integration of Immigrants in Vancouver, Canada* (Dordrecht: Springer).

Sandercock, L. and Attili, G. (Eds.) 2010. *Multimedia Explorations in Urban Policy and Planning: Beyond the Flatlands* (Dordrecht: Springer).
Sandercock, L. and Attili, G. 2010. *Finding our way. 90-Minute Documentary* (Vancouver: Moving Images).
Sarkissian, W. and Hurford, D. with Wenman, C. 2010. *Creative Community Planning: Transformative Engagement Methods for Working at the Edge* (London: Earthscan Publications).
Stretton, H. 1969. *The Political Sciences* (London: Routledge and Kegan Paul).
Trinh, M.-H. T. 1989. *Woman Native Other* (Bloomington: Indiana University Press).
Weisner, D. 1992. Media for the people: the Canadian experiments with film and video in community development, *American Review of Canadian Studies* 2(1), 65–99.

9

CHANGING THE LENS

Film as Action Research and Therapeutic Planning Practice

Introduction

In 2008, the Canadian government apologized to the victims of the Indian Residential School system and established a Truth and Reconciliation Commission whose goal was to mend the deep rifts between Indigenous Peoples and the settler society that engineered the system. The residential schools (along with the Indian Act and the Reserve system of Indian Lands) represented one of three ways in which the Canadian government has systematically marginalized and reduced to dependency status formerly sovereign peoples, that is, Indigenous Nations. There is an emerging scholarly literature within planning which examines planning's role in this ongoing process of colonization (Jackson 1998; Jojola 1998, 2008; Porter 2010; Dorries 2012). There is also a parallel literature calling for the decolonization of planning practices (Hibbard and Lane 2004; Hibbard, Lane and Rasmussen 2008; Jojola 2008; Porter 2010; Walker, Jojola and Natcher 2013). This essay is situated within this critique and also within the normative orientation that seeks forms of planning intervention that can contribute to a larger societal project of justice and healing. We draw on an emerging orientation to a therapeutic mode of planning (Sandercock 2000; Forester 2009; Erfan 2013), which begins with a recognition that the past is often very present in planning conflicts and that past injustices shape what is possible in the present and often prevent constructive solutions from emerging, until and unless there is a process of recognition and working through these historic wounds and memories. Further, we situate ourselves in what has been variously labeled the communicative turn in planning (Healey 1997; Sandercock 2003; Forester 2009; Innes and Booher 2010) acknowledging the limits of technical rationality and exploring new tools for accessing other ways of knowing, learning, researching, and practicing. Specifically, we draw on an emerging literature and practice that explores multimedia interventions in urban policy and community engagement and planning (Sandercock and Attili 2010a, b) in designing an action research project in which film becomes the central planning tool.

This essay asks whether and what kind of planning intervention might contribute to a process of healing and reconciliation, helping a deeply divided community 'to move toward a relationship of power sharing, mutual respect and joint problem solving' (*Report of the Royal*

DOI: 10.4324/9781003325451-13

Commission on Aboriginal Peoples, vol 1, part 1, 1996). Further, it tests a new methodology in using film as a catalyst for this planning intervention. We begin with a brief sketch of the political and historical context of Indigenous–state relations in Canada. We then situate ourselves theoretically and methodologically in the literatures that address decolonizing and transformative and therapeutic planning as well as digital ethnographic techniques of intervention. The main part of the e describes and discusses our five-year action research project in north-central British Columbia, a project which began with the collaborative making of a film about the struggle between a small First Nation[1] and the adjacent municipality; continued with the collaborative design of a community screening and dialogue around the film; and concluded with a tentative evaluation process. The essay discusses why we chose the making of a film as our mode of research, how we approached this as a collaborative planning exercise, and the ways in which this was conceived as action research in a therapeutic and transformative planning mode. We conclude with reflections on the effectiveness of this therapeutic intervention and its broader applicability, as well as the challenges inherent in this kind of work.

Historical Context

The displacement, dispossession, and disempowering of Native peoples in what is now known as Canada began in the 17th century but was systematized in the latter part of the 19th century through the federal policy of assimilation (*Report of the Royal Commission on Aboriginal Peoples*, vol 1, 1996). This policy was enforced through three 'technologies of power' (Jacobs 1996): the Indian Act, the Reserve system of Indian Lands, and the Indian Residential School system. The Indian Act of 1876 determined that 'Indians' were now wards of the state, deprived of their sovereignty, and dependent on the federal government for virtually every aspect of their lives, lands, finances, and governance, from cradle to grave. In Canada's western-most province, British Columbia (BC), dispossession occurred by forcing Natives onto reserves while the province refused to make treaties (with a handful of exceptions) and refused to acknowledge or enter into any debate about Native title. By 1930 there were 1,536 small reserves scattered across the province, and Natives were now confined to a third of 1% of the land base of the province (Harris 2002, 263). The third technology of power was the Indian residential school system, which forcibly removed Native children from their families, forbade them to speak their language or engage in any of their traditional cultural practices and, according to official documents, was intended to destroy the culture by 'killing the Indian in the Native child' (Molloy 1999). The cruelty with which children were treated, in addition to the direct attempt to destroy their cultural inheritance and family ties, has had long -term, inter-generational impacts such as high levels of addictions, violence, and suicide rates (*Report of the Royal Commission on Aboriginal Peoples*, vol 1, 1996).

Our research took place in the traditional Carrier (Dakelh) territory of north-central BC. By the early 21st century, there were 203 First Nations recorded in BC under the Indian Act, of which 17 of them are Carrier bands ('band' being the Indian Act terminology for formerly sovereign Nations and their subtribes). The Carrier peoples were forced onto reserves in the first two decades of the 20th century. The Ts'il Kaz Koh Nation (more widely known as the Burns Lake Band), a subtribe of the Carrier, were allocated 400 acres of reserve land in 1914, but lost one-third of it the following year when the town site for what is now the Village (that is, municipality) of Burns Lake was laid out. That expropriation marked the beginning of a century-long battle between the Band and the Village, a battle that culminated in a confrontation in 1999–2000 that ended up in the Supreme Court of BC. What had begun apparently as a municipal water and sewer issue was transformed into an Indigenous/human rights issue when the Village shut off water and sewer services to the reserve over a taxation dispute that exposed the ongoing operations of colonialism at the municipal level (Stanger Ross 2008). And that was the trigger for our involvement in this community. But before we embark on that story, we present a brief overview of our theoretical and methodological lenses.

Theories and Methods

We bring a social justice orientation to our work, and we are provoked by the challenge to mainstream planning practice expressed in the concept of 'decolonizing planning' (Jackson 1998; Jojola 2008; Porter 2010). If, as these authors argue, planning has been deeply implicated in the colonization process through its spatial technologies of power (such as mapping, surveying, land use, and strategic planning) as well as its more recent processes of consultation or inclusion, which fail to acknowledge Aboriginal rights, title, or sovereignty, how might planning become a more ethical and socially just practice? How might it contribute to the process of healing and reconciliation identified as a national challenge in the prime minister's apology in 2008? We agree with Jojola (2008) that this amounts to no less than a 'long-overdue paradigm shift' for mainstream planning. But this still begs the question of *how* to do such work, where to begin, and at what scale. And here we are drawn in two new directions: first to an emergent approach that some have called 'therapeutic planning' (Sandercock 2000; Erfan 2013), and second to the use of more creative tools of engagement.

From Peter Marris's *Loss and Change* (1974) to the more recent works of Baum (1997), Forester (1999, 2009), and Erfan (2013), some scholars have been drawing attention to the role of emotions in planning practice and arguing for the importance of working with and through people's hopes, fears, memories, wounds, indelible experiences of loss and trauma. Sandercock initially developed what she called a 'therapeutic planning process' in which a dialogic space is created for the unspeakable, for talk of fear and loathing as well as of hope and transformation (Chapter 5). This involves the design of a safe space in which conflicting parties can meet and speak without fear of being dismissed, attacked, or humiliated, a new space of recognition in which historic injustices are acknowledged as a necessary prelude to addressing contemporary conflicts. This kind of planning work requires exceptional interpersonal, relationship-building, and facilitation skills. It requires patience and the ability to establish trust. It involves difficult dialogue and negotiation across the gulf of cultural difference and historic injustice. When done well, it can open up new spaces of possibility, enable conversations that have previously been unthinkable, and lead to commitments to co-existence and collaboration hitherto deemed impossible. In response to the challenges of this difficult work, practitioners and researchers have been exploring the role of the arts in this kind of community engagement and conflict resolution (Sarkissian and Hurford 2010). We have extended this exploration into the applications of multimedia, including film and video (Sandercock and Attili 2010a, b).

Our community-based action research approach is intended as a contribution to transformative planning, helping to bring about a shift towards more equitable economic, social and political relations between Native and non-Native residents of Burns Lake (and possibly beyond). We became involved in a local struggle for both reconciliation and the decolonizing of planning, and we brought with us these conceptual tools and some planning skills, along with technology and artistry (film as an expression of the power of story), and an ethical orientation to collaborative and empowering modes of planning. Thus began a three-stage, five-year action research project using film as a therapeutic planning intervention.

High Noon in Burns Lake

We begin with the story of the research, filming, and editing of the documentary *Finding Our Way* (Attili and Sandercock 2010), explaining the 'who, what, and why' of our collaborative film-making, our partners and purposes. We then describe the action piece of the action research: how, with help from colleagues, we took the finished film back to the communities whose story it tells, organizing screenings followed by dialogue circles, and then evaluating that process. But first we provide a sketch of the research site and the questions that intrigued us about local power dynamics.

Burns Lake, a small town on Highway 16 (the northern transcontinental highway) midway between Prince Rupert on the coast and Prince George to the east (Figure 9.1), has a population of almost 5,000 people, almost equally divided between First Nations and non-Indigenous Canadians. While Native people have been living in the region for at least 5,000 years (Tobey 1981; Muckle 2007), European settlers were beginning to arrive only after the completion of the transcontinental railway in 1913. What drew us to the town initially was the story of a conflict – ostensibly over taxation and payment for municipal services – between the Burns Lake Band and the Village, which erupted into a Supreme Court case in 1999.

FIGURE 9.1 Caption: Screen shot from *Finding Our Way* showing location of Burns Lake and historic photo of Carrier people.

Although the Band won this case, the Village (ie municipality) then shut off water, sewer, and fire services to the reserve in the winter of 2000, ignoring the Supreme Court ruling. What followed was a stand-off between the antagonists, which drew national media attention and allegations of human rights violations against the Village.

This illegal exercise of power by the Village demonstrated their lack of recognition of Aboriginal rights and reflected a darker history of discrimination and segregation: a history in which Native people walked on one side of the street and non-Native people on the other; Native people were not served in local restaurants and stores; and nor were they employed in local businesses and industries. While efforts were made after the Supreme Court case to open a dialogue between the antagonists, these efforts soon fell apart through lack of trust and the ongoing failure to respect Native rights. For example, when a new high school was built in 2005, the Burns Lake Band asked for the return of the land on which the old high school had been built, which had been appropriated from their reserve. Instead, the School District starting tearing down the buildings, prompting the Band to occupy the site. Eventually the Band was given the land on a 99-year lease, and they wanted to turn it into a Gathering Place for all the clans. This new conflict rubbed the old wounds, and the ongoing planning process of trying to get development permits for the new premises through the Village's planning bureaucracy seemed to be a further illustration of the persistence of a colonial planning process and mentality.

But there were local forces for change, seen in a small group of anti-racism activists as well as a group of high school youth, Native and non-Native, who had collaborated on the writing and performing of an anti-racism song, 'Leave it Behind'. The demographics were also shifting in favor of the Native population, which will soon be a numerical majority. And 20 years of federal legislation since the new Canadian Constitution of 1982, which had established Aboriginal rights and title, was creating a larger context of change, one that the Village appeared to be resisting. Our interest was, first, in understanding how such a deeply divided community had been created – through what relations of power; and second, whether there was a role for us as planners, engaging in a therapeutic planning intervention through the making of a film.

We were invited to Burns Lake by one activist – known locally as the 'anti-racism lady' – as potential storytellers, and we set about talking with as many people as possible, Native and non-Native, to get a sense of the scope and complexity of the story. We had close to a hundred conversations over a 12-month period, while simultaneously pursuing library and document-based research. The more we learned about the operations of power and privilege in Burns Lake, the more compelling we found the story to be, and we began to suspect that it was a microcosm of Native/non-Native relations in Canada. Our perception was that this 21st-century conflict was deeply rooted in the history of colonization. On the part of most non-Native residents, there was a lack of awareness of First Nations' lived experience of colonization, and particularly of the traumatic intergenerational effects of the Indian residential schools. And on the part of First Nations, there was anger and frustration, since historic injustices had never been acknowledged, were ongoing, and were not being addressed. In other words, the past was very much still present, distorting what contemporary goodwill existed and blocking a path forward. The more we talked with Native people, the more we heard about healing, about the kinds of healing they had undertaken both as individuals and as a community. What struck us was that there had been very little healing between Native and non-Native folks, even though a Community Healing Project had been initiated after the Supreme Court case. We wondered how community planning issues such as economic and social development could be dealt with in the absence of some real work across this cultural divide.

Thus, we conceived the film as a way of opening and encouraging a difficult dialogue, of changing the lens on the past, and of necessarily 'unsettling' the settler society. Because we wanted to encourage dialogue, but a dialogue starting from a fresh perspective (a new lens), we envisaged a collaborative film-making process[2] in which we would be creating the dialogic space for new stories to be told, by voices hitherto unheard. The film would begin with an investigation of the causes of the events of 1999–2000, the Supreme Court case and the shutting off of services to the reserve, revealing not only a contemporary quarrel over the taxation of a mill on Band land, but also the preceding 80 years of conflict over expropriated land. Then we would excavate even further, to uncover the story of colonization and its technologies of power, which had resulted in First Nations being forced onto reserves, and the century-long onslaught on their culture. And finally, we would look at attempts within the community to begin to shift these toxic relationships, and pose the question, 'is there a way forward?'

We approached both the Band and the Village seeking their collaboration. We explained our ethical protocol, which was to bring back rough cuts of the film to the community (Native and non-Native) at every step of the editing process to ask for their input; to offer every individual whom we interviewed the right to withdraw from the film if they didn't like how we used their words; and to bring the final cut back to the community and arrange a community-wide dialogue. The actual editing process took two and a half years and involved eight trips back to the community to share the footage with them, sometimes one-on-one (for the more 'sensitive' interviews), other times in group screenings.

We asked our collaborators what they would like to see come out of such a process. The then mayor, Bernice Magee, guardedly agreed to cooperate after we made the case that the film could potentially contribute to a shift in understanding within the Village, between Native and non-Native residents, that might result in a less confrontational stance and more willingness to cooperate on joint projects for social and economic development. As she says in the film, 'working with First Nations is a challenge'. But she also says that 'it's a learning process', and that certainly turned out to be true even during the making of the film, as each 'side' in this conflict began to learn more about each other during the rough-cut screenings. The Band was much more enthusiastic. They wanted their story to be told, and they trusted us to tell it. This trust, Chief Rob Charlie explained to us much later, came about because 'you took the time to get to know us'.[3] The Band was not sure what might change in the town, if anything, as a result of the telling. But they recognized, as the chief explains in the film, that they could not survive alone, and that they were willing to forgive, although not to forget, in order to move forward as a people.

The Action Part of the Action Research: Designing Community Dialogues

Initially we had conceived the screening and dialogues as the *action* part of our project, and the making, editing, and rough-cut screenings of the film as the *research* part. In hindsight, every stage of the film-making was part of the 'action', that is, a therapeutic planning intervention, in that it cumulatively brought community members, hitherto antagonists, into new relationships with each other simply through watching and listening to each other in the rough cuts. In other words, the therapeutic work that we had imagined beginning through the formal dialogue process that we were about to design actually began in the course of making the film, during which a lot of meaningful relationship-building work across the cultural divide began to happen.

Four years after we entered this community, once we had the final cut of the film, we returned and formed an Advisory Committee – comprising the mayor; the chief of the Burns Lake Band; the high school principal; the high school drama and dance teacher, who has a prominent role in the film through her work with youth; the staff sergeant of the Royal Canadian Mounted Police; two young Village councilors; and the 'anti-racism lady'– to work with us to design the film screening and community dialogues.[4] We envisaged two screenings and dialogue events, one for youth (with whom we had done specific engagement work during the filming stage) and one for the community at large. These community partners, after watching the film together, all agreed that this would not be an easy film for the community to watch, that it was not 'sugar-coated' and would likely provoke a range of reactions, from surprise and anger to guilt and denial, empathy, and hope. We collectively pondered how we might encourage the community to watch with an open mind and heart, and with some mindfulness about their own emotions as they watched. The chief suggested that perhaps community leaders should introduce the film and convey this message and, somewhat to our surprise, the mayor readily agreed.

We then collectively brainstormed what the community dialogue process might look like, and decided to break the audience into dialogue circles of eight to ten people, each of which would have two facilitators, a youth and an Elder. Our Advisory Group helped us to recruit locals whom we then trained as facilitators. We also did a preview screening for the high school teachers, anticipating that the high school youth would be eager to discuss the film with them after the screening for the youth. The teachers were so engaged by the film – one said it 'lifted the veil of silence about what's been going on in this town' – that they offered assistance as facilitators. The high school principal acknowledged that the school had been part of the problem, and he was eager to get the film into the classrooms. The training sessions were in themselves profoundly moving events as this small sample of the local audience expressed their emotions and their support for using the film as an educational tool. The youth expressed anger that they had not been exposed to these stories in school. Some of the Native youth talked of growing up being ashamed of being Native, but now, with the film's existence, they felt pride and wanted to share the film with their friends and family, immediately! Others spoke of growing up not wanting to hear their parents' stories of pain, especially the pain of the residential school experience, but now they understood why it was important to hear and discuss those stories and understand the intergenerational impacts. Each person spoke in these deeply felt ways, and it was clear, during this preparatory stage, that further healing work was beginning.

Each of the two screenings began with food and live music played by the local youths who had written the anti-racism song a few years earlier. As planned, the chief and mayor then introduced the film. They stood on stage together, the mayor acknowledging that the film told an important story and that the Village had made mistakes that needed to be acknowledged, while the chief spoke of how, before the film process, he had given up hope in the town, but now his hope and vision of working together was revived. We then launched into the 90-minute screening, followed by the dialogue circles. These facilitated dialogues ran for 90 minutes, focusing on three questions. What was the most significant impact of the film on you? Is the past still present in this town? And if so, how might this community move forward together now, and what would your role be? One of the facilitators in each group took detailed notes, and the mayor and the chief, who moved about the room as witnesses during the dialogue circles, each made closing summaries of what they had heard and felt during this time. Both of them reflected on the spirit of hope for moving forward that they heard in the circles and saw on the flip charts in the form of visions for the future, including a vision for an intercultural Gathering Place for the town, and the desire of the youth to be involved in intercultural community activities.

Fifty high school youth (half of them Native) attended the youth screening and dialogue and 150 people (at least one-third Native) attended the community-wide screening, half of whom remained for the dialogue and then stayed an extra half hour (after a five-hour session) to fill out our evaluation questionnaire.

Reflections on Therapeutic Planning

The primary question that framed our action research was whether there was some kind of therapeutic planning intervention that could address the history of colonization in this deeply divided community and thereby open up a *space of possibility* for healing. The second, more methodological, question was whether film could serve as the catalyst to create this new space for dialogue and healing, by changing the lens through which people viewed each other. We have three ways of addressing these questions. The first (and least satisfactory) is through a grant-mandated survey questionnaire administered at the end of the two evenings of dialogues. The second is the extensive notes taken by facilitators for each dialogue circle, as well as our own observations as participants and witnesses. The third is more anecdotal and medium term, based on ongoing relationships in and with the community. But before we discuss each of these, we re-visit our earlier discussion of 'therapeutic planning', now framing this in an Indigenous–settler context.

Our point of departure was Leonie's definition of therapeutic planning 'as the process of bringing people together not only to share their experiences and work in solidarity, but also to work through their differences … to talk of fear and loathing as well as of hope and transformation' (Sandercock 2000, 24). Beyond planning, though, there is a much broader literature addressing healing and reconciliation, and specific literature in which the historical context is the trauma of colonization that affects ongoing Indigenous–settler relations (Ross 1992, Duran and Duran 1995, Waldram et al. 2008, Regan 2011). This was the complex setting in which we found ourselves. We learned from this literature that there are at least three kinds of healing that need to happen in the context of colonization. The first is the internal psychological healing of individuals traumatized by the experiences of colonization, and specifically the Indian residential school experience. The second is a broader collective healing of Indigenous communities whose culture has been under siege for centuries. And the third is healing and reconciliation between colonizers and colonized, settlers and Indigenous Peoples. As non-Indigenous planners engaging with a decolonizing methodology (Smith 1999, Kovach 2009), we did not see it as our business or our competence to address the first and second kinds of healing. Rather, drawing inspiration from Regan (2011) we focused on the third, especially on the notion of 'unsettling the settler'.

Doing an evaluation right on the heels of an event is a common practice and is often required by funders, but it also has its limitations, likely yielding responses from those most enthusiastic about the event as well as those who hated it. Not only did our provincial government funders require such an evaluation but they also dictated specific questions and even format. We were able to negotiate a little wiggle room, but not much, so some of what follows must be understood in the light of these restrictions.

In response to the first question, 'How well did this screening and dialogue help your community address racism?', 17 of 21 replies among the youths gave the highest possible score for this answer. In response to the second question, 'How well did this event help your community identify pathways to working together across cultural differences?', 20 of the 21 replies circled 'very good'. There were similar responses in the community-wide screening. Additionally, there were very positive answers to a question inquiring about people's overall awareness of historical and current day relations between Native and non-Native peoples after (compared with before) the dialogues. Qualitative answers to questions we had added, from both youth and adults, to the question 'How will you act on what you have learned through the screening and dialogue?' contained many expressions of the desire to volunteer to work on community projects such as the Gathering Place, which was re-imagined during the dialogues as an intercultural facility for the whole community, rather than as a meeting place only for First Nations.

Questionnaires like these are extremely limited in what they can reveal about the qualitative experience of an event, or its longer-term effects. All of us in the project team took away from the dialogue circles profound and visceral memories of levels of honesty and empathy, of the sharing of tears, of an energized and enthusiastic youth, of the opening of a space of hope. A high level of energy and commitment was released, especially among the youth, some of whom have since gone on to become local leaders, including one of the non-Native lads who, at age 21, was elected mayor of the town in 2011. Others started a Facebook page to inspire other youth to get involved in social change. Three days after the dialogues, the local newspaper ran an editorial titled 'Inspiring' (*Lakes District News* 9 June 2010, p. 4), which referred specifically to this commitment to social change evident among the youth who were there.

On reflection, too, the comments that the mayor made immediately before and after the screening were a kind of local equivalent to the prime minister's apology. While she did not apologize per se, she did admit some responsibility on the part of the settler society for 'mistakes' in the past. We suggest that 'apology' at a local level is an important complement to the apology at the national level – perhaps even in some ways more meaningful, since most of what affects the lives of people on a day-to-day basis happens at the local level. This could then be seen as a very important part of the ongoing healing process.

It would be reasonable to say that our community dialogues served as a voluntary public setting for opening up and cathartic sharing of some very personal stories about the past and the present, stories of both pain and shame. Some of the Native youth admitted to and talked through some of the personal and family struggles that they had previously kept private. Some older Native folks divulged, for the first time, a lifetime of pain as a result of their own or other family members' experiences of residential school. Some members of the settler society admitted shame at what had transpired both historically, and in the very recent past, right under their noses. The quality of listening and empathic engagement among participants was intense. It seemed at the time, and was later confirmed to be a turning point, both for the tellers of these stories and for the folks hearing and receiving them, the beginning of new relationships in which older biases were now dropped and different and better relationships were now possible. This is what Bush and Folger (2004) call 'recognition', one of two pillars of conflict transformation. Recognition of others in their full humanity rather than as 'other' is significant in the context of our project, which called for a re-setting of the social web of relationships, something that is acknowledged to be central to the meaning of healing both within Indigenous communities and in conflict resolution between communities (Waldram et al. 2008).

After the dialogues, the mayor and the chief as well as various community members took public responsibility for working on the Gathering Place as an intercultural project. To us, this affirmed that the film was effective as a catalyst for cross-cultural conversation, had created the space for that conversation, and was instrumental in transforming the historic conflict. While we have not been able to monitor, from close quarters, the ways in which social relations have been changing in the three years since the community dialogues, our ongoing personal connections in the town continue to reveal relational as well as institutional and political changes.

What did we hope to set in motion with our presence and with our film? The project was always conceived as an intervention into an already highly charged political force field. We aimed to provide some fresh eyes through which a polarized community might be able to stand back and see itself in a different light, seeing both its dark past and the potential now being demonstrated by some leaders and in some local institutions to move towards a future based on mutual respect, recognition, and partnership. We acknowledge that there were already forces for healing in the community, alongside their opposite. Our intent was never neutral. It was to align with the forces for healing and reconciliation and give whatever little push to that project that we could. We cannot *prove* any direct impacts from our film work, but we can talk about what has emerged in the past three years, and we can talk about the learning that took place, both our own and that of the divided community.

We have noted the energy and sense of possibility that emerged from the dialogues.

In August of 2010, three months after the dialogues, the Village of Burns Lake adopted a motion of support in principle for one of the ideas that had emerged from the dialogues: the renovation of the old high school as a Gathering Place for the entire community. The mayor personally drafted a motion that the Village Council write to the School District asking that they transfer ownership of the land back to the Band so that the Band could qualify for federal funds for capital improvements. And she urged the Chamber of Commerce and the Regional District Planning Office to write similar letters. This action on her part would have been unimaginable four years earlier.

Nathan Edelson, a Vancouver-based community economic development and land use planner who had helped us conceive the dialogues, has been working for the past three years as a liaison between the Band and the Village in getting architectural drawings for the Gathering Place as well as seeking capital funding. Bernice Magee retired as mayor in 2011 but continued to serve as a staunch supporter on the committee that is pushing the Gathering Place project. Progress is necessarily slow and rarely linear, but the materialization of the Gathering Place, when it happens, will be a significant symbolic as well as social and economic advance for this community.

Is there something special about film as a tool for therapeutic planning? Why did we choose the process of film-making as the catalyst for our planning intervention? We argued earlier that some practitioners working in deeply divided communities have begun to experiment with the potential of the arts to provide creative approaches that go beyond the usual planning toolkit, to enable conversations that had previously been unthinkable. In this particular community, our intuition, informed by our initial research in the community, was that there was a need for a new lens, a way for people to hear and see each other differently than the ingrained patterns of colonization had determined, to see the past with different eyes, and to provoke people to moral reflection on the past. We knew that film could reach and disturb people by accessing emotions as well as intellect. That is the well-known power of storytelling in general, and of the visual medium of film (Sandercock and Attili 2010). We also reasoned to ourselves that if we 'built' this film mindfully over a period of time, in a collaborative and inclusive way, people would be more likely to watch it with an open mind, knowing that they had had the chance to tell their story, as had others.

What we do know from the detailed facilitators' notes and our own witnessing of comments in the dialogue circles in response to the first dialogue question, 'what moved you most/what stood out for you in the film?' was that the film had an incredibly powerful, even searing impact, and there were many comments along the lines of 'we had no idea this had happened' or 'now the veil of ignorance has been lifted' or 'why aren't we being taught this history in school?' Our conclusion is that it was a combination of the content of the film, the way we had edited it, and the collaborative process of making and editing it that produced such a profound response in the community. It is reasonable to conclude that there is something about the public nature of film that gives the type of acknowledgment and recognition that would otherwise not have taken place.

In the intervening years since the community dialogues, other communities have used our film for similar purposes of addressing the same deep-seated historic conflicts, often asking us and our community partners to facilitate the screening, which has created some Nation-to-Nation learning. This leads to two final questions concerning the applicability of this film in other communities and the broader applicability of therapeutic planning beyond the Indigenous/non-Indigenous realm of conflict.

Although our film tells the story of a specific conflict in a particular place, it is also a story with a timeless theme: a people dispossessed, deep historic wounds, and ongoing and unresolved injustices. It is both the specificity and the universality that makes it resonate beyond north-central BC. At a rough-cut screening at the annual conference of the Canadian Institute of Planners in 2008, dozens of people lined up afterwards to tell us that this historic conflict is exactly what is playing out in *their* home community (planners from all across Canada), except that in their communities 'nobody is talking about it, it's the elephant in the room'. Screenings in Australia, New Zealand, Hawaii, and New Mexico have equally resonated, especially with Indigenous audiences, who say 'this is our story!' Palestinians, too, have asked us to bring the film to their homeland and introduce the same therapeutic process. We conclude, then, that the film offers a microcosm of a much larger part of Canadian (and others') history, which explains why many School Districts across Canada and colleges and universities in the USA have purchased the film for classroom use.[5]

But what of the therapeutic planning approach? We propose that such an approach is essential in deeply divided communities precisely because the past is so present, so deeply felt, so visceral, that it is impossible to break through and begin to do more traditional planning work without first acknowledging and addressing the past. When we first arrived in Burns Lake, there was such antagonism between the mayor and the chief that they literally could not be in the same room for more than five minutes without one of them (usually the chief) walking out in protest, anger, despair, or frustration. Is this approach only relevant in situations of Indigenous/non-Indigenous conflict where neighboring communities are 'stuck' in the past? Our tentative answer, supported by related research and literature, is that there are a wide variety of planning situations in which a therapeutic approach would be necessary and beneficial.

Marris (1974) drew attention to the grieving that communities experience when faced with the loss of something precious: typically, a place attachment that is being undermined by change. Such changes can be unexpected (natural disasters) or part of a planning process (infrastructure, housing, transportation), and typically planners are surprised and irritated by the emotional resistance they encounter, and tend to dismiss such resistance as selfish NIMBY behavior. Mindy Fullilove, an African American therapist, has written about the destruction and displacement of hundreds of African American inner-city communities and the devastating psychological impact and dysfunction that resulted from the urban redevelopment schemes of the 1950s and 1960s (Fullilove 2004). Contemporary top-down approaches to the densification of cities, albeit with good ecological intentions, are meeting emotional resistance as residents see a familiar way of life being threatened. Whether such changes are desirable is not our point here. Rather, since change is inevitable, we suggest that planners need to pay more attention to and be more respectful of the emotional impacts of change in order to understand the resulting conflicts, and to design better processes for recognition of and working through these emotions of loss and grieving, rather than trampling over them. Therapeutic planning, then, is relevant not only to Indigenous planning or social planning but also to most planning endeavors, whether developmentally oriented or striving for preservation, sustainability, and social justice. Marris (1974) has offered the most substantive social psychological argument as to why this is the case, and it is surprising that his work is not more widely cited in the planning field.

Caveats and Further Research

In this final section, we raise two issues as cautious caveats to the above discussion. The first is the issue of cause and effect, or proof. The second is the potentially vexing issue of the planner as therapist.

First, and simply put, we cannot *prove* a cause–effect relationship between the making and screening of our film and subsequent changes in social, political, and institutional relationships. The work of therapeutic planning is the work of social innovation, and it is not well or best served by metrics. There is an emerging literature on 'developmental evaluation' (Patton 2010) that seeks to trace the medium and longer-term impacts of the kind of policy/planning intervention that we attempted. We have sought research funding to pursue such an evaluation and regard this as a broader challenge facing social innovation through planning.

Second, our work necessarily raises the question of the planner in the role of therapist. Are we overstepping our professional boundaries by even using the term 'therapeutic planning' and thereby trespassing in the realm of psychologists? This is an important question, which would take another essay to address (see Erfan 2013 for an extended discussion of this issue). We do propose that there ought to be a broader debate about this. If planning is, by definition, not just about navigating a relationship between knowledge and action (Friedmann 1987) but also about negotiating a relationship between the past and the future, a relationship that is often freighted with tension, conflict, and even trauma, then perhaps planning is an ideal forum for talking about trauma and healing. As Erfan expresses this: 'planning is naturally where the past comes into conversation with the future. If, as planners, we appreciate, gently support, and make room for these kinds of conversations, we can be on our way to becoming healers' (Erfan 2013: 254). But Erfan also issues the caution that therapeutic planning practice requires the acquisition not only of specific skill sets and tool kits but also the developing of a way of being, 'attitudes, qualities, manner and mind-sets, and the personal, emotional, spiritual, physical, and imaginative resources that make this work possible' (Erfan 2013: 264).[6] This is both a necessary and daunting new direction for planning research and practice.

The therapeutic planning intervention that we designed began with a healing process in the context of historic trauma and injustice, catalyzed by dialogues that the film enabled. It proceeded through recognition that the past is still present, and moved on to a visioning process engaging with how things might be different. While this is very much a work in progress, we have presented some evidence that this small town, which had been characterized by disparities in power, violations of trust, and lingering unresolved disputes, is moving towards relationships of mutual respect and joint problem-solving.

Notes

1 There are three distinct Indigenous groups in Canada: Inuit, Metis, and First Nations, each of whom had different legal standing under the Indian Act of 1876. Since the revised Canadian Constitution of 1982, which acknowledged Native rights and title, official legal language uses the umbrella term 'Aboriginal' to describe all Native peoples, but Indigenous people never use this term. 'Indian' is a term used by the Indian Act but dismissed by Native people (except for their own internal/self-referential humorous purposes). 'Native' is more colloquial, and typically how Canada's Indigenous people refer to themselves, although with regional variations. 'First Nations' is a term most predominant in western Canada (Gray 2011).

2 Participatory film making, in which film makers train members of a community and then hand over equipment and control, has many virtues, but we did not have the time or the budget to do the necessary capacity building. Further, we chose a collaborative rather than participatory mode of film making because our goal was to deeply involve the community in the dialogues precipitated by the film, in which their stories would be presented through a fresh lens, rather than have them shoot and edit their own film.

3 Interview with Rob Charlie in the film's Epilogue.

4 At this point we partnered with Scott Graham (Social Planning and Research Council of BC) and Norma-Jean McLaren and Nathan Edelson (community development and planning consultants) in the design and facilitation of the dialogues and training of local facilitators.

5 The DVD contains a 32-page discussion booklet for schools and community groups to guide them in the use of the film, and a recommendation that counselors be present and available, given the potentially psychologically disturbing nature of some of the subject matter. Information about whom has purchased the film is provided by our distributor.

6 See Erfan (2013: 262–280) for an extended discussion of how to become a therapeutic planner.

References

Attili, G. and Sandercock, L. 2010. *Finding Our Way. 90 min Documentary* (Vancouver: Moving Images).

Baum, H. 1997. *The Organization of Hope* (Albany: State University of New York Press).

Bishop, A. 1994. *Becoming an Ally: Breaking the Cycle of Oppression* (Halifax: Fernwood).

Bush, R. A. and Folger, J. P. 2004. *Promise of Mediation: the Transformative Approach to Conflict* (San Francisco: Jossey Bass).

Dorries, H. 2012. Rejecting the "False Choice": foregrounding indigenous sovereignty in planning theory and practice. Unpublished PhD dissertation. University of Toronto, Department of Geography.

Duran, B. and Duran, E. 1995. *Native American Post-colonial Psychology*. New York: SUNY Press.

Erfan, A. 2013. An Experiment in Therapeutic Planning: learning with the Gwa'sala-'Nakwaxda'xw First Nations. Unpublished PhD dissertation, University of British Columbia, School of Community & Regional Planning.

Forester, J. 2009. *Dealing with Differences* (Oxford: Oxford University Press).

Friedmann, J. 1987. *Planning in the Public Domain: from Knowledge to Action* (Princeton: Princeton University Press).

Fullilove, M. 2004. *Root Shock: How Tearing up City Neighborhoods Hurts America, and What We Can Do about It* (New York: Random House).

Gray, L. 2011. *First Nations 101* (Vancouver: Adaawx Publishing).

Harris, C. 2002. *Making Native Space* (Vancouver: UBC Press).

Healey, P. 1997. *Collaborative Planning*. (2nd edn.) 2007) (London: Macmillan).

Hibbard, M. and Lane, M. B. 2004. By the seat of your pants: indigenous action and state response. *Planning Theory and Practice* 5(1), 97–104.

Hibbard, M., Lane, M. B. and Rasmussen, K. 2008. The split personality of planning: indigenous peoples and planning for land and resource management. *Journal of Planning Literature* 23(2), 136–151.

Innes, J. and Booher, D. 2010. *Planning with Complexity* (New York: Routledge).

Jackson, S. 1998. Geographies of co-existence: native title, cultural difference and the decolonization of planning in north Australia. Unpublished PhD dissertation. Sydney: School of Earth Sciences, Macquarie University.

Jacobs, J. M. 1996. *Edge of Empire: Postcolonialism and the City* (London: Routledge).

Jojola, T. 1998. Indigenous Planning: clans, intertribal confederations and the history of the All-Indian Pueblo Council. In: Sandercock, L. (Ed.) *Making the Invisible Visible: A Multicultural History of Planning* (Berkeley: University of California Press).

Jojola, T. 2008. Indigenous planning: an emerging paradigm. *Canadian Journal of Urban Research* 17, 1, Suppl.

Jojola, T., Natcher, D. and Walker, R. (Eds.) 2013. *Reclaiming Indigenous Planning* (Montreal: McGill Queens University Press).

Kovach, M. 2009. *Indigenous Methodologies: Characteristics, Conversations, Contexts* (Toronto: University of Toronto Press).

Marris, P. 1974. *Loss and Change* (London: Routledge and Kegan Paul).

Milloy, J. 1999. *A National Crime: the Canadian Government and the Residential School System, 1879–1986* (Winnipeg: University of Manitoba Press).

Muckle, R. 2007. *The First Nations of British Columbia: an Anthropological Survey* (2nd edn.) (Vancouver: UBC Press).

Patton, M. Q. 2010. *Developmental Evaluation* (Thousand Oaks: Sage).

Porter, L. 2010. *Unlearning the Colonial Cultures of Planning* (Aldershot: Ashgate).

Regan, P. 2011. *Unsettling the Settler Within: Indian Residential Schools, Truth Telling, and Reconciliation in Canada* (Vancouver: UBC Press).

Report of the Royal Commission on Aboriginal Peoples. 1996 5 vols (Ottawa: Government of Canada).

Ross, R. 1992. *Dancing with a Ghost: Exploring Indian Reality* (Markham, Ontario: Octopus Publishing Group).

Sandercock, L. 2000. When strangers become neighbors: managing cities of difference. *Planning Theory and Practice* 1(1), 13–30.

Sandercock, L. 2003. *Cosmopolis 2: Mongrel Cities of the 21st Century* (London: Continuum).

Sandercock, L. and Attili, G. (Eds.) 2010. *Multimedia Explorations in Urban Policy and Planning: Beyond the Flatlands* (London: Springer).

Sandercock, L. and Attili, G. 2010a. Digital ethnography as planning praxis: an experiment with film as social research, community engagement and policy dialogue. *Planning Theory and Practice* 11(1), 23–45.

Sarkissian, W. and Hurford, D. 2010. *Creative Community Engagement* (London: Earthscan Publications).

Smith, L. T. 1999. *Decolonizing Methodologies. Research and Indigenous Peoples* (London: Zed Books).

Stanger Ross, J. 2008. Municipal colonialism in Vancouver: city planning and the conflict over Indian reserves, 1928–1950s. *Canadian Historical Review* 89(4), 541–580.

Tobey, M. 1981. Carrier. In: Helm, J. (Ed.) *Handbook of North American Indians: Subarctic* (vol. 6) (Washington, DC: Smithsonian Institution).

Waldram, J. B., et al. 2008. *Aboriginal Healing in Canada: Studies in Therapeutic Meaning and Practice* (Ottawa: Aboriginal Healing Foundation).

Walker, R., Jojola, T. and Natcher, D. (Eds.) 2013. *Reclaiming Indigenous Planning* (Montreal: McGill-Queens University Press).

10

EDGE OF THE KNIFE

Film as a Catalyst for Indigenous Cultural Revitalization? (2022)

> This culture never died, and will never die … This film is for the rest of the world to see how deeply we're connected to land, sea, and sky. Everything is connected to everything else.
>
> *(Diane Brown, matriarch, language teacher, actress,* Edge of the Knife*)*

> Language is healing. The film will change lives because of that, inspiring new generations of language learners.
>
> *(Erica Ryan-Gagne, actress,* Edge of the Knife*)*[1]

Introduction

The previous three chapters in this section have detailed the evolution of my film making praxis as a community planning intervention, from its conceptual origins in theorizing the power of story and storytelling (Chapter 7) to the first two documentaries (Chapters 8 and 9), which were conceived as experiments in community-based action research. The learning from those two multi-year partnership projects became both the inspiration and the foundation for a third, much more ambitious film project: a three-way partnership between myself (activist/filmmaker/planner), the Council of the Haida Nation (the governing body of the Haida people), and the legendary Inuit film making collective Isuma, whose first feature film, *Atanarjuat: The Fast Runner*, in 2001, had revolutionized Indigenous film making and won numerous national and international awards.

This was such a complex project, politically, ethically, and emotionally, that it's impossible to tell as a single narrative. The complete story can never be told – or not, at least, by one person. From my necessarily partial perspective there are two inter-connected stories, which I'll awkwardly call the external and the internal. By 'external' I mean the series of observable, documented activities involved in the planning and implementation of an idea: the idea of film as a catalyst for community development and cultural revitalization for the Haida Nation. And by 'internal' I'm referring to what it felt like (as a White female settler) to be a part of this process, the actual lived experience, and particularly the learning involved, both about myself

DOI: 10.4324/9781003325451-14

and about working in partnership, over a period of five years. To make all of this manageable, I'm devoting two chapters to this project, the first telling the external story, the second the internal, although sometimes it is simply not possible to separate the two stories. In this chapter I describe the conceiving, planning, making, and distributing of *Edge of the Knife*, the first feature film made in the Haida language, as a success story – as the successful outcome of a community-based action research project that experimented with the use of film as a catalyst for Indigenous cultural/language revitalization and community economic development. As such, it was the culmination, for me, of 30 or so years of working to expand the 'languages' of planning through multimedia work, as well as attempting to work in genuine community-based partnership. That, though, is just my White-centric perspective. More importantly, it was exhilarating not only for those Haida who were directly involved in the production but also for the wider community, who derived enormous pride from this depiction of their culture, along with many more unquantifiable benefits. I will draw on post-production interviews with community members, later in this chapter, to sketch some of what success meant to the Haida Nation and community.[2]

There are many more layers to this story, not only, but yes, also partly because this was a partnership between myself, a non-Indigenous White settler and privileged senior scholar, and a proud, resurgent, and powerful Indigenous community: the Haida Nation. Chapter 12 will tell the more personal, internal story of the experience of being both inside and outside this project;[3] of what it meant and felt like to be consciously attempting (and not just theorizing about) unsettling the colonial culture of planning; about potentially different meanings of reconciliation; about the ultimate importance of the *ways of being* that are central to this kind of partnership work with Indigenous communities; and whether and what kind of future there is for such partnerships. These are all themes central to Part 4 of this book, 'Navigating Indigenous Worlds'. But first, the 'success story'.

Prelude

The experience I describe in Chapter 9, of working for the first time with Indigenous Nations, changed me as a person, bringing insights about my inescapably colonial self (explored further in Chapter 11), and it changed my intellectual and political project in planning. Up until this time I had been focused on social justice *in the city*, asking questions about who got what, where, and why. Thereafter, I started to ask what justice might mean for Indigenous Peoples in Canada (Chapter 13), and I committed to unsettling planning in whatever ways I could: through pedagogy, research, and practice.

My relationship with the Haida Nation, which began in 2013, was a direct result of the new Master's program in Indigenous Community Planning (ICP) at the University of British Columbia, UBC, (see Commentary Four). In 2012, we began what grew into a four-year relationship of mutual learning with the Skidegate Band Council (SBC). Skidegate is one of the two Haida villages on the island archipelago of Haida Gwaii, which was colonized by the British in the mid-19th century and re-named the Queen Charlotte Islands. In 2012, the Skidegate Band Council became one of a growing number of First Nations in British Columbia (BC) who were seeking to replace top-down Indian Act planning with a community-based model that goes by the name of Comprehensive Community Planning (CCP), with the term 'comprehensive', in this context, signifying the desire to return to ancestral traditions of planning that are inherently holistic (Skidegate Band Council, 2017). Through this partnership, two of our students were embedded in the Skidegate community, under the mentorship of faculty member and community planning practitioner Jeff Cook. The top three priorities identified at the end of the first year, through an extensive community engagement process led by our students, were language revitalization, the creation of jobs that would keep youth on Haida Gwaii, and protecting the lands and waters of Haida Gwaii through sustainable development (ibid).

The Haida people have inhabited the remote island archipelago of Haida Gwaii, 100 kilometers offshore from Prince Rupert in northern British Columbia, since time immemorial, and their language is an isolate, that is, unrelated to any other known language. By the early 20th century, the impact of European diseases and government policies of assimilation had reduced the Haida Nation from an estimated 25,000 people, pre-contact, to just 580 survivors. Once residing in over 300 villages around the extensive coastline, the Haida were confined by 1900 to just two small reserves, today known as the villages of Skidegate and Old Massett, while the European settlers lived a few miles away in the adjacent towns of Queen Charlotte and Massett, in a classic pattern of spatial apartheid. Haida culture and language faced the threat of annihilation by the mid-20th century thanks to the genocidal intent of the Indian Residential Schools policy in place from the late 19th century to 1996, which forcibly removed 150,000 Indigenous children across Canada from their families and communities, punished them severely for speaking their native languages, and subjected them to horrific physical, emotional, and sexual abuses (Molloy 1999; TRC 2015). For the past 50 years or so the Haida Nation has been politically resurgent. But the struggle for economic, cultural, and language revitalization is ongoing. Geographic isolation, lack of opportunities for youth, and inundation of Western media have compounded the challenges of revitalization (Simpson 2017; http://www.isuma.tv/haida/film-development).

The Haida Nation regards their language status as in crisis, given the diminishing number of fluent speakers. By 2014, less than 1% of the community was fluent, fewer than 20 people, most of whom were over 70 years old. The Haida were concerned that there is not enough interaction between these few fluent Elders and younger community members to organically support long-term language revitalization goals. Intergenerational transmission of the language is critically endangered, and opportunities for Elder teachings are increasingly limited, so the Haida Nation welcomed the concept of the film project for this reason. As Dana Moraes, community planner for the SBC, explained: 'This film came from our community. It came from community planning. In our community they said they wanted to, number one, they wanted to revitalize our language. It was in a crisis and still is in a crisis. So we wanted to revitalize our language. We wanted to protect our land and waters'.[4] They also regarded the digital media and film industry as a sustainable approach to support a green economy, aligning with traditional Haida values to protect their lands and waters. For the past 150 years, European settlers had turned Haida Gwaii into a resource-extraction-based economy, dependent on logging and commercial fishing. The Haida Nation aspires to self-sufficiency and wants to support community initiatives that provide work experience and new opportunities for youth. The film was seen by the executive of the Council of the Haida Nation (CHN) to support the three core Haida aspirations noted above, as part of a transition to self-determination.[5]

I visited Haida Gwaii at the end of that first year of the new ICP program to witness our two students delivering to chief and council the results of their work with the Skidegate community. During that visit I began to develop a personal connection with two members of the community: Dana Moraes, who had just been hired as their community planner, and 'Babs' Stevens, the chief administrative officer for the Band. On that first trip, in the typical way of Haida hospitality, Dana had taken me wading for crabs at dawn on a bitterly cold day. We filled several buckets with huge Dungeness crabs, which we immediately cooked and took into the Band office to share. That was the beginning of many more foraging, cooking, and eating adventures that Dana and I were to share. With Babs, I connected through a mutual interest in healing, both individual and collective, personal and planetary. She and I share a belief that the work of community planning can and should be the work of healing, that there are deep historic wounds that need to be healed, as well as repairing the damage we humans have done to Mother Earth.

A conversation between the three of us about my previous documentary with the two Carrier Nations in north central BC led to Babs' suggestion of making a film about the Haida. Before I had time to give this more than a moment's thought, Babs pulled me into a meeting with the chief counselor, Billy Yovanovich. Getting straight to the point, he asked me first 'what would such a film be about?' To which I replied, 'that would be entirely up to your community'. Then, 'who would own this film?' And my response, 'You (the Haida) would'. 'OK, so how do we get started?' was Billy's follow-up. I said I'd need to raise some seed money. He gave me his blessing and offered letters of support. And so it began!

I was thinking in terms of a documentary. I turned to a new friend, Norman Cohn, to see if he had any ideas about sourcing funds for such a documentary, and his response surprised me. He explained that in the current funding landscape, it was easier to get financing for feature length, dramatic films than for documentaries. Norman was a co-founder, with Zacharias Kunuk, of the now-legendary Inuit film collective Isuma, which had started making videos of knowledge keepers in the 1980s to preserve both knowledge and language. Their first feature film, *Atanarjuat: The Fast Runner* (2001), made in the Inuktitut language, was a breakthrough Indigenous film, winning the Palme D'Or for Zacharias as director at Cannes and Best Film at the Toronto International Film Festival. Isuma, through Cohn, had evolved a low-budget way of making films using community members for every aspect of production (Evans 2008).

A few weeks after that conversation, Dana and I were meeting with the president and five-man executive of the Council of the Haida Nation (CHN) to 'sell' this idea to the CHN. The idea had quickly grown in three directions. First, it would be a dramatic feature, like *Atanarjuat*, rather than a documentary. Second, the entire community would be involved in every aspect of the production, from writing to acting to costumes to the building of sets, based on Isuma's community-driven model of Indigenous film making. Third, it had grown into the idea of a three-way partnership, between myself, Isuma, and the Haida Nation. What was potentially transformative about this approach was that there would be Nation-to-Nation mentoring and transfer of knowledge from the Inuit to the Haida.

The Haida president declared this to be a 'no-brainer'. It fit with the CHN's goal of bringing the Haida into the 21st-century digital economy, providing jobs for youth, protecting their lands and waters, and showcasing their culture. The only question was, 'How do we proceed?' I offered to write a grant application for development money to the Social Sciences and Humanities Research Council's Partnership Development Fund. The president noted that the Haida admired the Inuit film *Atanarjuat*, believed that a similar kind of film *needed* to be made about the Haida, and were thrilled with the idea of being mentored by the Inuit in the art and financing of film making. Dana and I left the meeting floating on air.

The subsequent successful research grant application asked: 'Could a film be used in the context of community development to serve these Haida aspirations?' We received $200,000 for the first stage of the project, the community-driven development of a film script. In May 2014, I set to work with my Haida and Isuma partners on the 'process design' of the project – that is, how we might come up with a film script in the next 12 months. With a sabbatical about to start, I was able to spend much of the next year living on Haida Gwaii, but my Isuma colleague, Jon Frantz, was based in Igloolik, thousands of miles and thousands of dollars in airfares away, which presented immediate and ongoing challenges. The three-way partnership was based on 'in-kind contributions' from all the partners, which translates as *volunteer work*. Fine for me, with an academic salary. Not so fine for Jon, on a much smaller salary from Isuma, which depended on him working on Isuma's films in Nunavut. And definitely not fine for getting cooperation from the Haida, who never appreciated being asked to work for nothing on a project that may or may not ever materialize. Naively, 'we' did not realize in May 2014 just how much of a problem this was going to be.

By 'we', I am here referring to myself and Jon Frantz, as we carried most of the workload over this first year. Jon was a graduate of our Master's program at UBC with an initial background in banking (which was to prove very helpful), but once he graduated, Jon set up his own video production company with the intent of using video in planning work. Jon brought special personal gifts to this work: exceptional patience; a natural ease with folks from all walks of life; a love of fishing, hunting, and all things outdoors; and no ego that I have ever noticed. He was a great listener. And, since he had moved to Igloolik and been hired by the Isuma collective, he had learned a great deal about their model of film making and had been mentored by Norman Cohn in cinematography, and all of the financial aspects of film making. I billeted with a Haida family on the Skidegate reserve (in order to channel the grant funds covering my accommodation into the Haida community and get better embedded in that community). That decision about where to live, I later learned, became a factor in the emerging politics of the project. By always staying in the southern village, I became identified as being in the 'Skidegate camp' in some political sense, which did not help in the subsequent task of building relationships in the Old Massett community. This requires some explanation in terms of formal and informal structures of governance in the Haida world.

People, Relationships, and Governance

The Indian Act of 1876 had stripped Indigenous Nations of their power and authority, if not their sovereignty (Borrows 2016; McNeil 2017),[6] but it turned them into 'wards of the state'. Subsequent amendments to the Act attempted to outlaw traditional governance practices in which power had rested formally with hereditary chiefs (part of a clan system), but in actuality involved collective decision making and extensive talking. The Indian Act had driven these practices underground and replaced them with a system of Band Council elections every two years, at the same time installing 'Indian agents' appointed by the federal government in every community to oversee virtually every aspect of life on the reserves. The two reserves onto which the few remaining Haida had been forced to live at the end of the 19th century each have an elected Band Council (the Skidegate Band Council, SBC; and the Old Massett Village Council, OMVC) and an administrative structure similar to municipalities, with a staff organized along functional lines: housing, education, health and social development, infrastructure, and so on. But this entire system has been dependent on 'handouts' from the federal government, which in turn involved learning how to play the game of grant writing and reporting and meeting the demands of the Department of Indian Affairs in Ottawa. Indigenous dissatisfaction with this system has been seething for 100 years, and many Nations continued traditional governance practices surreptitiously. But the Haida are among the first in recent times to create a parallel system of governance of their own devising, the Council of the Haida Nation. The CHN was designed to take the big picture view of the wellbeing of the Nation as a whole; to devise policies to address that, including legal challenges to the authority of the federal government through the court system; and to include hereditary chiefs in this process aimed at self-determination and sovereignty.[7]

Those who sit on the Council of the Haida Nation are also elected, and there is equal representation of councilors from the two villages. Every effort is made to treat the two villages equally. This concern is not only all-pervasive in Haida governance. It was something we had to consider every step of the way, from the script development process right through to hiring folks as cast and crew, and to considerations of where the film would be shot – all the way down, in fact, to the number of lines of dialogue in each dialect in the script yet to be written. (The two villages each have their own dialect of the Haida language, and there's a third dialect, spoken in Hydaburg, Alaska, by Haida who had initially fled there to escape smallpox back in the 1860s.) And the fact that I chose to have a home base in Skidegate and to set up our project office in the Haida Heritage Museum, located on the Skidegate reserve, was read by some in Old Massett as an indication of bias from the outset.[8]

This was one aspect of the governance structure to be dealt with. And there were also relationships to be built. Among non-Indigenous planners who work with Indigenous communities, there is a shared understanding that establishing trust is the first and foremost task, and this involves building relationships, which takes time. Indigenous community planning work is, above all, relational, not transactional – in other words, the opposite of the typical Western way of getting things done. It sounds straightforward enough, in theory, but in practice it's not so simple for those of us who've worked in colonial institutions for many years and internalized a different way of being. Unlearning that colonial/Western way of being was/is crucial to working successfully with Indigenous communities. I thought I had learned this lesson in the years spent working on the previous documentary, where Giovanni and I were given the ultimate compliment by the chief of the Burns Lake Band, who said 'we trusted you because you took the time to get to know us'.[9] But there's an inherent contradiction in, on the one hand, depending on funds from Western/colonial institutions like the Social Sciences and Humanities Research Council (SSHRC) and the Canada Media Fund, which run on timelines, deadlines, accounting, and reporting, and on the other hand, the need to just hang out and get to know people and develop that trust. This tension ran through the entire project.

We had conceived this project as a community-based and community-driven action research project in support of the three core Haida aspirations noted above. It sounded good in theory. And it must have been persuasive in the grant application, since we received the maximum funding available under that grant. And by most measures of success, the results were spectacular, as I describe in the next section. Reflections on my always awkward and uncomfortable insider/outsider role in what started out as a three-way partnership are the subject of Chapter 12.

The Success Story

To start the project in a good way and establish the ethical protocols of the project, Jon Frantz and myself formed a four-person Haida Advisory Group: Diane Brown, a highly respected matriarch and fluent Haida speaker, who always opened our meetings with a prayer in Haida; Jason Alsop, a Skidegate community member who chaired the Cultural Committee of CHN and was also a star basketball player with the Skidegate Saints; Simon Davies, the non-Indigenous Communications Manager for CHN, who had been working with the Haida for several decades; and Cynthia Samuels, from Old Massett, also a CHN councilor. Jason and Cynthia gave us, we hoped, a direct line into the CHN; Simon's role as Communications manager would be helpful in spreading the word about the project, and his political savvy was already valuable; and Diane had been a leader in language revitalization for decades as well as holding immense cultural knowledge. She was a living legend. She could be a sweet *nanaay* (grandmother) one moment and fierce warrior the next. You didn't want to get on the wrong side of Diane. The relationship I was able to build with her was one of the most personally rewarding features of this entire project. We were to share many laughs, quite a bit of teasing, and some tears, of sadness and joy. Diane believed in this project from the outset and once, early on, confided that perhaps the reason she had narrowly escaped death recently was because she was meant to be involved in the film. Ultimately, she had a starring role as one of the *nanaays* in the film, alongside her daughter and granddaughter, and also helped with the language training of community cast members.

This founding group met over lunch, brainstormed the first steps of the project, and set out the ethical protocols for the partnership. The foundational principles of the partnership were that the Haida would have the decision-making power on all aspects of the project and would have majority ownership of any film that might eventuate. Jon Frantz and myself were essentially in advisory and mentoring roles, myself specifically on the creation of a film script, and Jon on all other matters relating to the financing and producing of a film. The Haida readily acknowledged that they had no experience with film making and were eager to learn about all aspects of the process. We intended to put a lot of emphasis on capacity-building within the Haida community, in hopes that if we were able to get a film project completed, it would not be a one-off project but rather the precursor to a film-based economic enterprise on Haida Gwaii.

Jon committed to working with the Haida to set up a Haida film production company in which the Haida would be the majority owner, Isuma a minority owner. And I committed to working closely within the Haida Advisory Group in the script creation process. I explained how I would go about this, hosting movie nights in the summer to get people talking about Indigenous films, followed by community story-gathering workshops and scriptwriting workshops; and a writing contest to choose the scriptwriting team, whom I would mentor and with whom I would co-write the script. We also discussed making a documentary about the whole process as it unfolded, and it was understood that Giovanni Attili and I would work together on that, given our previous experience. The Advisory Group was happy with this process and these principles.

Haida colleagues suggested that it might be difficult to get community involvement during summer when many people are out fishing and laying in supplies for winter, as well as visiting relatives, and so on. So, we abandoned plans to hold workshops during summer and instead simply 'hung out' and tried to get to know people and get the word out about the film project. I left Haida Gwaii at the end of July feeling very insecure about the community engagement process still to come, and questioning whether we could actually get Haida community buy-in.

When I returned in September (2014), Giovanni joined me (from Rome) with the intent of filming the upcoming community workshops as part of the proposed documentary about the making of the film, while I started working with two Haida community members: Dana, the Skidegate community planner, and Lucy Bell, the leader of Xaad Kihlgaa Hi Suu.u (Speak Haida Society: the northern Language Society in Old Massett). Together we co-designed and co-facilitated two community story-gathering workshops in which community members were asked to suggest what Haida story or stories they would most like to see portrayed in a feature film. Around 50 folks attended the Skidegate workshop, 40 in Old Massett, and we broke people into small groups at 'work tables' where we recorded conversations as well as using flip charts to capture story ideas. It was thanks to Dana and Lucy that we had such good attendance at these workshops, from elders to youth and everyone in between. They phoned individuals and urged them to attend, arranged for transportation for those who needed it, and arranged for community caterers to provide local food favorites. There was a high energy level and a lot of excitement among those who attended, and at the same time an undercurrent of disbelief that all this effort would amount to anything. They were all too familiar with White folks from the city coming to the community with big ideas for economic development, only to have them vanish into thin air after spending grant moneys that the community members never saw any benefit from. Nevertheless, we came away from the workshops encouraged by the participation and interest, and my spirits rose a notch from where they'd been at the end of the summer.

The story-gathering workshops were quickly followed by scriptwriting workshops that I offered in both villages, again with the invaluable organizational help of Dana and Lucy. My intent was to generate interest among Haida of all ages in being part of a small scriptwriting team, ideally two people plus myself, who would work with the story ideas already generated and come up with a script by May of 2015, the deadline for annual submissions for production funding to the Canada Media Fund (CMF), a semi-government agency with funds specifically for Indigenous film making. Attendance was much smaller at these writing workshops (15 in Skidegate, about the same in Old Massett), but those who came were keen to understand what was involved in writing a film script, and I used my training from the scriptwriting program in the Film School at UCLA (Commentary 1) to set out basic dramatic principles, while at the same time discussing what was different about the approach of some Indigenous film makers, including our Inuit partner.

At these same workshops I launched a writing contest, previously discussed with the Advisory Group, in which members of the Haida community were invited to submit a short story/film idea of between three and five pages. The jury comprised two Haida women (one from Skidegate the other from Old Massett) who were self-described film buffs, plus Jon Frantz and myself. We already had two guidelines from our Advisory Group, which reflected the often-fraught political and cultural relationship between Skidegate and Old Massett, which was that the writing team must have equal representation from each village, and that any script created must reflect in more or less equal proportions the two dialects of the Haida language, the 'northern/Old Massett' and the 'southern/Skidegate' (Xaad Kil and Xaayda Kil). The jury read the submissions 'blind' and reached a unanimous decision about the two best stories. One was by Skidegate community member Graham Richard, the other a joint submission by brothers Gwaai and Jaalen Edenshaw from Old Massett, both well-known Haida artists/carvers from a high caste Haida lineage, who had grown up with cultural teachings.

Now, mid-November 2014, we had a writing team, and the four of us gathered over brunch in the café at Tlell (halfway between Skidegate and Old Massett), where we mapped out how we would work together over the next six months to come up with a script by May 2015. We then worked together three days a week for six months in the development of the film script *Edge of the Knife*.[10]

During this process we set up another Advisory Group with whom we could discuss aspects of our evolving story to make sure everything was culturally accurate and that we were not trespassing on any protocols such as specific family ownership of certain stories and songs. There were three constraints/guidelines for the script from the outset. One was from Jon, as acting producer, in terms of the scope of production that would be affordable, given his best guess about what our budget might be. Based on Inuit experience, he believed that we could access between C$1–2 million if we had a good script and some Inuit participation. That meant that we could not even re-create an entire traditional village, let alone move between time periods – such as from pre-contact to post-contact to contemporary – because of the cost of sets and costumes. A second constraint was the language one already mentioned, which meant we had to invent a story that would 'naturally' involve more or less equal numbers of dialogue lines in northern and southern dialects. And finally, our Advisory Group hoped that we might come up with a story that reflected the strength and health of Haida culture, technologies, and ways of living, pre-contact.

Given these guidelines, by the end of our first writing weekend we had settled on what was known as the 'Wildman' or 'Gaagiixid/Gaagiit' story, for one very practical, cost-related reason that it would involve basically one location (somewhere remote on the coastline of Haida Gwaii), but also because it was the most-mentioned story in our community story-gathering workshops. We then had to invent a whole dramatic structure around the variety of stories that had been passed down through time of someone who is lost at sea in a canoe accident, is the lone survivor, makes it to land, but then experiences guilt as the only survivor, and descends into madness as he struggles to survive in the forest with uncertain prospects of ever being rescued. And if he is rescued, can he be brought back to his humanity after a descent into madness? Can he be brought back into community? We were excited too about this story for its potential contemporary parallels, notably the ways in which some community members are losing themselves to drugs and other forms of addiction.

Each weekend, we worked long days, interspersed with pit stops at the handful of fast food joints in both villages. By Christmas we had a 10-page story outline; by early March we had 50 pages shaping up; and by April we were able to get the script to our cultural advisors in plenty of time for them to review it. They gave us some great advice about specifics, such as songs, details for a welcoming ceremony, and gratitude for the first salmon of the season. Jon worked through April to prepare the detailed, 1,300 line-item budget to accompany the script for submission to the potential funder.

Jon had been trying to interest various Haida in being mentored as producer, but no-one had shown any interest. At this point we were very much dependent on Jon's financial expertise, acquired through his work with Isuma, in getting a budget prepared. And in doing that work, Jon had to secure a guarantee on paper from the CHN that they could provide 10% of the estimated budget, as forewarned. He was having trouble reaching any of the CHN executive. In fact, it seemed to us that the Haida leadership were being very hands-off/wait and see, as though they didn't really believe the project would ever materialize. We were a little frustrated at this apparent lack of involvement, and anxious about missing the funder's deadline.

But we made it. The script and budget were submitted for production funding to the Canada Media Fund and Telefilm Canada, and a few months later we learned that we were successful and would receive $1.89m in production funds. While the script was a crucial component of this success, so too was Jon's financial expertise, along with the track record of our Inuit partner, Zacharias Kunuk, whom we listed as executive producer. This news really galvanized support in the Haida community. Now that there was money on the table, the project was real. But all sorts of political complications now entered the picture, such as choice of location, choice of director/s, questions of sourcing authentic 'props' such as canoes and longhouses, and translating the script into the two dialects.

Fifteen Elders from SHIP (the Skidegate Haida Immersion Program, established by Diane Brown in the 1990s) translated the southern dialect portion of the script over the next year, while two elders, the sisters Dolores Churchill and Jane Adams, living in Hydaburg, Alaska, eventually translated the northern dialect. This had been included as an expense in the grant budget, so it was not volunteer work, but still, the extraordinary dedication to the task on the part of both translation teams told us how important this potential film was to them. As Dolores said, subsequently, 'I'm 88, and I still want to help people learn the language. We must protect our language. It's the vital part of our culture'.

Jon worked with the CHN executive in creating a majority Haida-owned film production company, building on the Nation-to-Nation learning and knowledge transfer that had been foundational to this project. And pre-production unofficially began in early 2016, when Jon and I sat in a coffee shop in Massett and discussed whom we thought should/could direct the film. We were committed to hiring an Indigenous director, as well as Indigenous folks in as many key creative roles as possible, but there were no Haida we were aware of with feature film experience. There was, however, a Ts'ilhqotin director, Helen Haig-Brown, who had made an award-winning long-form documentary as well as some stunning short features, and was now living on Haida Gwaii with her Haida partner and their baby. We speculated about pairing Helen with Gwaai, who a few years earlier had co-directed with his brother a play in the Haida language and had also produced animated short videos. I had developed enormous respect for his cross-cultural fluency, his film savvy, and his artistic sensibility. We interviewed them both to see, among other things, how they each felt about the prospect of working together. And that was the beginning of an outstanding creative partnership between those two.

They then ran casting auditions and acting workshops during Spring and Summer of 2016 in the two Haida villages, following the Inuit model of working with untrained community members. According to CMF funding rules, production was supposed to start in Fall of 2016, no more than 12 months after receiving the funding. Here was a classic dilemma for Indigenous film making: funding from colonial institutions such as the Canada Media Fund, with its rigid rules and deadlines, did not seem to consider the challenges facing small Indigenous Nations (with no previous experience) in gearing up for such a major project. In our case, the cast hadn't even begun the difficult task of learning their lines in the Haida dialects. And there were ongoing arguments over the choice of location, because there was a lot of money at stake, depending on whether the chosen location was closer to Old Massett or Skidegate. There were ongoing delays over whether to build new canoes and longhouses or whether to rent existing ones from community members. There were crucial production positions, such as cinematographer and sound and lighting technicians, which hadn't been filled. Capacity issues were an ongoing concern. Finding Haida folks to work on the project often didn't happen until the last minute. All of these were major headaches for Jon, as the now official producer. But the language learning was perhaps the biggest challenge of all, and something that was central to the whole purpose of making the film.

The directors decided on a two-week 'language boot camp', during which the cast were isolated from their families and housed in log cabins at Hiellen on the north coast of the north island, where there was one-on-one coaching from the fluent Elders and a couple of proficient young speakers.[11] Erica Ryan-Gagne, one of the actors, had been studying the language at the University of Northern British Columbia (UNBC) and was attracted to the film project because of the language goals. 'At the bootcamp', she said later, 'I was living my dream, learning more about dialect differences … these powerhouse speakers in one place. I went as an actress but then got thrown into teaching and facilitating … Gwaaganad (Diane Brown) coached me on intonation, and she told me, if I pray in our language, the ancestors will help me'. Erica's plan was to become a language teacher in the schools, and she dreamed of the kids 'bringing it home and making their parents learn. The young people will inspire'.

Willy Russ was a youth worker in Skidegate when we began the project. He heard about the writing contest and came along to the workshop to hear more. For him, being cast as one of the two lead male roles was 'two dreams in one coming true: learning to be an actor and learning the language'. Now he is teaching language to youth through food and other activities. 'I felt a big responsibility to learn. I started to think and dream in Haida. Hopefully the film makes it more appealing for everyone to learn the language'.

I was able to visit the boot camp for a day and witness the actors struggling to get their vocal cords and face muscles familiar with the 35 consonants, two tones, and 20 sounds unique to the Haida language. Willy shared with me at that time just how difficult it was, but added that it was 'the best kind of challenge'. This boot camp took place in April 2017, two months before filming was due to start, and my impression (voiced to nobody at the time) was that the film was going to be a challenge because very few of the actors were managing the sounds, let alone being able to acquire emotional expression and fluency with sentences. A year after the film had successfully premiered, I dared to share that opinion with matriarch Diane Brown, one of the language coaches. And to my surprise, she said she had felt exactly the same anxiety at that time. But in spite of the daunting challenge, or perhaps because of it, the boot camp itself was helping to create a dedicated community among the cast, already feeling a huge sense of responsibility (to the ancestors as well as to the future) to 'get this right'.

Pre-production proceeded apace, with two Haida community members hired as production managers (one from Skidegate, one from Old Massett), and more as set and costume designers and set builders. A location was finally chosen. Filming took place in the remote northwest wilderness of the archipelago, close to Old Massett, in June–July 2017. Editing then took another year. By June 2018, the film was ready for release and was submitted to various festivals. Exceeding our own expectations, the film was selected to premiere at the prestigious Toronto International Film Festival in September 2018 and then at the Vancouver International Film Festival a couple of weeks later, where it won Best Canadian Film. It went on to win Best Film at the ImagiNative Film Festival in Toronto in October, and the following year, 2019, it won Best Film, Best Actor, and Best Supporting Actor at the American Indian Film Institute's annual Film Festival. There was a special screening for 300 parliamentarians at the Museum of Civilization in Ottawa in Fall of 2018. It was a delirious time for the cast and directors.

But before any of these prestigious public events, we arranged screenings for the Skidegate and Old Massett communities on Labor Day weekend, where the community halls in both villages were packed (beyond fire regulations capacity) as 50% of each community showed up, and cheered for 20 minutes after the final titles. There were tears of joy as the aunties (matriarchs) and Elders gave their nods and hugs of approval.

*

So far, then, this is a story of a two-fold success. Most importantly, the film has been a huge hit among the Haida community and has fueled enormous cultural pride as well as interest in the language itself. Especially among the Elders, hearing their language spoken throughout a 100-minute riveting film was something they had never expected to experience in their lifetimes. When I asked Diane Brown about her reaction to the first screening, she told me she cried all the way through it. Tears of joy! According to Willie Russ, one of the two male leads in the film:

> After this movie was made, and after it was screened, there's been a huge growth spurt in people wanting to reconnect with the language and signing up for language programs … because this movie was made, people are feeling and realizing the importance of our language … And so I feel the elders are very proud. They feel heard, they feel cared for and loved … For a long time they felt their knowledge was going to pass on with them but I don't feel like they feel like that way anymore.[12]

Gwaai had said many times during the writing, filming, and editing that the most important audience was the aunties and Elders. His wish came true.

But beyond the Haida community there was the astonishing national and international reception of the film among Indigenous and non-Indigenous audiences alike. The film sold out all of the scheduled three screenings at the Vancouver International Film Festival (VIFF) and extra screenings were arranged, one of them in an 1,800-seat auditorium. It too sold out, with lines snaking for blocks in downtown Vancouver. Another week of screenings later that year in the VanCity cinema also sold out and led to extra screenings. The Canadian Broadcasting Corporation (CBC) picked up television rights for five years. Air Canada picked up rights to screen the film on its flights for two years. As one of TIFF's Top Ten Films, it toured all over Canada in 2019.

It had been my hope, when I wrote the initial grant application, that the film might inspire other Indigenous Nations to see the potential of film as a medium for language revitalization. The Gwaii Trust (a non-profit on Haida Gwaii) did provide $10,000 for a Haida community member to take the film on a tour of Vancouver Island and Lower Mainland communities. Graham Richard, Communications manager with CHN and one of the three Haida script writers, took on that project and reported ecstatic audiences everywhere, Indigenous and non-Indigenous.[13] But there was no funding for further screenings and workshops with other Indigenous communities in the north or the interior of the province, or beyond.

The extent of attention the film has attracted beyond Haida Gwaii, nationally and internationally, exceeded our wildest expectations. The reception of the film by a Maori audience on the Orakei marae in Auckland in June 2019 was especially gratifying. Helen Haig-Brown, Dana Moraes, and I were able to travel there on my SSHRC funding for a screening at a

NAISA conference (Native American and Indigenous Studies Association) as well as on the marae. It was the latter screening that was deeply satisfying, as the Maori Elders immediately understood it as a story of healing through reconnection with community, and declared that the film should be seen by Indigenous youth everywhere, as a way of countering the epidemic of Indigenous youth suicides.

Beyond the acclamation, another more tangible impact was that $800,000 of the budget went directly to Haida community members employed during the production. Some capacity-building happened (but nowhere near enough), thanks to the UBC and Inuit partners training the Haida in various aspects of film making, including financing, management, and logistics as well as scriptwriting, directing, and cinematography. No-one has yet quantified the impact on language revitalization or other less tangible impacts, but our post-production interviews indicate many other less quantifiable benefits such as cultural pride, community-building, healing between and within families and villages, and connections between youth and Elders.

Erica Ryan-Gagne talked about how special it was being on set: 'everyone on the land, being Haida, like the old days. It was very healing.' Diane Brown shared that she had been 'in a sad dark place before this film. We were losing elders, and there's not enough happening in the schools. But … film lasts forever. It's the best way for children to learn … It takes a whole family to learn the language. It's not easy'. I was able to spend a week on the set (at the remote location of Yaan, northwest of Massett), sharing a tent with Diane; her husband, Dullskin; and her daughter Lauren and granddaughter Emma, both of whom were in the film. While we were there, Diane shared with me that she felt the presence of the ancestors at Yaan, and that it was a wonderful experience to be all living together in the camp. 'It took me back to feeling like it was pre-contact … Our love for our place is the same as it always was'. And four years later, Diane reports that granddaughter Emma hopes to train as an actress, while her older sister is enrolling in film school in Vancouver to study scriptwriting[14].

Leo Gagnon, a carver and blacksmith who played one of the Elders in the film, talked about how he wants to be involved in language through knife-making and paddle-making and weaving. 'It was like the good old days being out at Yaan'. Like most of the cast, Leo got a traditional tattoo from artist Corey Bullpit before the film shoot (which he proudly displayed on his torso during a scene in the film where he catches an octopus). Leo was excited at 'how many silent speakers were coming out of the woodwork' during and after the production. (Silent speakers are people who once knew the language but had repressed it as a result of colonization.) James MacGuire, a young musician who became the props manager for the production, 'fell into this after helping out at the language bootcamp … I learned a lot, in the process, about our history. I consulted with the elders on set about the use of the artifacts we'd borrowed (from the Haida Heritage Museum) such as Dolores' weaving set-up. The film pumped me up to get more involved in being a cultural guide'. His biggest hope for the film was that it would be a 'a teaching tool for language'. Tyler York, a carver and exceptional athlete who had the lead role in the film as 'The Wildman', commented that he had started learning Haida in elementary school from the elders 'but in high school I was told I had to learn French if I wanted to go to Uni. So my language was taken from me again'. At the language boot camp, he was especially inspired by Ben Young, a Haida living in Hydaburg, Alaska, who is the only functionally fluent young speaker and was one of the coaches for the film. Ben now teaches immersion classes in elementary school in Hydaburg. Tyler chuckled as he noted that Ben's youth and language fluency made his 'competitive nature kick in'. Tyler had spent 8 hours during language bootcamp having a tattoo hand-poked on his chest by Corey Bullpit. That activity seems to have been another factor in the community-building and pride that was happening during pre-production. Tyler believed that 'the film will show the world we're still here. It will inspire youth and get them off their phones'. And Dolores Churchill also noted 'we came to feel like a real community when we were on the set. I had a strong connection with the past, especially when I saw ancestral remains in the woods ... The children interacted with all of us. It's a great way to learn. I saw the children speaking Haida afterwards. Wonderful ... We're now practicing more language and I'm coaching people over the phone'. Likewise for Sherry Dick, who played one of the *nonnis/nanaays* (grandmothers): 'My nonni spoke Haida all the time when I was a child. To hear it again makes your heart warm ... The bootcamp experience was awesome. So beautiful. It's put a fire under me about learning Haida. My son is fifteen now and I want him to learn'.

We haven't been able to reach out to members of the Haida community who weren't directly involved in the production, but in a post-production interview with Graham Richard in 2018, I asked if he had any sense of the impact on the wider community.

> I feel like everyone kind of bumped up a level. People who wanted to act but were too shy, have been bumped up to a level where they're inspired to the level where they're like, 'oh, maybe I could act, maybe I could do all kinds of things'. Right. Yeah. I think this is very affirming, overall, for the whole community. It brought so much dignity to everybody's heart right. Yes, it was beautiful. Beautiful project.

The sisters Dolores and Jane, who had translated the northern dialect for the project, should have the last word. 'We spoke to each other all the time on the set in Haida. There was so much love on the set. A community. All wanting to learn'.

The Haida defined their own criteria for the success of this project, and the quotes shared above indicate how far the film went towards achieving that success, extending far beyond tangible, quantifiable factors to the less tangible but arguably more precious dimensions of community-building and healing from the wounds of colonization.

*

There were two major medium and longer-term intentions behind this project. The first was that the film itself, and the fact that it is made in the Haida dialects with English subtitles, would generate interest – especially among youth – in learning the language, and that teaching and learning relationships between Elders and youth would be sustained beyond the end of the film production. Anecdotal evidence (some of it embedded in the quotes used in this chapter), as opposed to systematic data collection, suggests that this has been happening. The second hope was that there would be sufficient training during this project for the Haida themselves to be able to continue to generate film projects under their newly created film production company, energized and inspired by this project. Four years beyond the premiere of *Edge of the Knife*, this does not yet appear to be happening.

In retrospect, the whole project was perhaps too ambitious for an Indigenous Nation with virtually no experience in film making. The CMF and Telefilm have created a system that enables experienced production teams to access professional financing. It is not designed to be a capacity-building fund to teach and develop the skills required to make a professional film. We had assumed that we could develop the skills required within CMF's time frame. But

CMF and Telefilm operate on a colonial industrial model of film making, with tight timelines and complicated financial accounting and reporting. It's near impossible, conceptually and practically, to transplant or impose that model on an Indigenous Nation used to operating in more traditional ways, and still suffering the intergenerational traumas caused by the residential schools. Those who were hired to work on the project found themselves in an intense environment that demanded long days on the job, day after day, at a relentless pace. The clock was always ticking, but community members have other priorities, such as seasonal harvesting, taking care of extended family, and so on. There was immense dedication to the success of this project, illustrated in some of the quotes from cast and crew above. But the time pressures of the pre-production period, and of the film shoot itself, were an unwelcome surprise to most people. In retrospect, it may have been wiser to start with a smaller, documentary project while building capacity, and gradually build to the scale of a feature film. The Isuma folks had 15 or so years of that kind of small-scale experience before taking on their first feature film project, and they had evolved a financial model and support infrastructure that enabled them to make a film every couple of years. The constraints of the CMF budget meant that too little capacity-building was possible during the project, perhaps resulting in a burnout among those who were initially fired up about getting a 'film industry' going for the Haida Nation.[15] An even more fundamental mistake in the conception of this project may have been in expecting that the Isuma 'model' could be transplanted to another Nation. Every Nation has its own priorities, and film making has not yet emerged as one of the Haida Nation's top priorities. Only time will tell whether these two medium and longer-term aspirations will ever be realized. In the meantime, the Haida community still basks in the success of *Edge of the Knife* and all that it says about their culture.

This story continues in Chapter 12, where I reflect on my own learning about partnerships, and so much more, in the contemporary Canadian political context of reconciliation.

Notes

1 All quotes from Haida community members in this chapter, unless otherwise referenced, are taken from the video tapes of interviews with them conducted during post-production. See footnote 2.

2 These interviews were conducted by Haida community members Dana Moraes and Graham Richard as part of a Social Sciences and Humanities Research Council grant I received in 2017, the purpose of which was to evaluate the impact of the film on the Haida community, and beyond. Dana and Graham were both part of the production crew: Dana as production manager for Skidegate, Graham as script supervisor. Dana had previously been the community planner with Skidegate Band Council, and Graham works in communications for the Council of the Haida Nation. The video interviews cited can be viewed in their entirety on http://www.isuma.tv/haida/language

3 While this obviously appears to center Whiteness when this is really a Haida story, my intent is to explore and share the learning involved in working in such partnerships, for the potential benefit of those non-Indigenous folks contemplating this kind of work in the future.

4 *Edge of the Knife* Cast & Crew Q and A, Sept 9, 2018, Toronto International Film Festival, You Tube video, posted by TIFF Talks, Sept 8, 2018: https://www.youtube.com/watch?v=eT7pbYGWwnA

5 Meeting with the CHN Executive, September 10, 2013, in which Dana Moraes and myself propose and explain the film idea, seeking CHN support.

6 Mainstream accounts of the Indian Act, even by (non-Indigenous) scholars sympathetic to Indigenous rights and title, have typically described the Act as stripping formerly sovereign nations of their sovereignty. Increasingly, Indigenous (Borrows 2016) and non-Indigenous legal scholars (McNeil 2017) dispute that claim, arguing that nothing can strip Indigenous peoples of their sovereignty: that their sovereignty is innate, embodied, and resides in practice and law. Thanks to Scott Graham for this clarification.

7 See https://www.haidanation.ca/mandate/
8 This account of the rivalry between the two reserves/villages is my own interpretation, based on many conversations in both communities over a 5-year period.
9 Chief Rob Charlie, speaking in the Epilogue to *Finding our Way* (Moving Images, 2010)
10 The film draws its name from a Haida saying, 'the world is as sharp as a knife' – that as we walk along we have to be careful not to fall off one side or the other. This is the world we live in – so close to nature, our lives always hanging in balance – one slip can change everything.
11 An interactive language platform was also created as part of the CMF funding, and Elders recorded tapes on MP3 players of all the lines for each of the characters, which the actors could then study in their own time.
12 Ibid, Q and A, TIFF, fn 5.
13 Interview with Graham Richard, Oct 31, 2019.
14 Phone conversation with Diane, May 23, 2022.
15 These reflections are based on a series of conversations with Jon Frantz in May-–June 2022 and necessarily reflect our assessment, which may differ from those of the Haida participants.

References

Borrows, J. 2016. Unextinguished: rights and the Indian act, *University of New Brunswick Law Journal* 67, 3.

Evans, M. R. 2008. *Isuma. Inuit Video Art* (Montreal: McGill-Queens University Press).

McNeil, K. 2017. *'The Inherent Right of Indigenous Governance', Osgoode Hall Law School, Osgoode Digital Commons, All Papers* (vol. 319). Available at: http://digitalcommons.osgoode.yorku.ca/all_papers/319.

Simpson, L. B. 2017. *As We Have Always Done. Indigenous Freedom through Radical Resistance* (Minneapolis: University of Minnesota Press).

Skidegate Band Council 2017. Gud Ga is (being together to talk), comprehensive community plan Available at: http://www.skidegate.ca/wp-content/uploads/2022/02/ccp2012_2017.pdf.

PART IV

Navigating Indigenous Worlds

Praxis and Pedagogy

FIGURE 11.1 *Musqueam Foreshore* by kind permission of Musqueam artist Susan Point, who owns the copyright to the artwork. Image Copyright: Coastal Peoples Fine Art Gallery, Vancouver, BC, Canada.

Commentary: The Inner Journey: 2007–2022

The three chapters comprising Part IV chart my inner journey over the past 15 years since I began to work with Indigenous communities, first in praxis and then in pedagogy. They are deeply personal as well as political accounts of my ongoing attempt to 'unlearn the colonial cultures of planning' (Porter 2010). It is not a flattering story. I suppose it affirms the truism that we learn most when we are in the most discomfort: unsettled, unmoored, uncertain of the ground we stand on. Clueless about how to move forward. We learn from our missteps and mistakes; from retrospective reflection on our self-delusion; from questioning our moments of self-righteous indignation. In this Commentary, as in the previous Commentaries, I want to pay particular attention to the relationships that have underpinned this journey: to pay tribute to my teachers, for their generosity and humanity.

This journey into Indigenous worlds began, unintentionally, in 1995. I had been asked to curate a special issue of the journal *Planning Theory*, and I was using the occasion to put together a set of essays that would challenge mainstream planning history ('the Official Story') by introducing new theoretical lenses and new positionalities. I brought together a group of young scholars for a paper session at the annual Association of Collegiate Schools of Planning (ACSP) conference, most of them doctoral students at UCLA, to get feedback for their contributions to the journal issue (Commentary, Part I). These were the beginnings of what I was envisaging as new historiographies for planning, making the invisible visible. After we'd all presented and the floor was open to questions, someone at the back of the room stood up and, in a very soft-spoken, non-confrontational way asked, had I considered including an Indigenous perspective in this collection? This was one of those moments that surely all scholars dread: the question we have not thought about and have no idea how to respond to. Suddenly intellectually naked, I said that I hadn't, but would be keen to discuss. And that was how I met Theodore (Ted) Jojola, a member of the Isleta Pueblo, who was then and still is teaching community planning at the University of New Mexico.

The chapter Ted subsequently wrote, for what a few years later became my edited book, *Making the Invisible Visible: A Multicultural History of Planning* (1998), rocked my world in as profound a way as my meeting with Giovanni Attili did in 2001 (Commentary, Part III). I was expecting that Ted would write about the colonization and dispossession of Native American peoples, and he did. But there was more. From him I learned of ancestral Indigenous traditions of planning stretching back to time immemorial, traditions equivalent to what Western planning would call community and regional planning. Today it's an honor to call Ted a friend, and we've worked together and supported each other's efforts at decolonizing planning education and practice over the subsequent decades. Ted has devoted a lifetime to educating the planning profession about its complicity in the colonial project, and was successful in 2016 in getting substantial funding to establish an Indigenous Design and Planning Institute at the University of New Mexico. He is a gentle, generous, and tireless teacher of us settlers, and I am indebted to him for his teachings, his kindness, and his support in numerous ways.

During my five years back in Australia (1996–2001), the relationship between First Peoples and the Australian state, as the colonizing power, had risen to political and legal prominence since the historic 1992 'Mabo decision' by the High Court. A land rights movement had been gathering momentum since the 1970s. Eddie Mabo, representing the Murray Islands in Torres Strait, made the case to the High Court that his people's entitlement to the Murray Islands had not been extinguished by the settlement of Australia by the British. In handing down the decision in favor of Eddie Mabo's claim, the Court found that common law in Australia recognizes a form of title which is also applicable to the mainland. This decision dismissed forever the fallacy of *terra nullius*, of land unoccupied prior to British settlement. It opened the way for the passage of the Native Title Act of 1993, and implied that all lands, including urban lands, were once legitimately native lands and potentially open to claim (Butt and Eagleson 1996). Indigenous activists had also been drawing attention to the high rates of incarceration and deaths in custody of Aboriginal people and to the lost generations of Aboriginal children, stolen from their parents and removed to urban centers, adopted by White families, and forced to attend church-run schools. The refusal of Prime Minister John Howard, elected in 1996, to offer any sort of apology or reconciliation process had led to 'Sorry Books' being placed in public libraries throughout the country for anyone who felt moved to sign, and a series of 'Sorry Day' marches were organized in every capital city in 2001. Half a million people had participated in the march in Sydney. I was one of 300,000 folks who marched in Melbourne.

I was tangentially aware of an emerging scholarship, mostly coming out of history (Reynolds 1989), geography (Jackson 1998), and postcolonial studies (Jacobs 1996), that had begun to rewrite the foundational story of Australia. No longer a glorious story celebrating the brave pioneers, now a dark story of dispossession and genocide. It is sobering, and very revealing of the depth of the challenge of seeing through our White eyes, that the three scholars I cite here (Reynolds, Jackson, and Jacobs) are all non-Indigenous. I was unaware of Indigenous scholarship that had already rewritten that foundational story, countless times, and it didn't occur to me, then, to look for Indigenous scholarship that could inform my thinking.[1] In fact, I had not yet begun to think about how this work might 'spill over' into planning and require a/nother complete rethinking of how we told the history of the profession and, above all, of how we thought about and taught planning practice.

That all changed when Libby Porter came into my life, as a PhD student, in 1999. Libby had practiced planning, and was frustrated with that practice, before returning as a grad student, and she now wanted to think about how planning systems might better accommodate Aboriginal rights and aspirations. At least, that's where she started. We grappled together with ways of re-theorizing planning, and I was no help at all. But Libby went on, in the next decade and beyond, to rewrite the story of planning as a colonial project, a major theoretical and historical contribution (Porter 2004, 2007, 2010; Porter and Barry 2016; Jackson, Porter and Johnson 2018).

Libby had succeeded at least in planting new questions in my mind about the role of planning in Australia's history of colonization. That led me, in my role as an editor of the journal *Planning Theory & Practice*, to curate a group of essays in 2004 on 'Planning and Indigenous Communities'. Libby contributed an important piece, 'Unlearning One's Privilege: Reflections on Cross-cultural Research with Indigenous Peoples in South-East Australia', while I editorialized about issues of sovereignty, knowledge, and ethics. Geographer Sue Jackson had broken new ground in 1998 with her studies of the role of planning in the northern Australian towns of Darwin and Broome, demonstrating how planning had been and continues to be a spatial technology of cultural dominance, repressing Indigenous People's perspectives and concepts of place (Jackson 1998). Libby's own PhD, eventually to become the path-breaking book *Unlearning the Colonial Cultures of Planning* (2010), went much further, articulating theoretically and practically how planning was and still is a colonial project. In my acceptance in Houston in 2015 of the ACSP Distinguished Planning Educator award, I said the following:

> 'Mentors and apprentices are partners in an ancient human dance: the dance of the spiraling generations, in which the old first lead, then get to follow. I was blessed with

a great mentor. And now I have inspiring apprentices to follow'. Libby was one of a handful of such inspiring apprentices who have become my teachers.

By the time Libby's seminal first book was published, I had started to work with Indigenous communities, as Chapter 9 has described. The ways in which that work challenges one's very *way of being* is the often-painful subject of Chapters 11 and 12. I have explained (Commentary, Part III) the serendipitous way in which this opportunity first came about, in Burns Lake in northern BC, and how I thought about it at the time as an experiment in using film as a therapeutic planning intervention. But by the end of that project, I realized that the experience and the learning I was forced to undertake had changed me, as a person, as well as how I thought about planning education and practice. That inner transformation is best described in Chapter 11. I won't re-hash it here. I would, though, like to pay homage to those who taught me from those communities during that project, as described in the poem: specifically, (Chief) Corrina Leween and senior policy advisor Mike Robertson, from the Cheslatta Carrier Nation; (Chief) Rob Charlie from the Burns Lake Band; as well as the town's non-Indigenous anti-racism activists, especially Pauline Goertzen, Rayanne Charlie, and Doug Phair.

But for all of the existential power of that experience of working in partnership with the Cheslatta and the Burns Lake Band, for the unforgettable experience of *activating the heart*, for the lessons learned about how to work in a good way with First Nations, the ultimate and hardest lesson was in returning to my own 'community', the university, and asking some tough questions about what I/we were teaching and in what ways we were deeply and blindly entrenched in ongoing colonial practices. I began to ask about the decolonizing work that needed to be done in my own backyard.

⋆

My backyard is the University of British Columbia, which sits on the traditional, ancestral, and unceded territory of the Musqueam, the people of the river grass. The question staring me in the face in 2010 was this: why isn't my School (of Community & Regional Planning, SCARP) talking about any of this, this spatial apartheid and cultural genocide, and planning's role in it all? I started to talk about it and was encouraged by the then Director of SCARP, Penny Gurstein, to make some recommendations about curriculum change. I had learned just enough to know that I should consult the Nation whose unceded land UBC is occupying. I opened a conversation with our Musqueam neighbors, inquiring about their possible interest in such a program, hoping they would want to be involved, although I had no idea what that might look like.

Several years and many difficult as well as productive conversations later, SCARP entered a formal partnership with the Musqueam Nation in the design and delivery of the Indigenous Community Planning (ICP) program, and in the Fall of 2012, six students entered that two-year Master's program, four of them Indigenous. I've co-written, with my Musqueam and UBC colleagues, several essays on the progress and challenges of that program (Sandercock et al. 2013, 2016; Chapter 13), which has just completed its tenth year, is now admitting ten students each year, and has graduates doing outstanding work throughout British Columbia and beyond.

Even as we teach this program and are proud of it, we've come to understand that there is so much more learning work to be done within UBC, within my own school, and within each of us, especially those of us who are White settlers. For example, one of the hardest things I've had to learn, in the process of uncluttering my colonial mind, is to let go of some of the very attributes that contribute to success in academia. My way of charging ahead and getting done what needs to be done, respecting and meeting institutional, funding, and other deadlines, had been *my way of being* since I was a junior academic. This was how I'd been socialized. Deadlines. Results. Articles. Books. Films. The more you do, the more you get asked to do. I was on a behavioral treadmill I seemingly couldn't get off. I had always been in motion, never at rest. I was a punctuality fetishist. I couldn't bear to waste a moment. The whole concept of 'hanging out' was completely foreign to me. I had adapted unquestioningly to the institutional deadlines without which a university can't function. These were the worst possible qualities to bring to working with an Indigenous community. It took me a while to grasp, for example, that our Musqueam partner did not appreciate being asked to work on UBC's or potential funders' timelines, and that to expect that they would revealed the depth of my still-colonial self. Once I did grasp this, relationships improved, but then I was at odds with half of UBC, from our Department Head to our admin staff at the time, as well as the folks in the Development Office of our Faculty, resulting in a few unpleasant conflicts with them all.

But one of the many joys of immersion in the ICP program these past dozen years has been the hiring of Indigenous colleagues. We were fortunate to be able to hire Jessie Hemphill in 2018 as coordinator of the Practicum part of the program (experiential, community- and land-based learning). Jessie, whose family lineage is of the 'Nakwaxda'xw, Ligwitlda'xw, and Métis Nations, had recently graduated from the Community Planning program at Vancouver Island University after working as the community planner for her own community, as well as serving as a councilor for the District of Port Hardy. She had established an Indigenous community planning consulting practice, Alderhill Planning, with two Indigenous colleagues.[2] We had initially brought her in as a mentor to one pair of students in the Practicum, and such was her impact on all of us that we were eager for more of Jessie. Thanks to her willingness to spend a day each semester with my own ICP class, I've been able to learn from Jessie three important concepts central to ICP work, but surely relevant to all planning work. The first is the concept of *nested systems*, which expresses an Indigenous worldview of the interconnectedness of all of the webs of life, from the individual to the family to the community, then to the land, and ultimately to the cosmos. The second is (among other things) an expression of Jessie's personality: she talks about '*planning for joy*' and asks us to center this as an approach, as a way of being as well as a way of doing planning work. Imagine the difference this could make to professional practice. And finally – bearing in mind Canada's history of cultural genocide through the Residential Schools policy for over 100 years, a policy violence that has caused ongoing intergenerational trauma for all Indigenous communities – Jessie teaches about the necessity of a *trauma-informed planning*. This is a recognition not only of the need for individual as well as collective healing within Indigenous communities, but also a warning to those working with Indigenous communities that trauma is ever-present, can manifest in unexpected ways at unexpected times, and needs to be addressed for what it is, a deep psychic wound. There is no 'How to' manual for this kind of work, although Aftab Erfan's essay on how therapeutic planning can address collective trauma is a good starting point (Erfan 2017)[3]. As our Practicum Instructor, Jessie is able to prepare our students as much as is possible for this kind of work.[4]

At the same time as we hired Jessie half-time as our Practicum Coordinator, we were able to bring in Dr. Maggie Low as our first full-time tenure track Indigenous appointment. Maggie, a status member of Wikwemikoong Unceded Territory, had just completed her PhD, working with the Heiltsuk Nation. She brings knowledge of resource management and climate change challenges into the program, along with an innate generosity of spirit and an understanding that this work is all about process. Maggie has now stepped into leadership of the program. Equally delightful and gratifying has been the hiring of Shauna Johnson, whose family lineage is Tsawout on her mother's side and Tsimshian from Laxkwala'ams on her father's side. Shauna is a graduate of the ICP program, and now serves as a mentor to one of the Practicum placements each year.

★

In parallel with my developing pedagogy in ICP, I was developing the praxis of film as a planning intervention in the service of Indigenous cultural revitalization in my work with the Haida Nation. In Chapter 12, I reflect on a different kind of learning, at once personal and political. There is a contested conversation in Canada about the concept of reconciliation, its meaning and the implications for those wanting to work in partnership with Indigenous communities. My personal conclusion is that context is everything when it comes to the potential of such partnerships. I found myself receiving mixed signals from different members of the Haida community over the years that it took to complete the film. At the time, this was very unsettling and confusing. In Chapter 12, I take a deeper look at that personal unsettling, because it is a not uncommon experience for White folks working with Indigenous communities. With the benefit of some distance from the experience, the insight for me becomes that a feeling of powerlessness is only possible when one is used to feeling powerful – so used to it, in fact, that we are not aware of inhabiting the world in this way, in our taken-for-granted White bodies. I draw on Sara Ahmed's work to probe this notion of Whiteness – and the associated, taken-for-granted White power to act in this world – as invisible to those who inhabit it (Ahmed 2007).

My reflections about partnership are three-fold. First, as a non-Indigenous partner, it's important to see oneself at the service of the (specific) Nation, supporting Nation-building and self-determination. That means seeing ourselves as the partner who perhaps brings money and/or access to resources, and perhaps some relevant expertise or experience to the project, while expecting nothing in return. We should prepare to write ourselves out of or be written out of a project at any moment, then perhaps be invited back in at another moment. Learning when and how to let go, and not taking it personally, is a crucial sensibility to cultivate. Second, be aware that there may be conflicting paradigms at work in the partnership. For some Indigenous Nations, reconciliation may mean reparations, a taking back. For others, it means collaboration, partnership, at least for now. Third, no community is monolithic in its outlook and approach, including Indigenous communities. Be prepared to be confused, used, honored, discarded, erased. Just be open to being in loving relationship.

I have come to understand that working in a good way with Indigenous communities requires a *different way of being*: one that is more about process than results, one that is above all about building relationships, developing trust, working at the pace of the community, approaching things with a beginner's mind, and listening to and taking one's cues from the community. This work has changed me. How it has done so is in part reflected in Chapter 11's poem, in part in my self-critique in Chapter 12, and again but differently in Chapter 13, where I try to apply what I've learned in working with the Musqueam and Haida Nations to the broader challenges of how a city/region like Vancouver might move forward to a more just co-existence with the Indigenous Nations who have inhabited these lands (which we now call the Lower Mainland) since time immemorial.

The inner journey continues, in these chapters as well as in real life.

Notes

1 Thanks to conversations with Libby Porter in April 2022 for this retrospective insight.
2 Alderhill has rapidly grown into one of the leading consultancies in Canada, not only in Indigenous community planning but also more broadly, doing anti-racism and diversity training across a wide range of institutions and jurisdictions. Alderhill now has 11 full-time staff, some of whom are graduates of our ICP program.
3 Aftab is another of those 'apprentices' from whom I have been able to learn so much: first from her dissertation work, which had started out as an exploration of my own writings about the potential of a therapeutic approach to planning; and subsequently as a practitioner, now working as the Chief Equity Officer with the City of Vancouver. The ICP program was fortunate to have Aftab teach a core course on Facilitation, Negotiation, and Conflict Engagement for almost a decade.
4 This account of Jessie's teachings is based on five or six years of having her in my classroom delivering one-day intensive sessions, and on a follow-up conversation in February 2022 to confirm the accuracy of my account.

References

Ahmed, S. 2007. A phenomenology of whiteness, *Feminist Theory* 8(2), 149–168.

Butt, P. and Eagleson, R. 1996. *Mabo: What the High Court Said and What the Government Did* (2nd edn.) (Sydney: Federation Press).

Erfan, A. 2017. Confronting collective traumas: an exploration of therapeutic planning, *Planning Theory and Practice* 18(1), 34–50.

Jackson, S. 1998. Geographies of Co-existence: native title, cultural difference, and the decolonization of planning in northern Australia (Unpublished PhD dissertation (Sydney: School of Earth Sciences, Macquarie University)).

Jackson, S., Porter, L. and Johnson, L. (Eds.) 2018. *Planning in Indigenous Australia. From Imperial Foundations to Postcolonial Futures* (London: Routledge).

Jacobs, J. 1996. *Edge of Empire. Postcolonialism and the City* (London: Routledge).

Porter, L. 2004. Unlearning One's privilege: reflections on cross-cultural research with indigenous peoples in South-East Australia, *Planning Theory and Practice* 5(1), 104–109.

Porter, L. 2007. Producing forests: A colonial genealogy of Environmental Planning in Victoria, Australia, *Journal of Planning Education and Research* 26(4), 466–477.

Porter, L. 2010. *Unlearning the Colonial Cultures of Planning* (Aldershot: Ashgate).

Porter, L. and Barry, J. 2016. *Planning for Coexistence? Recognizing Indigenous Rights through Land-Use Planning in Canada and Australia* (London: Routledge).

Reynolds, H. 1989. *Dispossession: Black Australia and White Invaders* (Sydney: Allen and Unwin).

Sandercock, L., Sparrow, L., Grant, L. and Cook, J. 2013. Indigenizing planning education, decolonizing planning practice: first steps', *Plan Canada* 53(2), 48–52.

Sandercock, L., Sparrow, L. and Cook, J. 2016. Ways of knowing: engaging in partnership for indigenous planning education, *Plan Canada, Special Issue*, December, 49–52.

11

FINDING MY WAY

Emotions and Ethics in Community-Based Action Research with Indigenous Communities (2018)

Prologue

In 2006, I was invited to the town of Burns Lake in north central British Columbia by a non-Indigenous anti-racism activist who'd seen a documentary I'd just finished about the challenges of racism and xenophobia in a culturally diverse neighborhood in Vancouver (Chapter 8). I was invited as a film maker, as a potential storyteller. But what was the story or stories that needed to be told? After a year of local research, I thought the story was about Canada's history of apartheid, expressed through the local history of conflict between Indigenous Carrier people and White settlers. This was a story that posed a major question, not just for Burns Lake but also for Canada: given this toxic history, is there a way forward? We conceived the making of a film in and with this deeply divided community as an action research project that would culminate in the careful design of cross-cultural community dialogues that could potentially create a space for healing and reconciliation – what I've since come to describe as a therapeutic planning intervention (Chapter 9).

During and beyond this five-year community engagement, it gradually became clear that there was an equally important (to me) parallel story, about educating the heart, about what I've come to call 'loving attachment' (Erfan and Sandercock 2012). That is a story in which the researcher's gaze turns inwards and asks an unsettling question: in addition to what may have changed in Indigenous/non-Indigenous relations in Burns Lake as a result of the action research, how did the experience change me? Here are those intersecting stories.

1.
Let's start with the Dark Side.
My profession vocation mission,
Planning:
that child of Enlightenment
'linking knowledge to action',
but whose knowledge, for what actions?

DOI: 10.4324/9781003325451-16

In the public interest, they say,
'managing our co-existence in shared space',
but for whom?
An apartheid kind of co-existence?

The unpalatable truth:
for all its good intentions
progressive motivations
stories of heroes and heroic projects,
Robert Moses' Mephistophelean
transformations of New York and environs;
Harlan Bartholomew's Plan for Vancouver
removing the Indians from the city,
they don't belong here in 1929,
they belong to the past
beyond the city beyond modernity.

For all its good intentions,
the dirty little secret of the historic record
is self-serving, justice-defying,
status quo-protecting:
not just good intentions gone wrong,
through lack of knowledge
or greed or corruption,
although plenty of that for sure,
but also deeply implicated,
facilitators of social exclusion
economic deprivation
marginalizing minorities
colonizing Indigenes
perpetrators of structural violence
while 'just doing our jobs'
implementing legislation
within a colonial system.
Land stolen children stolen
people forced to relocate
for the 'administrative convenience'
of the state:
infamous DIA (Dept of Indian Affairs)
or for economic development:
in the national interest, of course.

Land-use policies, by-laws, legal
and bureaucratic exercises of power,
structural violence authorized
by the state:

the ordering and policing
of certain bodies
in space:
certain bodies marked as different,
as Other,
as lesser, as 'not persons'
as a problem to be solved.

Like the 'Indian Problem':
that is spatial planning, that is the
Reserve System of Indian Lands
confining Natives to a tiny portion –
less than half of one percent
of their former territories –
of what was becoming the province
of **British** Columbia.

Like the Indian Residential Schools:
that is social planning,
a hundred years of planning
to destroy community and culture
'to kill the Indian in the Native child',
to break all bonds of attachment.

Like the Indian Act of 1876:
reducing hitherto sovereign Nations
to 'wards of the state',
that is governance planning.

An official policy of assimilation,
using three technologies of power
that together spell
GENOCIDE.

And the worst of it is,
this is not just history.
This is the colonial present.

2.
And I still call myself a planner?

3.
Can planning/planners redeem ourselves
repair the wounds of the past
become an instrument

of healing and justice?
Or is this another in a long line
of professional delusions
about so-called
'progressive planning'?

Can new forms of practice lead to healing
this tragic past?
What do we need to learn?
What do we have to unlearn?
Who do we have to become,
to play this decolonizing role,
of the profession, its practices,
and
most challenging of all,
of our non-Indigenous selves?

4.
And me, specifically,
as opposed to the convenient
yet confusing academic 'we':
immigrant Australian working class
White girl, socialized
into Anglo-American academia.
Professor at an early age,
yet still the sense of the impostor.
Class does that, as does Race,
and so many other dividers
that tell us our place.

When only 4% of Australian professors
were women, and I became one of them,
ambivalent academic from the get-go
always thinking I understood what it was
to be a minority, blind
to the contradictions of my own privilege
in the upwardly mobile meritocracy
of the baby boom generation,
always taking the part of the underdog
always the social justice orientation
always the sense of fighting the good fight,
suspicious of accusations of White privilege
when I had never 'felt' my privilege.

Identity stories are just
so much more complicated,

I told myself,
and conveniently left it at that.

And/yet now in Canada
and beneficiary of research grants,
and an invitation to visit a (to me) remote
northern interior town,
in the depth of my immigrant ignorance and
still unacknowledged White privilege,
all my radical planning baggage intact,
I begin a journey of confronting
Indigenous Canada and,
reluctantly but inevitably,
also myself.

5.
Encountering 'the north',
stereotypes pave the way:
the people
suspicious of outsiders
city folks university folks government folks,
maybe all for good reasons.
The landscape, God's Own Country,
lakes rivers mountains wildlife
hunters' paradise no place for sissies
in minus 40 winters and bug-infested summers.
Cowboy mentality and
plenty of Indians to kick around.
You don't have to be in Canada long
to absorb these stereotypes.
Could I leave them behind
in the closet of the urban townhouse
or would they shadow me
hiding in my back pocket?
Could I find and occupy
a space of neutrality?
Should I?

What about my outsiderness
as urban and academic,
which is also an insiderness
in power relations and resources?
Would *anyone* talk openly with me,
Native or non-Native?

What about my ignorance?
How to inhabit a 'contact zone'

always already contaminated
by history:
teachers and priests
social workers and cops
hellbent on saving or improving,
civilizing or assimilating?

How else but with a 'beginner's mind'.
Go with humility, prepare, ask,
be prepared to look foolish,
expect to step right in the shit
and to stink some of the time
and be uncomfortable
most of the time.

6.
Driving Highway 16 the first time
before I came to know it
as locals do
as The Highway of Tears,
for the number of Native women
who've died or disappeared
hitchhiking it:
driving those three hours
from Prince George,
it wasn't the beauty
but the devastation of the landscape
that struck me. Swathe after swathe
of pine forest
turned red or grey, dead.
Pine beetle infestation
piercing the wooden heart
of the regional economy.

No big surprise then to see so many
boarded storefronts along the main street;
a bleakness about this town
no obvious sense of civic pride
no beautification
just a few half-dead flower baskets
hanging from light poles.
Highway 16 suddenly becoming Main Street
for a mile through the Village of Burns Lake,
Supermarket and pub
framing the eastern entrance
cemetery and motel

bookending the western exit,
and it felt like most travelers just
went right on through,
nothing visibly attractive or enticing
unless Pioneer Museums are your thing.

My first thought, coffee?
I'm not going to find a decent cup here.
I was right about that (urban brat),
but not about most of the other thoughts
that came with me from Vancouver.
Like doubting
whether healing could be possible
in such an uninspiring setting.
My metropolitan baggage
took time to unpack.

The Village of Burns Lake:
what you see at first seems
just barely enough to make a town,
some cheap cafes dilapidated hotel
a bank or two
drugstore tacky motels post office
with the Queen's photo,
newspaper office RCMP fortress
community college churches library;
bowling alley closed down
movie house closed down,
where do the kids go?
Skating rink and skateboard bowl,
spiffy new high school elbowing
the dilapidated old one, transformed
into Burns Lake Band offices
flying the Mohawk flag (!)
right in your face
on the Highway.

What you don't see at first
are the social and spatial relations
off the Highway.
Two Reserves within the town,
their stories a microcosm
of Canada's race relations:
Ts'il Kaz Koh's Rez on the swamp land
between railroad and lake,
allocated before the Village itself

was laid out,
gobbling up a third of the Rez
in one bite,
after the Transcontinental was built,
same time as the Great War,
and more of their land appropriated
after the next world war
for the high school the hospital
some shady housing deals,
and then a huge chunk to accommodate
another forced relocation,
Lake Babine Nation moved
from their territory into Burns Lake
for the administrative convenience
of the almighty DIA.

Passing through, you don't see any of this
and you don't see that, until recently,
Natives walked on one side
of the street and White folks on the other,
or that there were race riots on this same
Main Street,
hundreds of riled up whites
laying into Natives with baseball bats,
or that Natives
weren't allowed in certain stores and
certainly weren't hired.

Apartheid,
more poetically 'the two solitudes',
invisible to the untrained eye.
Invisible to me
until my local guides
opened my White metropolitan eyes.

7.
The project begins
with an invitation
to collaboration,
proceeds with co-production
of ethical protocol
mutual give-and-take, building
to mutual learning
negotiating suspicions
mediating mistakes
scoping out the extent

of the welcome the extent
of the project itself.

Film as planning intervention
privileging the past
over the future
creating a space where healing
might begin
if denial could end.
That is the vision.
Telling a story that might shift
the balance between the polarities
of a deeply divided community.

Looking for possibility.

8.
I met him in the Band office
this legendary Chief,
long braids, flint eyes,
said to 'eat White folks for breakfast'
warrior reputation
known for confrontation
no hostages taken.

He'd been in a running battle
with Village, Mayor, Settlers,
for twenty years:
a battle for stolen land,
recognition, rights, dignity.

The moment of introduction
a decisive moment
words incidental
he didn't look into my eyes
but through them
and saw whatever is within.
An undressing of my being.
My heart rate surged
the desire to be somewhere else
was overwhelming.

We went outside for a smoke
some incidental talk
then an invitation.
Come with me.

He took me down to the railway tracks
dividing the Rez from the town
swamp land from high ground.
There's our story.
What do you see?

And so began a fieldtrip
into our mutual history
he the first among many
willing to share story
as teaching, guiding me to see
to feel my hitherto unexplored
White privilege.

His willingness to share
his invitation to family sweat lodge
ceremonies, his jokey emails,
kept me returning because
and in spite of the bitter taste
of the learning I was experiencing.
Absorbing the guilt of my culture.

9.
What did I see at the railway tracks?
On one side, the Town side,
a pretty municipal sign
carefully painted, proudly announcing
'Welcome to Burns Lake:
Gateway to Tweedsmuir Park'.

Across the tracks, on the Rez side,
The Wall.
Words of fury splashed across it
words screaming at a passing train
screaming at town residents
if they cared to look, (but didn't)
INDIANS HAVE NO RIGHTS IN THIS TOWN!
END APARTHEID NOW
INJUSTICE ANYWHERE IS INJUSTICE
EVERYWHERE!
YOU CAN CONTROL OUR WATER
BUT
YOU CAN'T CONTROL OUR SOULS.

The Wall,
separating the homes of those whose

water and sewerage had been turned off
mid-winter year 2000
by the Village,
from the homes of those
who had turned it off.

A flashpoint burned into the history
of this town that is not a community
but two solitudes:
Native and non-Native.
A flashpoint that is both
ancient history and
21st century.

10.
Thus began my education,
other teachers following.
The women Elders who cried
as they shared what they perceived
as their failures as parents.
And I cried with them
as they told of their humiliations
and deprivations
and separations from family and
community
in Residential School;
the attempts to escape, the attempts
to protect siblings, the punishments
cruel and brutalizing,
intended to destroy
all bonds of attachment
intended to destroy
community and culture.

The Elder
whose mother refused to teach her
Carrier language because
she would only be punished
for speaking it.
And the terrible shame
of not speaking it.

The Ojibway mental health counselor,
migrated east to west
body pierced green mohawk hair,
shared a shocking story

of his crime and punishment,
confession and self-criticism,
but above all of healing himself
through connecting to culture,
and now there for the youth, 24/7.

The senior policy advisor
who'd come with his rancher dad
from California via Wyoming
and was changed forever on reading
the Indian Act
of 1876,
and insisted I read it,
who introduced me to the tragic story
of the Cheslatta Carrier people,
evicted from their homelands for
a hydro-aluminum project, they
in the way
of our 'development'.
A half-day in the telling,
my listener's tears flowing,
no way to walk away, forget,
do nothing, feel nothing.

The Youth who told me of their shame
being born Native,
and the non-Native music teacher
working with those youth to create
their anti-racism song,
and set them on a different path.

The non-Native high school dance teacher
married to the Chief
defying the unspoken rules
broaching the difficult topics
through performance:
Residential Schools!
Homosexuality!
Hair!
'Baby step by baby step
so much work to be done
I'm just a small part of it
but I have to do my part'.

And the two (ex)Mennonites:
one with a business degree

running the Community Forest,
hanging with the anti-racism crowd,
challenged by the Mayor to explain
his choice of friends,
his answer simply
'I agree with their values'.

The other, the Anti-racism Lady,
who'd invited me here to witness
the stories, introduced me to Right
Left and Centre,
Native and non-Native,
believer and non-believer,
never told me what to think
and above all taught me
the true meaning
of partnership
between Indigenous and non-Indigenous Canada.

The dance of getting to know
taking the time many stumbles
missed steps missed cues
not listening not watching
being made fun of kindly
observing other ways of teaching
discarding preconceptions
discovering three dimensions.

11.
A process unfolds by mutual consent.
I am outsider interpreter ally
and critical friend
turning their stories into film
potentially healing
potentially explosive
addressing apartheid
not sugar-coated.

Film as inquiry
exploration meaning making
film with a purpose, a process
moving toward possibility –
toward dialogue
the process itself advancing
the purpose because and despite
the conflict.

A year of research
four months filming
three years editing, returning
screenings for feedback
corrections additions suggestions.

Four months designing
the space of community dialogue
advised by the Main Characters
heroes villains everywo/man
Mayor Chiefs Elders
Youth Teachers Police.

Training local youth and Elders
as facilitators,
organizing music and food and
counselors, just in case,
so much psychological wreckage
in colonization's wake.

Structuring the dialogues:
What moved you most?
Is this your story?
Is the past still present?
What should be done?
What will you do?

Own your emotions but
own your history too
and be prepared to see it
through Others' eyes
and emotions.
Leave all defenses at the door
and enter the screening
your lives on display
your hearts and minds open.

12.
The dialogues begin:
talking circles spirit feather
soft voices angry voices
silences tears shock shame
surprise relief catharsis
hope:
cautious and feisty visions
of moving forward together.

13.
Was there some healing
across the cultural divide
through this planning intervention?

What does transformation look like?

There is no metric,
just the stories
the voices of hope.
Post-dialogue questionnaires,
for what they're worth,
showed changes in attitude:
a desire to get to know 'the Other'
shock at living in ignorance so long
about the lives of 'the Other'
shame at believing the dominant narrative
about 'the Other'.
The quality of listening and engagement,
remarkable.
The energy for change,
palpable.
Some of the Native youth started
a Facebook page
for change in their community.

There was recognition,
there was empowerment,
there is the beginning of a new story
about the need to move forward
together
in full recognition of past injustice,
the need to build a new web
of social relationships,
and an understanding that the hope
is with the youth.

14.
What did I learn about planning
in deeply divided communities?

That to address collective
trauma amnesia denial,
we need collective forums,
public spaces for working through
historic and ongoing injustices

for bearing witness
for recognition
for taking responsibility
for bringing the past into conversation
with the present and future
in spaces carefully designed
to protect the vulnerable
and unsettle the powerful.

Colonial planning knew what was best
for its 'subjects' and
reduced them to dependency.
Progressive planning sought to abolish
cultural difference in the name of equality.
Postmodern planning sought inclusion
thru recognition of difference
but didn't recognize
claims to rights, title, sovereignty.

How then to imagine
a decolonizing planning project?

As *a way of being*,
starting from a depth
of not knowing,
humble open-hearted caring
listening more than talking
not helping or solving
but supporting,
always and only when invited.

Reversing the gaze,
interrogating our own institutions
observing our own daily actions
for the guaranteed
evidence of paternalism
disguised as expertise.

Moving beyond guilt
beyond craving for legitimacy
to a politics and praxis
of love and justice:
knowing love is not enough
that justice means making noise
when I see injustice,
a research/practice

respectful and fearless
caring and courageous
standing alongside
walking together, working
face-to-face.

That's what being an ally means.
But where does this politics begin?

Surely at home.

15.
What did I learn
about myself?
Myself in relation
myself and my profession
myself and my privilege?

How/was I transformed?

Drawn to this way of working
from a career-long feeling of
Lack:
in those critical arguments
and assured pronouncements
about who and what matters,
those devastating judgments
that crush or boost,
intellectual power games
gate-keeping the establishment.
I've done all of that and
for some of it,
am ashamed
for being good at
playing that game.
The Lack that has no name
but resides in the body and psyche
as wound,
whose only healing
is *activating the heart.*

Working in partnership
in community
shows me/us a different way
of being in the world.
You/I cannot be in community

without loving attachment
humble open-hearted vulnerable
patient critical hopeful
listening thru silence
walking lightly speaking softly:
without these and more
I will always be suspect,
undeserving of trust.

The Lack now lives in me
as a vow
to honor this way of being
to make the connections
to all my relations,
everywhere.
I don't always keep this vow
and at such times
I disappoint myself and know
I am ethically suspect.

How to practice loving attachment?
The chemistry of attachment
is relationship.
The ethics of attachment
is reciprocity.
The work of attachment
starts from learning
the meaning and practice of respect.

And I keep stumbling upon
new ways of being disrespectful,
in spite of good intentions.
Those good intentions are
the biggest trap, leading to
self-righteous indignation
instead of humble self-questioning.

My great good fortune
are the teachers
who don't give up on me,
with varying degrees
of gentle patience
and tough love
they are showing me the way.

Inching my way out of ignorance.
Heart activated.
Full of gratitude.
A work in progress.

A Note on Sources

This poetic essay/prose poem has been informed by many scholars whose works are cited below, and whose words and thought fragments have found their way into my essay, as well as by the people of Burns Lake, the Ts'il Kaz Koh Nation, and the Cheslatta Carrier Nation.

Bibliography

Attili, G. and Sandercock, L. 2010. *Finding Our Way. 90 min Documentary* (Vancouver: Moving Images).

Barry, J. and Porter, L. 2012. Indigenous Recognition in state-based planning systems: understanding Textual Mediation in the contact zone, *Planning Theory* 11(2), 170.

De Leeuw, S. 2004. *Unmarked. Landscapes along Highway 16* (Edmonton: NeWest Press).

Deloria, V. 1999. A flock of anthros. In: *Spirit and Reason. The Vine Deloria Jr. Reader* (Golden: Fulcrum Publishing).

Erfan, A. 2013. An Experiment in Therapeutic Planning: learning with the Gwa'sala-'Nakwaxda'xw First Nations. (Unpublished PhD dissertation (University of British Columbia, School of Community & Regional Planning)).

Erfan, A. and Sandercock, L. 2012. Plato's lacunae: on the value of loving attachment in community-based planning research and practice, *Planning Theory and Practice* 13, 4.

Friedmann, J. 1987. *Planning in the Public Domain: From Knowledge to Action* (Princeton: Princeton University Press).

Healey, P. 2007. *Collaborative Planning* (2nd edn.) (London: Macmillan).

hooks, b. 2000. *All About Love: New Visions* (New York: Harper Collins).

Jojola, T. 2008. Indigenous planning: an emerging paradigm, *Canadian Journal of Urban Research* 17, 1, Suppl.

Kovach, M. 2009. *Indigenous Methodologies: Characteristics, Conversations, Contexts* (Toronto: University of Toronto Press,).

Marris, P. 1974. *Loss and Change* (London: Routledge & Kegan Paul).

Pelias, R. 2004. *A Methodology of the Heart* (New York: Altamira Press).

Porter, L. 2010. *Unlearning the Colonial Cultures of Planning* (London: Ashgate).

Pratt, M. L. 1992. *Imperial Eyes: Travel Writing and Transculturation* (London: Routledge).

Regan, P. 2011. *Unsettling the Settler Within* (Vancouver: UBC Press).

Smith, L.T. 1999. *Decolonizing Methodologies* (London: Zed Books).

Stanger-Ross, J. 2008. Municipal colonialism in Vancouver: city Planning and the conflict over Indian reserves, 1928–1950s, *Canadian Historical Review* 89(4), 541.

Wilson, S. 2008. *Research Is Ceremony* (Halifax: Fernwood Publishing).

12

PARTNERSHIP PRAXIS IN A 'RECONCILIATION' CONTEXT

What Is Mine to Do? (2022)

> Whiteness could be described as an ongoing and unfinished history, which orientates bodies in specific directions, affecting how they 'take up' space.
>
> *(Ahmed 2007, 150)*

Introduction

As I noted in the Commentary to Part IV, I've been involved for the past dozen years in a decolonizing and indigenizing educational experiment at the University of British Columbia, in partnership with the Musqueam Nation, with whom we co-designed and now co-deliver a Master's program in Indigenous Community Planning (ICP). Our initial goal was that at least 50% of those entering the program each year would be Indigenous. Some years, we've had more than that, many years less, which means that often at least half of the students are non-Indigenous. The most frequently asked question by these non-Indigenous students – who are increasingly sensitive about the space they are taking up, as per Sara Ahmed's opening quote – is an existential one: is there really a role for me in working with Indigenous communities, or am I just perpetuating the colonization of those communities by my very presence? The unambiguous answer given by Musqueam knowledge holders, each year, is 'Yes, we welcome you, as long as you understand the following …' (to which I shall return).

In spite of those initial supportive words, these non-Indigenous students typically struggle for the full two years, sometimes more, and not all succeed in finding a place of relative comfort in this work. I emphasize *relative* comfort because I don't think we (White settlers) *should* ever be completely comfortable (Robinson 2020). Rather, we should always be questioning our roles and ethical responsibilities, as the outsiders that we are, and always will be. As a planner (as well as an educator) who has been working in partnership with Indigenous communities for the past 15 years, I want to speak from my White settler positionality and experience and use this chapter as an opportunity to unpack and reflect on four things: first, the range of motivations for wanting to do this work; second, navigating conflicting paradigms of reconciliation; third, some thoughts about 'being unsettled'; and finally, some personal and political fundamentals of successful partnership. In doing so, I am drawing on some of my own

DOI: 10.4324/9781003325451-17

missteps in the evolving partnership with Musqueam; the challenges I faced in my five-year partnership with the Haida Nation on the film project *Edge of the Knife* (Chapter 10); learning from my Indigenous as well as non-Indigenous colleagues; and learning, through our students' eyes, from the partnerships our ICP program has engaged in for ten years with twenty First Nations in British Columbia (BC).

Motivations: The Fluid Context of Indigenous–State Relations

In the face of four centuries of colonization, Indigenous Peoples have adopted many forms of resistance to their ongoing dispossession (Simpson 2014, 2017). There are a number of important landmarks in a growing Indigenous political and cultural resurgence in the past 50 years, among them, opposition to the infamous White paper of 1969 which would have abolished the Indian Act, and along with it, the very status of 'Indian'; opposition to the limited terms of the proposed repatriated Constitution in 1982, which led to the incorporation of Indigenous rights and title; Supreme Court challenges such as Calder (1973), Sparrow (1990), Delgamukw (1997), Haida (2004), and Ts'ilqotin (2014); and the numerous forms of direct action, such as blockades, that led to the Royal Commission on Aboriginal Peoples in 1991. Each form of resistance has produced concessions from the Canadian state, along with evolving normative understandings on the part of the state and settler society of what justice might look like for Indigenous Peoples and how best to re-frame Indigenous/state relations. Along the way, there have been divisions within Indigenous leadership about both means and ends, as well as divisions between the leadership of Indigenous organizations and grassroots Indigenous activism (Coulthard 2014).

In the early 2000s, when I first became involved with two Carrier Nations in north central BC, a significant part of my self-education was the five-volume report of the Royal Commission on Aboriginal Peoples (RCAP), delivered in 1996. My own approach to working with Indigenous communities and Nations was shaped by the RCAP recommendation of 'moving toward a relationship of power sharing, mutual respect and joint problem solving' (*Report of the Royal Commission on Aboriginal Peoples*, vol 1, part 1, 1996). This recommendation fitted neatly into my own collaborative planning practice, and formed the basis of the work that Giovanni Attili and I did with the Cheslatta Carrier Nation and the Burns Lake Band from 2006 to 2010 (Chapter 9). And it was that approach that carried over into the Haida film project, conceived as a three-way partnership (Chapter 10).

But the political and legal landscape has been changing rapidly in the past decade in response to increasing Indigenous litigation and activism. Prime Minister Stephen Harper's formal apology in 2008 to the survivors of the Indian Residential Schools ushered in a new official discourse couched in the language of 'reconciliation', and established a Truth and Reconciliation Commission whose Final Report in 2015 contained 94 Calls to Action. The City of Vancouver declared itself to be a City of Reconciliation and announced a Year of Reconciliation in 2013–2014. In 2020, the province of British Columbia passed legislation to implement the United Nations Declaration on the Rights of Indigenous Peoples (UNDRIP). Much of this has been regarded as a 'symbolic politics of recognition' rather than anything approaching justice for Indigenous Peoples (Coulthard 2014; Simpson 2011, 2017). In Michi Saagiig Nishnaabeg activist/scholar Leanne Betasamosake Simpson's view, the state's approach to reconciliation serves to neutralize the legitimacy of Indigenous justice claims by offering statements of regret and apology for harms narrowly conceived as having occurred in the past, thus offloading Canada's responsibility to address structural injustices that continue to inform our settler-colonial present. In doing so, the state can claim that 'the historical wrong has been righted and further transformation is not needed' (Simpson 2011, 23).

When Leanne Simpson and Yellowknives Dene scholar Glen Coulthard were formulating this opposition to a politics of recognition, I was formulating the three-way planning partnership for a film project with the Haida and the Inuit. And simultaneously there was an emerging Indigenous grassroots movements, *Idle No More*, supported by some non-Native allies, demanding land justice and an end to state violence against Indigenous Peoples. Some Indigenous leaders had shifted to a discourse of Nation-building, self-determination, and sovereignty (Coulthard 2014, chapter 6; Simpson 2014), while emerging Indigenous planners were formulating a 'trauma-informed' approach to working with Indigenous communities (Hemphill 2022).[1] The moderate language of RCAP has now been superseded by demands that social, climate, and land justice for Indigenous Peoples be addressed. The Haida film project came into being amidst this always evolving Indigenous political and cultural resurgence, and it is within this turbulent force-field of resurgence that Indigenous Community Planning must now learn to navigate (Robinson and Martin 2016; Robinson 2020).

Navigating Conflicting Paradigms: Reconciliation or Reparation

When I wrote the research grant application to the Social Sciences and Humanities Research Council's (SSHRC) Partnership Development Fund in 2013, to obtain seed money to get the proposed film project off the ground, my assumption was that this would be a collaborative process among the three partners – the Council of the Haida Nation; Isuma, the Inuit film collective; and myself – in the service of Haida aspirations for cultural and language revitalization and community economic development. I saw the project as an 'experiment' of sorts, an action inquiry into whether film could serve as a catalyst in support of these aspirations. My understanding of a 'collaborative approach' had been forged experientially in my previous two documentary projects with Giovanni Attili (Chapters 8, 9), and I assumed I would be involved in the whole process, from conception to completion, as part of a creative team. I had also written into the grant that Giovanni and I would be documenting the whole process, as researchers, in the form of a documentary ('The Making of *Edge of the Knife*'), as part of my accountability to the initial $200k SSHRC grant.

I was committed to two concepts that I had elaborated in both the grant application and in the 'pitch' to the Council of the Haida Nation (CHN) in 2013: one was the importance of Nation-to-Nation knowledge transfer, which was why the Isuma Collective was involved in the partnership; the other was that the Haida would own this project in the fullest sense, not just legally and financially, but that they would take on a leadership role. I didn't ever feel that it was 'my' project, but I did feel that I would remain central to it. In those first two years of grant writing, community engagement, and script development, I longed for more Haida engagement and 'ownership' of the project, as did Jon Frantz (representing the Inuit partner) when he started to work on the budget submission to the Canada Media Fund (CMF). We were frustrated by the Council of the Haida Nation's inaccessibility at crucial times. It felt like all the work was being left to us while the CHN adopted a 'wait and see' stance. That was the situation until we received the $1.89m in production funding from CMF. Then things began to change.

When the co-directors (whom Jon had hired) started work in early 2016, Jon and I had a planning meeting with them to set out roles and responsibilities moving forward. Jon had not intended to take on the lead producer role, hoping that there would be Haida who wanted to learn by doing, but there was no apparent interest in this role, which was puzzling. Jon's experience working as a producer and cinematographer on Isuma's two previous feature films helped set out the necessary steps for the co-directors to think about. Initially they were preoccupied with the tasks of casting and training community members as actors, and sorting out between themselves how they would divide their energies, while Jon was going to be taking care of the budget but staying in the background and letting them make all the artistic decisions. I was to work with Giovanni on the evolving documentary project for which he had already done some filming of the 2014 story-gathering and scriptwriting workshops that I had convened with community partners.

But not long after I returned to Vancouver to start planning the documentary process, word reached me, via an intermediary, that Giovanni was not considered appropriate for the project, living in Italy and with no connection to the community. I was initially taken aback and disappointed by this, but recognized it was a choice the Haida were making, and I had to respect it. I started to think about who else I might work with, including Haida folks with documentary experience. But before long I was informed, again second-hand, that an Indigenous film maker from eastern Canada, a friend of one of the co-directors, had been invited to make the documentary and would be working with the Canadian Broadcasting Commission (CBC) on this. That too made sense to me, in terms of indigenizing the whole process. But now I experienced an internal conflict. By April 2016, around the time of the language boot camp, I suddenly felt cast as an outsider, rather than as a partner. Instead of celebrating the fact that the Haida had now truly taken ownership (and were making decisions I hadn't anticipated), my still-colonial self was dwelling in the place of being shut out, feeling hurt, and wondering what had happened to the three-way partnership.

A journalist from the *New York Times* had been invited to the language boot camp at this time, and given full access for a week, along with the film crew from CBC. A subsequent front-page article in the *New York Times* talked about this revolutionary film project, the first ever feature film in the Haida language, and interviewed several of the cast members as well as Vern Williams, a Haida artist, musician and composer, who was present at the boot camp to provide singing and drumming and to compose songs for the film. Vern revealed to Catherine Porter, the *New York Times* journalist, that he had been forced to spend seven miserable years in one of Canada's notorious Residential Schools. Asked by Porter whether he regarded the film project as part of Canada's 'reconciliation', in the wake of the report in 2015 from the Truth and Reconciliation Commission about those Residential Schools, Williams replied: 'I don't call this reconciliation. Something was taken. We are taking it back' (Porter 2017, 1).

Those words of Vern's have helped me better understand the unfolding situation from a Haida perspective, and thus better understand what I had been experiencing. I had been feeling the bitter taste of exclusion, whereas what was happening from a Haida perspective, or at least *from the perspective of the specific Haida who'd now taken charge of the project*, was a 'taking back'. This was reparation, not reconciliation.

In retrospect, my version of working in a collaborative way did not correspond with the Haida folks who were now in charge of the production. They wanted this to be a Haida project, not a collaboration. What I had needed to learn all along, had I been more 'in tune with the social field', in Patricia Wilson's insightful language (Wilson 2019), was simply that I needed to let go, to understand what had been mine to do, getting the project up and running, and what was no longer mine to do. Now I just needed to get out of the way. The project had never been 'mine', and was no longer 'ours'.

On Being Unsettled: Exploring Asymmetries of Power

Recent scholarship influenced by feminist and postcolonial perspectives makes much of the fact that in university–community partnerships there is always a power imbalance that favors the (usually) White, middle-class, city-based, privileged academic. That was not my experience working in north central BC (Chapter 9), where there was deep suspicion of city folks and academic folks, and a lot of work had to be done to build trusting relationships before the planning work could begin. Nor has it been my experience in the planning education partnership with Musqueam. And it certainly was not my experience working on the Haida project. On the contrary, and even though I had obtained and was 'in control' of the initial funding for the project, I always felt like I was walking on egg shells. This felt sense of powerlessness is surely a paradox that needs further unpacking.

To work in partnership means, for me, to actually do things together, to co-design, co-create. Yet almost every effort to involve the Haida Nation, initially, felt like an imposition (with a couple of important exceptions). I tip-toed around Band Offices, spent weeks at a time feeling apprehensive and *not* making phone calls or knocking on doors, wondering what I was doing there. I was *never* sure of my 'place'. In fact, in that first summer, I had been told quite directly by a former president of the Council of the Haida Nation that the Haida were suspicious of, if not downright hostile toward, academics/researchers from the city. 'But', I protested, 'I'm a community planner who makes films. We're hoping to making a film in the Haida language ...' Clearly seeking his approval, clearly thinking of myself as not one of those people he disliked, I was (in spite of my overt humility) making an implicitly even more offensive claim: I'm here to help. Beneath my temporary discomfort, I had a White person's certainty about my intentions, my mission. But that clarity of purpose was regularly undermined by responses of indifference, distance, doubt, and suspicion, co-existing with curiosity, cooperation, and just plain opportunism on the part of the Haida.

If I intellectualized this, as I tried to do, I could accept it as part of a necessary cultural humility and part of the larger reconciliation framework that informed our approach. But if I am honest about the emotional dimension of the experience, as a White settler planner, it was very unsettling. I thought I had understood and accepted the importance and necessity of being unsettled, having spent some time digesting Paulette Regan's excellent book on this subject, *Unsettling the Settler Within* (Regan 2010), but 'knowing' something intellectually is different than actually experiencing it. And this was a more or less continuous experience of disorientation, of being unmoored, totally unsure of myself. This was a feeling of powerlessness.

There are many good reasons why White settler planners should expect to experience this unsettling disorientation. There's the history of 400 years of colonization that suggests that a welcome mat might not be the first thing that happens when one begins to explore or enter a partnership. There's the above-described conflicting paradigms of reconciliation versus reparation. There's our 'white fragility' (DiAngelo 2018; Cole 2020; Robinson 2020), which may cause us to be over-sensitive to apparent or real slights or distancing. There are also cultural and professional roots of this unsettling. As Western-trained planners and academics, we are task oriented, productivity oriented, habituated to meeting funding and other deadlines. Indigenous Nations, though, have their own governance systems and timeframes, rhythms, and cycles that are often closely interconnected with land and ceremony, as well as different and pressing priorities. Taking the time to get to know people should be the most important priority for non-Indigenous planners. It is foundational. It's also important to remember that no community is monolithic, even when a worldview is shared. In my work with the Haida Nation, I experienced being embraced and included by some individuals and families and shut out by others, which was part of the confusion, and it's not always clear when this is personal and when it's political. Also, it's not helpful to be or appear to be in need of constant validation.

Another permutation of power asymmetries is resource-based. While we, non-Indigenous planners, may bring resources and some valued skills to the partnership, and idealistically hope/expect to work collaboratively, which implies shared responsibilities and workloads, our Indigenous partner is likely to be under-staffed and under-resourced, and thus to have different expectations about 'sharing' a workload. All of which is to say that power relationships are not a given; they are fluid and unpredictable and demand a cultivated ability to 'read the social field' (Wilson 2019). These relationships are always both context-dependent and character-dependent. Every community is different, as I have had to learn, and arriving at a good relationship with one community doesn't guarantee that the next one will start well or go the same way. Learning as much as one can about a community and/or Nation in advance is obviously important.

But perhaps there was something else going on here, something even deeper. From a distance in time, I can now see a double-layered and paradoxical learning. I was, after all, the (White) scholar who for decades had been 'making space' for the voices of those who had been historically marginalized, dispossessed, discriminated against. 'Making the invisible visible', the title of one of my books, had been my mission, exposing injustice in the systems in which I worked: academia and planning. My own early experiences of classism, of being made to feel that I didn't belong in a certain place – the institution of the university (Introduction) – had pre-disposed me to have empathy for those students and communities 'from the borderlands' (Sandercock 1995, 1998). But I know I never will understand what it feels like to be an Indigenous and/or racialized body (Cole 2020) in all those White spaces, from institutions of 'higher learning' to public space to shops, transit. There I was on Haida Gwaii, with a chance to reflect on that, now that the tables were turned and I was the one who was out of place, not 'at home', not comfortable (Ahmed 2007; Maynard 2017).

But that paradox of tables turned is not what I reflected on at the time. I was too caught up in the lived experience of powerlessness, too unsettled. The deeper learning, only available to me now, is that a sense of powerlessness was only possible because I was used to feeling powerful, of being able to make things happen. I hadn't read Sara Ahmed's work at the time, but my experience was surely akin to her notion of White bodies always already being expected in the spaces of the world: White bodies take for granted that they belong, and that they can act (Ahmed 2004, 2007). Whiteness is only invisible for those who inhabit it. 'To be at home in the world is to feel a certain comfort: we might only notice comfort as an affect when we lose it, when we become uncomfortable … To be comfortable is to be so at ease with one's environment that it is hard to distinguish where one's body ends and the world begins. One fits …' (Ahmed 2007, 158). Whiteness allows bodies to move with comfort through space, and to inhabit the world as if it were home (ibid, 159). The existential reality of racism is grounded in this lived expectation, for White bodies, of the right to 'sink into space' (Ahmed 2007) and thus the profound disorientation when that feeling (of being powerful) disappears. And it's perhaps especially challenging for White feminists like myself who have been schooled/socialized in the language of empowerment. With this insight, though, the question becomes, what is an everyday politics of praxis for working within this differently understood power imbalance?

Personal and Political Fundamentals of Successful Partnership

What can I share with White settler planners seeking to work in partnership with Indigenous Nations in order to support the aspirations of those communities? Is there a future for such partnerships in the current contested context of recognition, reconciliation, reparation?

Given the nature of Indigenous resurgence outlined above, and the necessarily fluid nature of Indigenous–state relations, my own motivation has shifted from prioritizing and valuing partnership *as such* to asking in what ways can non-Indigenous community planners support the aspirations of those Indigenous Nations who choose to work with us. Once a Nation indicates such a desire, it's vitally important to establish and respect the aspirations of that particular community, wherever they are on the spectrum of recognition/reconciliation/reparations. (And whatever our own activist views are in relation to this question, it is *not* our professional role to impose them, under any circumstances.) Is the community using the language of community development and cultural/language revitalization (as the Skidegate Haida were when I first started to work with them), or are they using the language of Nation-building and self-determination? It would be helpful to be able to answer this question at the start of a project, to know where one stands, so to speak. But in the flow of daily life, things are much messier. As I discovered with the Haida Nation, communities are not monolithic, and depending on who holds what kind of power at any one time, the answers may vary dramatically.

In the short and medium term, many Nations in BC, for example, seem to believe collaborations and partnerships are a necessary/inevitable part of the *journey towards* Indigenous self-determination and sovereignty. This is true for those Nations with whom we've been working in the ICP program. They seek us out, having heard by word of mouth, usually from neighboring Nations, that we have something to offer. But the answer is different from Nation to Nation, depending not just on the internal politics, but also on any Nation's self-perceived capacity. Some Nations with whom I've worked do see such collaborations as necessary/desirable, others as a straightjacket that confines them to Western ways of knowing and doing. The reality is, there's a spectrum, and it's entirely the choice of any Nation whether to adopt a collaborative approach or not, depending in part on their trust in specific 'outsiders', their sense of their own capacity, and more.

It may be worth reflecting that collaboration is not inherent in traditional Indigenous ways of doing/governing. Elders were regarded as having the knowledge and experience that gave them the ability to be good leaders and decision makers. Historically, lives, comfort, and generational success depended on it. For very practical reasons, an Elder would not concede to a young person in the name of collaboration. Their views were not regarded as equally valid. This is still evident today in the great respect I've seen shown by youth to their Elders. According to Pueblo scholar Theodore Jojola (1998), collaboration was a practice in inter-tribal relations with regard to matters of resource management and dispute resolution. But this is more of an instrumental approach to collaboration, not an elevation of collaboration as a good in itself. I suppose that my collaborative approach is driven by my Western way of thinking, embracing a paradigm with increasing currency in Western planning, whereas from an Indigenous perspective, collaboration or partnership is likely a more instrumental stance, appropriate when we non-Indigenous planners bring some kind of value to the situation in the form of skills and/or resources. And when that value has expired, so too does the collaborative relationship. It's not personal. It's practical. And that's the situation I found myself in at the completion of the script development process. I had no further skills to bring to the production stages, and my ongoing presence as a researcher, documenting the unfolding filmmaking process, was not welcome, for very understandable reasons.[2]

What's important for those hoping to work in this way is to establish an understanding up-front about one's role as a team member, and even then, it might be wise to expect to be dropped, then perhaps added back in at various stages, as the Nation sees fit. For an outsider to adopt this approach involves an important shift, and one I hadn't made in the first few years of the Haida film project: a shift from the collaborative paradigm to the 'support of Nation-building and self-determination' paradigm. Then, one would think of this work as being an invited participant in Nation-building work, working for the interests of the Nation, stopping when asked to stop, starting when asked to start, and giving one's all during these working sessions. Learning to (know when to) let go. And not taking it personally. Perhaps paradoxically, this can also mean that non-Indigenous planners occasionally need to take more of a leadership role (when the situation requires it) than perhaps we might feel comfortable with. This is a tricky path to tread. There has to be the trust and experience to be given the okay to do that, and the lack of ego to give it up when the time is right, or when asked to exit.[3]

If this is a necessary part of moving forward with these kinds of partnerships in planning, then I/we non-Indigenous folks need to learn how to prepare for writing ourselves out or being written out of these stories. How do we cultivate that humility in our ego-driven culture? Perhaps we begin by redefining our role, seeing ourselves as the partner who brings access to money and possibly relevant knowledge/experience to the project, while expecting nothing in return. If the balance of benefit and praise lies heavily in favor of the Indigenous partners we work with, then we've taken another baby step towards redressing past wrongs. That's what it means to support Nation-building and self-determination. Historically, though, this has been a problematic area for academia, where there is an embedded need to own data, author papers, and be the Principal Investigator (PI) on projects. That follows the old colonial ideology, against which some non-Indigenous scholars are finally pushing back.

When I reflected on what the Haida artist Vern Williams told the *New York Times* writer, that this has nothing to do with reconciliation and everything to do with 'taking back' what was taken away, I began to understand that some of the Haida people involved in this project as it went into pre-production were working within a 'reparations' paradigm, whereas I was still in a 'reconciliation as collaboration' paradigm. What was problematic was that we were never able to articulate that difference between us, in real time. So perhaps another thing to note for those wanting to do this kind of work is that there was no monolithic (Haida) 'community' acting here. Rather, there were some folks operating in a reparations paradigm who wanted to minimize the role of the outsiders, and other folks in the community who were and still are prepared to see the whole process as a collaboration and be in a loving relationship. Different community members simply had different expectations of the collaboration: different paradigms, different theories of justice and social transformation. This insight is vitally important to me in my processing of the experience, but may also be important to name here, as a way of countering the colonial/anthropological tradition of viewing 'community' as monolithic.

Clearly, *learning to let go* is a – perhaps *the* – major lesson of this story. Although we (White settler planners) may often feel powerless, this doesn't make us any less powerful. And that 'right to feel powerful' too easily turns into self-righteous indignation. Too often I fell into the 'good intentions' trap that I wrote about in the Chapter 11 poem. Good intentions can lead to self-righteous indignation instead of humble self-questioning about what is, and is not, mine/ours to do. That deeper self-questioning is the political praxis of partnership.

The inner journey continues – in the next two chapters as well as in real life.

Notes

1 Personal correspondence with my UBC/ICP colleague and Indigenous planner, Jessie Hemphill.
2 Thanks to Jon Frantz for our conversation about the relevance of 'collaboration' (31 May 2022)
3 Thanks to Scott Graham and to Jon Frantz for sharing this insight from their own work with Indigenous Nations.

References

Ahmed, S. 2004. Declarations of whiteness: the non-performativity of anti-racism, *Borderlands* 3, 2: Available at: http://www.borderlandsejournal.adelaide.edu.au/vol3no2_2004/ahmed_declarations.

Ahmed, S. 2007. A phenomenology of whiteness, *Feminist Theory* 8(2), 149–168.

Cole, D. 2020. *The Skin We're in: A Year of Black Resistance and Power* (Toronto: Doubleday Publishing).

Coulthard, G. S. 2014. *Red Skin White Masks. Rejecting the Colonial Politics of Recognition* (Minneapolis: University of Minnesota Press).

DiAngelo, R. 2018. *White Fragility: Why It's so Hard for White People to Talk about Racism* (Boston: Beacon Press).

Jojola, T. 1998. Indigenous planning: clans, intertribal confederations and the history of the all-Indian pueblo council. In: Sandercock, L. (Ed.) *Making the Invisible Visible: a Multicultural History of Planning* (Berkeley: University of California Press).

Maynard, R. 2017. *Policing Black Lives: State Violence in Canada from Slavery to the Present* (Winnipeg: Fernwood Publishing).

Porter, C. 2017. Reviving a lost language through film, 06/11/2017, *New York Times*, 1.

Regan, P. 2010. *Unsettling the Settler Within* (Vancouver: UBC Press).

Robinson, D. 2020. *Hungry Listening: Resonant Theory for Indigenous Sound Studies* (Minneapolis: University of Minnesota Press).

Robinson, D. and Martin, K. (Eds.) 2016. *Arts of Engagement: Taking Aesthetic Action in and beyond the Truth and Reconciliation Commission* (Ottawa: Wilfred Laurier Press).

Sandercock, L. 1995. Voices from the borderlands: a meditation on a metaphor, *Journal of Planning Education & Research* 14(2), 77–88.

Sandercock, L. 1998. Ed. *Making the Invisible Visible: A Multicultural History of Planning* (Berkeley: University of California Press).

Simpson, A. 2014. *Mohawk Interruptus: Political Life Across the Borders of Settler States* (Durham: Duke University Press).

Simpson, L. B. 2011. *Dancing on Our Turtle's Back: Stories of Nishnaabeg Re-creation, Resurgence, and a New Emergence* (Winnipeg: ARP Books).

Simpson, L. B. 2017. *As We Have Always Done: Indigenous Freedom Through Radical Resistance* (Minneapolis: University of Minnesota Press).

Wilson, P. 2019. *The Heart of Community Engagement: Practitioner Stories from across the Globe* (New York: Routledge).

13

BEYOND COSMOPOLIS

Dreaming Co-existence as Indigenous Justice (2019)

Twenty years ago, in *Towards Cosmopolis: Planning for Multicultural Cities*, I dreamed of a city

> where social justice is more prized than law and order, where I have a right to my surroundings, and so do all my fellow citizens; … where only after consultation with us could decisions be made about our neighborhoods; where scarcity does not build a barbed wire fence around our carefully guarded inequalities; where no-one flaunts their authority and no-one is without authority; where I don't have to translate my 'expertise' into jargon to impress officials and confuse citizens. I want a city where the community values and rewards those who are different; where a high school is run differently from an accounting firm, and a teacher teaches by sharing her interests with her students; where everyone can paint the sidewalks and address passersby without fear of being shot; where there are places of stimulus and places of meditation; where there is music in public squares, and street vendors, and people take pleasure in shaping and caring for their environment. I want a city where my profession contributes to all of the above, where city planning is a war of liberation fought against dumb, featureless public space as well as against the multiple sources of oppression and domination and exploitation and violence; where citizens wrest from space new possibilities, and immerse themselves in their cultures while respecting those of their neighbors, and collectively forging new hybrid spaces.
>
> *(Sandercock 1998a: 218–219)*

Looking back, my utopian dreaming of a good city and good society was a curious amalgam of concerns about urban design, conviviality, neighborliness, inequality, and respect for difference in its multiple manifestations. The main project of that 1998 book was an attempt to integrate a theory of difference with a theory of social justice and the city. Conspicuously absent was an awareness of Indigeneity in the city, and the challenge that presents to Western planning.

I have always believed in and conceived of planning as a social justice project, but my understanding of what that project involves has evolved over time. In my first book, *Cities for*

DOI: 10.4324/9781003325451-18

Sale (1975), social justice in planning meant, to me, addressing the spatial dimensions of class inequalities. Later, I incorporated an environmental critique, and then a gender-based analysis of urban inequalities. During my decade living in Los Angeles (1986–1996), I began to see the city, and the planning profession, through the eyes of the very diverse group of graduate students in the Graduate School of Architecture and Urban Planning at UCLA. In that age of insurgent identity politics, African Americans, Asian Americans, Latino/as, feminists, gay students, lesbian students, queer students, and students with disabilities were all struggling for their 'right to the city' and did not see themselves represented in either the planning profession or the mainstream planning texts we were reading. I needed to think beyond class and gender inequalities to theorize cities of difference and to ask what challenge this posed to the planning profession. *Making the Invisible Visible: A Multicultural History of Planning* (1998) and *Towards Cosmopolis: Planning for Multicultural Cities* (1998) were the immediate products of that exploration.

Today, my main concern is that planning in the three settler societies where I have lived – Australia, the United States, and now Canada – was and still is a *colonial* project, based on the foundational injustice of the dispossession of Indigenous Peoples (Sandercock 2004a; Porter 2010; Porter and Barry 2016). There were no Native Americans in my classes at UCLA to challenge me, but at a conference in 1994 I met Dr. Ted Jojola, a member of the Pueblo Nation in what is now New Mexico, who was and still is teaching community and regional planning at the University of New Mexico. Ted introduced me to the concept of Indigenous planning as long pre-existing the arrival of settlers in what is now the United States (Jojola 1998; 2008; 2013). And thus began my slow and continuing education about planning as a colonial project.

Some people are put in our paths for a reason. In 2006, I was introduced to Pauline Goertzen, a non-Indigenous anti-racism activist in the small town of Burns Lake in north-central British Columbia, who introduced me to the Carrier people, and together they began to open my eyes to what had previously been invisible to me in Vancouver and Canada: the shocking history of Canada's relationship with its First Peoples, now referred to as First Nations, Inuit, and Metis Peoples, or Indigenous Peoples. I spent the next four years working with and making a documentary about the Cheslatta Carrier Nation and the Burns Lake Band, two sub-tribes of the Carrier Nation, each of whose stories are microcosms of the history of colonization and of the ongoing injustices inflicted on Indigenous Peoples in this country (Attili and Sandercock 2010). This learning experience, their willingness to share their story and trust me to tell it through the medium of film (Chapters 9 and 11), changed my life and work.

I am a descendant of Swedish and Irish settlers who came to Australia in the 19th century in search of a better life, almost certainly oblivious to the fact that the cost of that better life was the displacement and dispossession of that continent's First Peoples. Since 2001, I have been a settler in and now citizen of Canada, an uninvited guest on the unceded territories of the xʷməθkwəy̓əm (Musqueam), Skwxwú7mesh (Squamish), and Səl̓ílwətaʔ/Selilwitulh (Tsleil-Waututh) Nations. I earn my living at the University of British Columbia, which sits on the ancestral and unceded lands of the xʷməθkwəy̓əm (Musqueam). I am clearly a beneficiary, on numerous levels, of this still-colonial history.

For the past decade, since that first visit to Burns Lake, I have dreamed of how this might change: I dream of a post-colonial Vancouver and a decolonized planning profession. I dream of a different kind of co-existence between settlers, newcomers, and Indigenous Peoples in

Vancouver than has been the reality for the past 150 years. What would this co-existence look like, feel like?

★

In the spirit of an Indigenous approach to planning that always begins with the past, ('walking backwards into the future', as the Maori saying has it), I ask 'where have we been?' as a settler society these past 150 years; 'where are we now?' in the year of the 'celebration' of this 150th anniversary of the nation; 'where do we want to be?' by, say 2050, in terms of a post-colonial project of co-existence; and 'how do we get there?' In other words, the intended focus of this chapter is on *relationships*: past, present, and future.

Where Have We Been?

The Musqueam people, people of the river grass, have inhabited a large territory, now known as the Lower Mainland of British Columbia, since time immemorial. The 2015 multi-site multimedia exhibit *c̓əsnaʔəm: the city before the city* tells a part of this history. The University of British Columbia (UBC) has occupied a small but significant part of Musqueam's traditional, ancestral, and unceded territory for the past 100 years. The School of Community and Regional Planning (SCARP) has been in existence for a mere 65 years and has regarded itself as a leader in progressive planning. Yet, for the first 60 years of its existence, SCARP had virtually no contact, let alone connection, with the neighboring Musqueam people, or with Musqueam governance. Until one day it did. And I will return to this small step forward in due course.

Cities, and city planning, have been and still are key mechanisms of colonial expansion. As nodes within wider networks of power, command centers within regional economies, and settings for the concentration of settler populations, cities and the planning of them have been vital instruments in colonization, which is also to say, as tools of dispossession of Indigenous Peoples (Stanger-Ross 2008). This is my and our undeniable legacy, both as settlers and as 'Western' and therefore colonial planners.

'Municipal colonialism' is a term used by historian Jordan Stanger-Ross (2008) to describe settler territorial claims, expressed through municipal governments, about the presumed requirements of urban development or city-building, which has always been one of the twin pillars of nation-building. And historically associated with this city-building project has been the denial of the 'right to the city' of Indigenous Peoples. This public discourse, which first surfaced in the 1880s, was clearly stated in a 1936 editorial in the *Province* newspaper: 'a city is no place for the primitive wards of the government ... An Indian reserve in the midst of a big city is an anachronism' (Stanger-Ross 2008, 546).

Vancouver as a city did not exist before 1886, when it became the western terminus of the transcontinental rail line. But 30 years before that, in the process of what historical geographer Cole Harris has called 'the resettlement of British Columbia', the colony, which became the province of BC, had been forcing Indigenous People onto tiny reserves, depriving them of the resources of their traditional territories (Harris 1997, 2002). From 1886 onwards, these urban reserves became coveted both by settlers, from aspiring home owners to businessmen, and by various branches of government, from Harbors to Highways to Parks.

Those who are Indigenous to this area (whom anthropologists refer to as Coast Salish) were first removed, and then saw their memory erased in the first three decades of the 20th century, through a public discourse by which the urban was equated with 'civilization' and 'modernity' and Indigenous Peoples were equated with the primitive, seen as needing protection from the temptations of the city, and not considered worthy of occupying urban land since they were not using it to its full 'productive potential'. This is the 'impossible contradiction' named by Evelyn Peters (1996) to describe the 'prolonged political project' of identifying 'Aboriginal' as completely out of place/incongruous with the 'urban' and thus justifying a history of policies intended to separate/segregate First Nations from urban society.

Speaking in favor of a 1911 amendment to the Indian Act, Prime Minister Laurier explained: 'where a reserve is in the vicinity of a growing town ... it becomes a source of nuisance and an impediment to progress'. Accordingly, residents of 'any Indian reserve which adjoins or is situated in an existing town or city having a population of not less than 8000' could be legally removed without their consent if it was in 'the interest of the public and of the Indians for whose use the reserve is held' (Stanger-Ross 2008, 549). When the McKenna-McBride Commission toured the province between 1912 and 1916, re-examining the size of existing reserves, shrinking many of them, and carving out new ones, municipality after municipality requested the appropriation of reserve lands for such 'public purposes' as parks and exhibition grounds, industrial and residential expansion.

Today there are 11 First Nations with reserves in Metro Vancouver, but extraordinary efforts have been made at various times to dispossess and remove them, with 'due process' usually playing no role. Here are just three examples.

Kitsilano Indian Reserve, established 1869, expanded to 80 acres in 1877

Not long after this reserve was expanded, the land began to be coveted by a variety of interests. An 1891 map shows a series of ocean docks on the English Bay side of the site. In 1911 the Great Northern Railway offered $2 million for the site. A scandal erupted in 1913 when the provincial government announced it had purchased the reserve land from 20 Squamish males who lived on the site for $225,000. At that time the federal government refused to let the province take over the site. The people resident there repeatedly refused to move in the face of more than a decade of attempts to get rid of them, but were finally forced off by the province, which told them they'd be driven off by police if they didn't consent to sell. When the Indigenous residents finally capitulated in 1913 and were put on a barge taking them north, their houses and remaining belongings were immediately burned to the ground by the government (Stanger-Ross 2008).

The federal government bought out the provincial government's interest in the site in 1928, and two years later gave the city a right-of-way through the reserve for the Burrard Bridge. While the land officially remained a reserve until 1947, parts of the site had already been chipped away for the Armory. In 1936 the Mayor of Vancouver announced a $2 million plan for the reserve, which involved a stadium, auditorium, tennis courts, a bowling green, a bandstand, a rowing club, an open-air swimming pool, a bathing beach, and a landing for small pleasure craft (Mackie 2016, 2). This plan did not materialize then, but its existence was indicative of the mindset of the time, and reflected in Harlan Bartholomew's 1929 Plan for Vancouver, that the city was no place for Indians or for Indian reserves. Except for the 8 acres used in the building of the Burrard Bridge in the 1930s, this land remained vacant for 30 years until 1947, when the Squamish officially 'surrendered' the land to the federal Department of Indian Affairs. Parts of the land were then sold, including the Molson's brewery site, and in

1966 the federal government handed what remained of the reserve to the City for Vanier Park – all without consultation with the Squamish or Musqueam Nations.

Stanley Park

The Indigenous residents of what is now known as Stanley Park had made use of various sites on this peninsula for thousands of years, although the federal government refused to allocate a reserve there. In 1908, the federal government leased the 1,000-acre peninsula to the City for 99 years for park purposes, and the City and Parks Board then proceeded to use various tactics to evict and dispossess the Indigenous and mixed-race families living there.

Residents of the south side of Brockton Point (facing downtown) were forced out by 1931 and their homes burned. The Parks Board then erected four Kwakwaka'wakw (people from what is now North Vancouver Island) totem poles, completing what historian Jean Barman describes as 'a double erasure' of the resident First Nations and 'making the province's Indigenous heritage fashionable in sanitized forms that did not affect the realities of newcomers' everyday lives. The passion to rehabilitate the imaginary Indian who existed prior to the arrival of outsiders was very different from *co-existing with real people*' (Barman 2007, my emphasis).

Harlan Bartholomew's 1929 Plan for Vancouver

This much-celebrated 'modern' (scientific) plan provided a detailed analysis of the place of the Kitsilano and Musqueam reserves in the development of Vancouver. In short, consistent with the 'impossible contradiction' discourse identified above, both were designated to be incorporated into a system of 'large parks and pleasure drives' (Stanger-Ross 2008). No mention was made of what might become of the people living on these reserves. The Musqueam reserve was seen in this plan as occupying an important site at the mouth of the Fraser, 'well suited to the construction of a sports arena and an ideal setting for scenic promenades' (ibid).

The Squamish, Musqueam, and other First Nations have directly confronted and rejected these claims of municipal colonizers and continue to seek justice and compensation for their unceded territories through the Courts, the BC Treaty process, and specific land claims. But in spite of the acknowledgement and inclusion of Aboriginal rights and title in the 1982 Constitution, governments as well as private developers continue to ignore these still unresolved land rights.

Where Are We Now?

For more than a century, Indigenous People have resisted the appropriation of their lands, treaty betrayals, and other dirty tricks, using various 'weapons', from the Courts to politics to direct action. They have begun to win significant victories, involving acknowledgement, finally, of Aboriginal title (to land) and rights (to various uses of land and sea) in the Canadian Constitution in 1982 and Charter of Rights and Freedoms, although there is a huge legal vacuum still in the interpretation of rights and title and ongoing legal battles.

With 50% of Indigenous People in Canada now living in cities, the 'new frontier' of contestation for the *recognition* of Indigenous rights and title, the *reclaiming* of Indigenous spaces, and the *reconciliation* of Indigenous and non-Indigenous lifeways and desires is the urban

(Porter 2013). What do Indigenous rights and title mean in an urban planning context? How might contemporary Indigenous title, governance, and aspirations to economic development and cultural revitalization be engaged in urban planning? Planners are unavoidably involved in these struggles at the points of intersection between Indigenous claims and land use plans and regulations, what Porter and Barry (2016) call 'the contact zone'. Urban planners have yet to come to grips with what contemporary *co-existence* might look like in urban settings in this new political landscape of Indigenous rights and title.

Can planning have a positive role in this struggle for recognition, reclaiming and reconciliation, *or* will it continue to be essentially a colonial practice? Can we 'unlearn' the colonial culture of planning (Porter 2010)? Can planners redress the history of Western planning and transition to practices that are more culturally respectful and responsive to the co-creation of new institutions and forms of decision-making that fully acknowledge Indigenous rights and title? How can a still-colonial system be re-shaped, and by whom (Porter and Barry 2016; Livesey 2017)?

The challenges to Western planning theory, history, education, and practice are significant. Accelerating levels of conflict over the destruction of Musqueam and Tsleil-Waututh burial and sacred sites in Vancouver, North Vancouver, and other municipalities for the building of condos, housing projects, golf courses, parks, bridges, and freeways, and the failure to dutifully and meaningfully consult on any of these development proposals, do not guarantee a peaceful path to recognition and reconciliation. Ethically, planners cannot continue to hide behind colonial policy documents and legislation. We (settler-planners, settler-planning scholars) have to learn how to work through these challenges in constructive and inclusive rather than defensive or evasive ways.

In other words, I too have to advance my previous thinking about planning as a social justice project, to engage in what Porter and Barry (2016, 187) aptly describe as 'rethinking planning for Indigenous justice', and to apply that thinking where I live, in the City of Vancouver. But where to start?

There is no more than a fledgling literature in Canada addressing the challenge of community/urban/regional planning for Indigenous co-existence in cities (Porter 2013; Walker and Belanger 2013; Porter and Barry 2016). Walker and Belanger (2013), focusing on Canada's large prairie cities, begin by noting the social distance between Indigenous and non-Indigenous residents of these cities, a gesture of acknowledgement as to what *today's* form of co-existence actually *feels like*: living side by side, yet with no real contact or interaction. The same surely holds for Vancouver today. For two weeks in February 2010 during the Winter Olympics, the four host First Nations were 'showcased' to the world and had a central presence in the cultural pavilion in downtown Vancouver. Those two weeks aside, our urban reserves, diverse urban Aboriginal population, and history of Indigenous occupation of this region are not on most people's mental or experiential map of the city, whose streets, parks, schools, libraries and so on are all named after colonial 'heroes'.

In a significant departure for settler-planning scholarship, Walker and Belanger discuss Indigeneity as a civic strength, and the importance of urban Indigenous self-determination. They emphasize the necessity of decentering Western/White authority over the procedural and substantive knowledge that currently dominates planning, and of challenging the presumed cultural neutrality in the production of physical, aesthetic, cultural, social, and political space. And they describe three recent court cases that, together, establish urban Indigenous communities as political communities, and in the process, establish a legal framework to guide federal, provincial, and municipal officials to better determine what an urban Indigenous community is and what it represents to various government and non-government agencies.

They then ask how to improve city planning at the municipal/Indigenous interface and identify, with current examples, five ways for municipalities to improve relationships with First Nations and urban Indigenous communities: Civic Declarations, Community Accords, sector-specific protocol agreements, proactive policy frameworks for urban reserve generation, and more nuanced Indigenous engagement techniques (ibid, 201-209). All of these have undeniable symbolic as well as practical importance. A civic declaration such as Edmonton's 'Strengthening Relationships between the City of Edmonton and Urban Aboriginal People' and the 'Edmonton Urban Aboriginal Accord' (2005) perhaps signify a new era, a statement of intent on the part of the City, of cooperation between itself and its First Nations and Metis communities. But, absent research, we don't know yet in what ways, from symbolic to processual to material, this is or is not making a difference to the everyday lives of Indigenous and non-Indigenous People living in Edmonton. Has it translated, for example, into the co-production of Official Community Plans? Has it changed relationships between city staff and Indigenous organizations? Has it lessened the social distance between Indigenous and non-Indigenous People?

There must be things that the City of Vancouver could learn, positive and negative, from this decade of experimentation in Edmonton. From a transformative planning perspective, the key question would be, are these 'improvements' transforming colonial structures of Indigenous political, social, cultural and economic oppression, or are they perpetuating the status quo of power asymmetry and of requiring urban Indigenous organizations and Nations to continue to play within the rules of the still-colonial planning system?

I recognize as soon as I write it that this formulation of the question is too simplistic, too either/or, rather than recognizing that through the creation of new spaces of engagement there arises new possibilities for (gradual rather than dramatic) transformation, depending on many local factors, from leadership to the day-to-day interactions between City staff and urban Indigenous agencies and residents. Sector-specific protocols, in housing or health, for example, can improve the nature of housing provision and health services for urban Indigenous residents. And more nuanced engagement techniques might not only give voice to more Indigenous residents, but also, more importantly, enhance their sense of belonging in the city, diminish social distance between Indigenous and non-Indigenous city dwellers, educate those same non-Indigenous folks in unpredictable ways, certainly expand the horizons of urban and social policy makers through being exposed to Indigenous knowledge systems, and possibly result in urban design ideas for making more visible the history of Indigenous occupation of these now urban places.

What Is the City of Vancouver Doing?

The pursuit of economic and residential development by Metro Vancouver's municipalities for the past 100 or more years has devastated the territories, lifeways, and opportunities for thriving of the 11 First Nations for whom this region has been home since time immemorial. For example, the Musqueam and other Lower Mainland marine-based Nations have a saying that, before the arrival of Europeans, you could go down to the beach at low tide, and the table was set. Musqueam have not been able to harvest shellfish at the mouth of the Fraser for more than 50 years thanks to the settler economic development projects that have disrupted the ecosystem.

In 2004, the Supreme Court of Canada case **Haida Nation v.** British Columbia (Minister of Forests), [2004] 3 S.C.R. 511, 2004 SCC 73 established that the Crown has a legal duty to consult and accommodate First Nations when their interests are at stake. Whether municipalities, as 'creatures of the province', owe a similar legal duty to consult is still being argued on a case-by-case basis in the courts, but *ethically*, and given the new political and legal landscape of recognition of Indigenous title and rights, surely municipalities ought to be consulting with First Nations as they develop urban policies that impact those Nations. And, I would argue, the parties ought to be consulting with each other if we are to move towards new and more just forms of co-existence. But what mechanisms are in place for such dialogues, let alone joint decision-making?[1]

The City of Vancouver and the Province of British Columbia disgraced themselves in 2012 in approving a condominium development on Musqueam's ancestral village site at c̓əsnaʔəm (Marpole Midden). Unresolved conflicts, deep historic wounds, misunderstandings, double standards, poor communication, and just plain ignorance concerning Indigenous rights and title seem to perpetuate today the 'impossible contradiction' (Peters 1996) of urban Indigeneity and opportunities for evolving new practices of urban Indigenous planning that are respectful of Indigenous worldviews.

Since that 2012 conflict, the City of Vancouver has made several gestures towards reconciliation and recognition. In 2013, the City endorsed the United Nations Declaration on the Rights of Indigenous Peoples (UNDRIP). After declaring 2013–2014 to be the 'Year of Reconciliation', the City announced in 2014 that it would become a 'City of Reconciliation', aspiring to 'a sustained relationship of mutual respect and understanding with local First Nations and the urban Aboriginal community', and aiming to 'incorporate a First Nations and Aboriginal perspective into our work and decisions and provide services that benefit members of First Nations and urban Aboriginal communities'.[2]

In 2016, the City invested $100,000 in cultural competency training for 350 staff, and created two new positions: a manager of Aboriginal Relations, who works from within the City Manager's office to develop protocols and improve relationships with local First Nations; and an Aboriginal community planner, who works within the Social Policy Department on policies and programs for the urban Indigenous population. And since the 2015 release of the Truth and Reconciliation Commission's (TRC) 94 'Calls to Action', the City has pored through these and asked itself where and how they can contribute, within the mandate and Charter of the City. After identifying 27 recommendations they could act on, City staff organized these under three themes: healthy communities and wellness; achieving Indigenous human rights and recognition; and advancing awareness, knowledge, and capacity.[3] On the theme of recognition, the City made a significant statement in 2015, acknowledging that the land on which Vancouver is built is unceded territory (City of Vancouver 2016).

All of these shifts in the last few years have a certain symbolic significance. But what is the substance, on the ground? Today's Vancouver is a city of newcomers, and these newcomers, alongside the old-timers/settlers, all need awareness training about the real history of this land and its resettlement/displacement. It's important for the City to be setting the pace, just as they did in the 1980s and 1990s with anti-racism and diversity training to combat emerging xenophobia in the face of the new wave of immigrants from non-Anglo/European countries (Sandercock 2009).

The City has committed $200,000 for 'Reconciliation Projects' over the next three years. There is a Mayor's Task Force on Mental Health and an Aboriginal Healing and Wellness Centres working group, who are looking to understand and identify existing activities and opportunities for traditional Indigenous methods of healing and wellness in the Downtown Eastside (DTES). In 2014–2015 two health and wellness centers operated through the Urban Native Youth Association and the Lu'ma Native Housing Society were supported by City granting programs. There are plans to address Indigenous housing needs in the City. Of the homeless population in Metro Vancouver, 31% is Indigenous, although Indigenous People represent 2% of the metro population. In 2016, an Indigenous housing and wellness approach became the basis for how housing projects in the City are conceived and delivered. New housing projects will expand to include a complementary focus on the wellness needs of the urban Indigenous population and integrate cultural activities, recreation space, and childcare, as determined *in partnership with* local Indigenous agencies and the Metro Vancouver Aboriginal Executive Council. An Urban Aboriginal Peoples Advisory Committee has provided guidance to the City since 2013, and the City has a government-to-government structure in place to negotiate with the three main urban First Nations.[4]

These are good things in themselves, and the move towards a partnership approach is promising. If a journey of a thousand miles begins with the first step, each of these small steps is significant on the path to decolonization.

Meanwhile, on the western edge of the City of Vancouver, the Musqueam Indian Band's (MIB) reserve is a miniscule parcel of their once vast territory. UBC sits on their ancestral and unceded territory, occupying more land than the size of the Musqueam reserve. The MIB is an exemplar of the 'new' Indigenous community planning, receiving UN Habitat recognition for their 2011 Comprehensive Community Plan. But that's a plan that only applies to their reserve, which is bursting at the seams and simply inadequate in terms of land base to accommodate a growing population and aspirations for economic development opportunities. The Musqueam people's parallel attempts to reclaim land through the BC Treaty process have stalled, and they spent most of 2012 battling the City and Province to protect c̓əsnaʔəm, one of their burial and village sites, from a condo development that both municipal and provincial levels of government had approved (as discussed above).

What were those planners at their desks in City Hall thinking when they approved this permit? What did they know, or more likely not know, about Musqueam's ancient history of occupation of these lands, of their sacred sites, of their contemporary aspirations? What choices did the City have, once the province had approved the development application? Could they have changed their own bylaws in recognition of Musqueam's interest in and right to protect the site? As a planning educator, that question is important to me. After the cultural competency training of City staff and all the other activities described above, especially the appointment of an Indigenous community planner, we may see fewer of those kinds of egregious decisions.

In recognition of the lack of awareness among the planning profession in general and City of Vancouver planners in particular regarding Aboriginal rights and title and the aspirations of Vancouver's Indigenous residents, my personal project since 2010 has been to develop a formal educational partnership with Musqueam in the design and delivery of a new Master's program in Indigenous Community Planning in the School of Community and Regional Planning at UBC (Sandercock, Sparrow, Cook and Grant 2013; Sandercock, Sparrow and Cook 2016). After a year of conversations, Musqueam's Chief and Council agreed to this partnership in February 2012 and the program was launched in Fall 2012. After 60 years of living alongside but having nothing to do with the Musqueam community, those 'strangers' are now becoming our neighbors, and helping to educate a new generation of planners in how to work in a good way with Indigenous communities, respecting the knowledge systems, worldviews, and governance traditions of those communities, and supporting them in achieving their own aspirations for self-determination, cultural revitalization, and economic and social development.

This educational partnership is a learning journey of its own for SCARP faculty towards mutual respect and understanding, and is still in its early stages. But its very existence is both a small step on the longer journey of decolonizing the planning profession and a tribute to the visionary thinking of the Musqueam people, who have little to gain as a community from their involvement but have a great deal to offer to aspiring Indigenous community planners through the sharing of their own Comprehensive Community Plan and planning process and the sharing of their ways of knowing, their culture, and their history.

Where Do We Want to Be?

There is something about Canada that gives me a measured hope for a transformative planning towards Indigenous co-existence and partnership. For the past 50 years – since the 1968 change to federal immigration legislation opened Canada's doors to immigrants from non-Anglo/European countries – Canada has undergone its own quiet revolution. Once a nation of predominantly Anglo/Franco/European ancestry, there's been an evolution to one that is ethnically pluralistic and multi-faith – developing levels of tolerance, resilience, and adaptability (Chapter 6).

I don't believe that multiculturalism, or to be more accurate in the Canadian context, 'interculturalism',[5] is a failed project, as is being suggested south of the border and in parts of Europe and the UK. But it is an incomplete one, at least in the context of settler societies. Its vision of *inclusion* was blind to the foundational injustice of those societies: stolen land, dispossession, displacement, and cultural genocide, a whole series of violent public policies to enforce settler desires for land.

Can we now take an even bigger step, towards re-thinking planning for Indigenous justice, for a genuine co-existence in the City? Who do I mean by the 'we' here? I (can only) speak as a White settler-planner, concerned with unsettling planning education and practice. Planning has historically demonstrated its capacity to be both a site of reform and liberation *and* a site of oppression and control (Sandercock 1998b; Reardon 2003; Lane and Hibbard 2005). Can we (Western planners who wish to be allies with Indigenous Peoples in their struggle for land justice) collectively imagine and help bring into being a mutual, respectful co-existence that is the product of ongoing 'negotiation between western and Indigenous perspectives in which each has equal weight, and neither is positioned as dominant' (Porter and Barry 2016, 188)?

What are the challenges for planning in still-colonial contexts, specifically in Metro Vancouver and the City of Vancouver, where local governments retain statutory planning authority, but urban First Nations such as Musqueam, Squamish, Tsleil-Waututh, and Ts'awassen are increasingly realizing their powers as decision makers, developers, landowners, and planners? These urban First Nations now have their own Capital Development Corporations working as 'intergenerational investors' (Livesey 2017, 142), seeking to accomplish economic development goals for their communities within the constraints of an already built-out urban area. Conflict is guaranteed. What is important is that Western planners understand the cultural perspectives and imperatives of contemporary Indigenous engagement in urban development. Economic self-sufficiency is one imperative. Doing it their own way is another, with the right to plan and be their own planning agents (Matunga 2000, 38).

Will there come a time when the City of Vancouver's Official Community Plan is a product of a partnership between the City and Musqueam and Squamish and Tsleil-Waututh Nations, when Metro Vancouver is governed by a coalition of mayors alongside the chiefs of all 11 Lower Mainland First Nations and one or more representatives of the urban Aboriginal population? Will there be joint governance and decision-making arrangements? How would such arrangements be determined?

The short answer is that such new ways of governing would have to be co-produced, co-constituted, through a practical politics of co-existence. Porter and Barry outline the core of such a practice, captured in their apt phrase 'flipping the tables', as follows:

> The spatial boundaries assumed by settler planning to be static and normal would be suspended, made subject to negotiation. Indigenous spacialities would be carefully taught and expressed … The political authority of settler planning would also be suspended and subject to negotiation, no longer the final arbiter on land use disputes: the political authority of Indigenous peoples would be afforded parity. There would be no presumed process for decision making, but instead a negotiation about what the right decision rules might be in any given context. Decisions would involve the presence of country, of ancestor spirits, of the non-humans relationally involved in caring for place as much as they would statistics, projections, and regulations. All of the dimensions of that caring and responsibility would be present and negotiated.
>
> *(Porter and Barry 2016, 195).*

All of this presupposes a new generation of settler-planners and city staff with sufficient humility and a learned ability to listen – planners who are learners more than talkers and deciders, controllers, and regulators.

And How Do We Get *There*?

Here is my dream.

Change starts now, today, in Planning Schools across Canada. Curricula are re-written, not overnight, but as a process of negotiation with the Indigenous Peoples whose territories correspond with the location of the specific School. Educational partnerships are formed in each location, with the goal of mutual learning. Students spend half their time in the classroom, half on the land, learning from Elders and knowledge holders, learning how to know and see the land differently. They develop a careful attentiveness to the lifeworlds and planning intelligence possessed by Indigenous Peoples. They learn to question Western norms of property regimes and individual rights. Indigenous communities learn and adopt Western tools and techniques they deem useful (such as Geographic Information Systems, GIS) to better represent their worldview, ways of seeing, ways of relating to place. Indigenous and non-Indigenous students learn, side by side, and begin to learn about each other, too. Social distance begins to break down. Whole families get to know each other and visit each other's homes and territories.[6] They eat together, go to each other's weddings and funerals …

Instead of talking about building the capacity of Indigenous communities to do their own planning, which they were already doing for millennia before European contact (Jojola 1998; Musqueam Indian Band 2011; Matunga 2013), we need to focus on building the capacity of non-Indigenous planning professionals, city staff, and the systems within which they work. We need to see *ourselves*, our colonial mindsets, and our still-colonial institutions, especially universities and professional institutes such as the Canadian Institute of Planners, *as the problem that needs fixing*.

The lineaments of a decolonizing planning imagination might build on four radical sensibilities: political, audacious, creative, and therapeutic.[7]

To practice a politics of co-existence we, Western planners, must first decolonize ourselves and our relationships. There are Indigenous and non-Indigenous guides for that journey. One of the best is Paulette Regan. In *Unsettling the Settler Within* (2010), she teaches us the importance of being turned inside out, constantly in discomfort as we begin to enter relationships with Indigenous neighbors, co-workers, activists, fellow professionals. As we first attend meetings, begin to develop projects then partnerships with Indigenous colleagues, the relationship (or lack of one) may be tense, uncomfortable, confounding. We may be yelled at, or may get no response or a very delayed response. Get used to it. Hang in there. The more the discomfort, the more we need to ask ourselves 'is there something I should be learning here?' instead of repeating the same mistakes and expecting different outcomes because we have authority, or choose to revert to the authoritative mode.

Be *Political*

Use our knowledge and experience and influence to create spaces and possibilities for co-existence. Work to create the spaces for local Indigenous knowledges and governance systems in our life space and workplace. To be political is to fight against the myth of the neutral, value-free planner. To be political is to challenge the current requirements for professional planning competencies dictated by the Canadian Institute of Planning.[8]

Be *Audacious*

Teach local Indigenous history and the history of colonization as a required part of the history of planning in our region. Get used to risk, uncertainty, change, as the new normal in the contact zone between Western state-based planning and Indigenous aspirations for land stewardship, self-determination, and cultural revitalization. Learn (at least some words in) the local Indigenous language. Aim to work ourselves out of a job by doing capacity building, when invited, in the Indigenous community we're working with. Deliberately unsettle professional authority, our own and others. Be aware of the modes of interaction (tightly time-bound and rule-bound) that privilege non-Indigenous ways of being, knowing, and doing, and seek to undermine them. Challenge taken-for-granted language and concepts (stakeholders, inclusion, rights, land, property, expertise). Focus on developing relationships rather than on applying predetermined (colonial) rules and procedures. Learn to think and speak 'with reference to Indigenous spatialities and governance systems' (Porter and Barry 2016, 182).

Be *Creative*

Find creative ways to engage with Indigenous communities. Use the widest range of arts practices at y/our disposal, but always be aware of the local arts and traditions, notably storytelling and visual ways of communicating as well as songs, dances. Find creative ways to develop a relationship outside the office/professional arena. Develop creative accounting skills, whereby we can channel the resources of our institution or research grant to assist Indigenous partners in collaborating when no other resources are at hand. Find creative ways to interpret existing planning practice, to interpret rules in ways responsive to Indigenous communities' aspirations, at the same time as we are working to change those rules by acting in the political arena. Creatively use existing language and discourses to make way for new approaches to land management and decision-making. And find ways of supporting Indigenous efforts to 'speak back' to planning systems (Porter and Barry 2016, 173) and to pry open new spaces for shared problem-solving, knowledge sharing, and decision-making.

Develop a *Therapeutic Sensibility*

Never forget that centuries of colonization and cultural genocide (TRC 2015) have created deep collective wounds and ongoing intergenerational trauma in Indigenous communities. Learn to 'read' the context: the person or the community we're engaging with. Sometimes the best, or only, place to start is with this history, and thus with the work of healing, before any planning can begin (Sandercock and Attili 2014; Erfan 2017).

★

There is no perfectibility in human affairs. There is only incremental progress. The history of human existence is a history of pain, brutality, venality, mass extinction. No land, culture, nation, is exempt from this, no tribe entirely innocent. But there is still incremental progress, perhaps invisible to those with apocalyptic perspectives, but it's there, and is expressed, in our context, in a re-written Constitution and a Charter of Rights and Freedoms, in apologies to the survivors of Indian Residential Schools, and in the workings and recommendations of the Truth and Reconciliation Commission (TRC), which has a lot to say about what public institutions like universities should be addressing.

But this has to be fought for, year by year, day by day, relationship by relationship, thought by thought, and policy by policy. The impossible becomes possible, sometimes almost imperceptibly, over centuries, sometimes in one's own lifetime. Progress is never permanent. It's always threatened, must be defended, re-stated, re-imagined, if it is to survive.

★

'I live at the western entrance of a haunted house called Canada whose hungry ghosts, windborn spirits, call us to conscience' (Wong 2015, 64)

In poet and activist Rita Wong's eloquent words, the house of Canada will continue to be haunted until its foundational injustice is undone. Demolished. Planners and planning can, must be a part of that demolition and part of a simultaneous reconstruction project, fulfilling an 'ethical responsibility to support the recovery of Indigenous communities and to facilitate the restitution of Indigenous materiality and memory' (Matunga 2013, 31).

Notes

1 The Federation of Canadian Municipalities, in partnership with the Assembly of First Nations, has pilot projects in seven cities across Canada right now, aimed simply at establishing such a dialogue.
2 (http://vancouver.ca/people-programs/city-of-reconciliation.aspx)
3 http://council.vancouver.ca/20170124/documents/rr2a.pdf
4 http://council.vancouver.ca/20170124/documents/rr2a.pdf
5 There is an important distinction. Multiculturalism has been defined in many ways but its major shortcoming is seen as celebrating ethnocultural diversity while not requiring different cultures to associate with and get to know each other. Interculturalism, on the other hand, calls for such essential political virtues as mutual respect and concern, tolerance, self-restraint, willingness to enter into unfamiliar worlds of thought and experience, and a commitment to accommodation among and interrelations with others (Sandercock 2009: 219–222).
6 Every year, for the past seven years, the Musqueam Nation has invited the entire incoming class of Master's students (40) along with faculty and staff and their partners, to a feast in their Cultural Centre, during which their knowledge holders speak about their history and way of life, their teachings, and their contemporary challenges as a land-starved urban reserve.
7 Here I am building on ideas I suggested back in 2004 for a 'planning imagination for the 21st century' (Sandercock 2004b)
8 See Chapter 11 of Porter and Barry (2016) for an excellent discussion of the limitations of these required competencies and the invisibility of Indigenous history, Indigenous traditions of planning, knowledge systems, and so on.

References

Attili, G. and Sandercock, L. 2010. *Finding Our Way. 90 min Documentary* (Vancouver: Moving Images).

Barman, J. 2007. Erasing indigenous indigeneity in Vancouver, *BC Studies* 155, 3–30.

City of Vancouver 2016. Available at: http://council.vancouver.ca/20170124/documents/rr2a.pdf.

Erfan, A. 2017. Confronting collective traumas: an exploration of therapeutic planning, *Planning Theory and Practice* 18(1), 34–50.

Harris, C. 1997. *The Resettlement of British Columbia: Essays on Colonialism and Geographical Change* (Vancouver: UBC Press).

Harris, C. 2002. *Making Native Space: Colonialism, Resistance, and Reserves in British Columbia* (Vancouver: UBC Press).

Jojola, T. 1998. Indigenous planning: clans, intertribal confederations and the history of the all-Indian pueblo council. In: Sandercock, L. (Ed.) *Making the Invisible Visible: a Multicultural History of Planning* (Berkeley: University of California Press).

Jojola, T. 2008. Indigenous planning: an emerging context, *Canadian Journal of Urban Research* 17(1), 37–47.

Jojola, T. 2013. Indigenous planning: Towards a seven generations model. In: Walker, R., Jojola, T. and Natcher, D. (Eds) *Reclaiming Indigenous Planning* (Montreal: McGill-Queens University Press).

Lane, M. B. and Hibbard, M. 2005. Doing it for Themselves: transformative Planning by Indigenous peoples, *Journal of Planning Education and Research* 25(2), 172–184.

Livesey, B. 2017. An institutional ethnographic inquiry into planning for developing land returned under Treaty settlement in Waikato, Aotearoa New Zealand (unpublished PhD thesis (Massey University)).

Mackie, J. 2016. Mayor had a plan for Kits reserve, *Vancouver Sun*, 17 September, 2.

Matunga, H. 2000. Decolonizing planning: the Treaty of Waitangi, the environment, and a dual planning tradition. In: Memon, P. A. and Perkins, C. H. (Eds) *Environmental Planning and Management in New Zealand* (Palmerston North: Dunmore Press).

Matunga, H. 2013. Theorizing indigenous planning. In: Walker, R., Jojola, T. and Natcher, D. (Eds) *Reclaiming Indigenous Planning* (Montreal: McGill-Queens University Press).

Musqueam Indian Band. 2011. *Comprehensive Community Plan*. Available at: https://www.musqueam.bc.ca/community-engagement/ccp/

Peters, E. 1996. 'Urban' and 'aboriginal': an impossible contradiction? In: Caulfield, J. and Peake, L. (Eds) *City Lives and City Forms: Critical Research and Canadian Urbanism* (Toronto: University of Toronto Press).

Porter, L. 2010. *Unlearning the Colonial Cultures of Planning* (Burlington: Ashgate).

Porter, L. 2013. Coexistence in cities: the challenge of indigenous urban planning in the twenty-first Century. In: Walker, R., Jojola, T. and Natcher, D. (Eds) *Reclaiming Indigenous Planning* (Montreal: McGill-Queens University Press).

Porter, L. and Barry, J. 2016. *Planning for Coexistence? Recognizing Indigenous Rights through Land Use Planning in Canada and Australia* (London: Routledge).

Reardon, K. 2003. Ceola's vision, our blessing: the story of an evolving community-university partnership in East St. Louis. In: Eckstein, B. and Throgmorton, J. (Eds) *Story and Sustainability* (Cambridge, MA: MIT Press).

Regan, P. 2010. *Unsettling the Settler Within* (Vancouver: University of British Columbia Press).

Sandercock, l. 1975. *Cities for Sale* (Melbourne: Melbourne University Press).

Sandercock, L. 1998a. *Towards Cosmopolis: Planning for Multicultural Cities* (London: Wiley).

Sandercock, L. 1998b. ed. *Making the Invisible Visible: a Multicultural History of Planning* (Berkeley: University of California Press).

Sandercock, L. 2004a. Interface: planning and indigenous communities, *Planning Theory and Practice* 5(1), 95–97, 118–124.

Sandercock, L. 2004b. Towards a planning imagination for the 21st century, *Journal of the American Planning Association* 70(2), 133–141.

Sandercock, L. 2009. Towards a Cosmopolitan Urbanism, in L. Sandercock and G. Attili. *Where Strangers Become Neighbours: Integrating Immigrants in Vancouver* (Heidelberg: Springer).

Sandercock, L. and Attili, G. 2014. Changing the lens: film as action research and therapeutic planning intervention, *Journal of Planning Education and Research* 34(1), 19–29.

Sandercock, L., Sparrow, L. M., Grant, L. and Cook, J. 2013. Indigenizing planning education, decolonizing planning practice: first steps', *Plan Canada, Special Issue*, Summer 53(2), 48–52.

Sandercock, L., Sparrow, L. M. and Cook, J. 2016. Ways of knowing: engaging in partnership for indigenous planning education, *Plan Canada, Special Issue*, Dec 2016, 49–52.

Stanger-Ross, J. 2008. Municipal colonialism in Vancouver: city planning and the conflict over Indian Reserves, 1928–1950s, *Canadian Historical Review* 89(4), 541–580.

Walker, R. and Belanger, Y. 2013. Aboriginality and planning in Canada's large prairie cities. In: Walker, R., Jojola, T. and Natcher, D. (Eds) *Reclaiming Indigenous Planning* (Montreal: McGill-Queens University Press).

Wong, R. 2015. *Undercurrent* (Gibsons: Nightwood Editions).

Conclusion

Mapping Possibility: The View from 2022

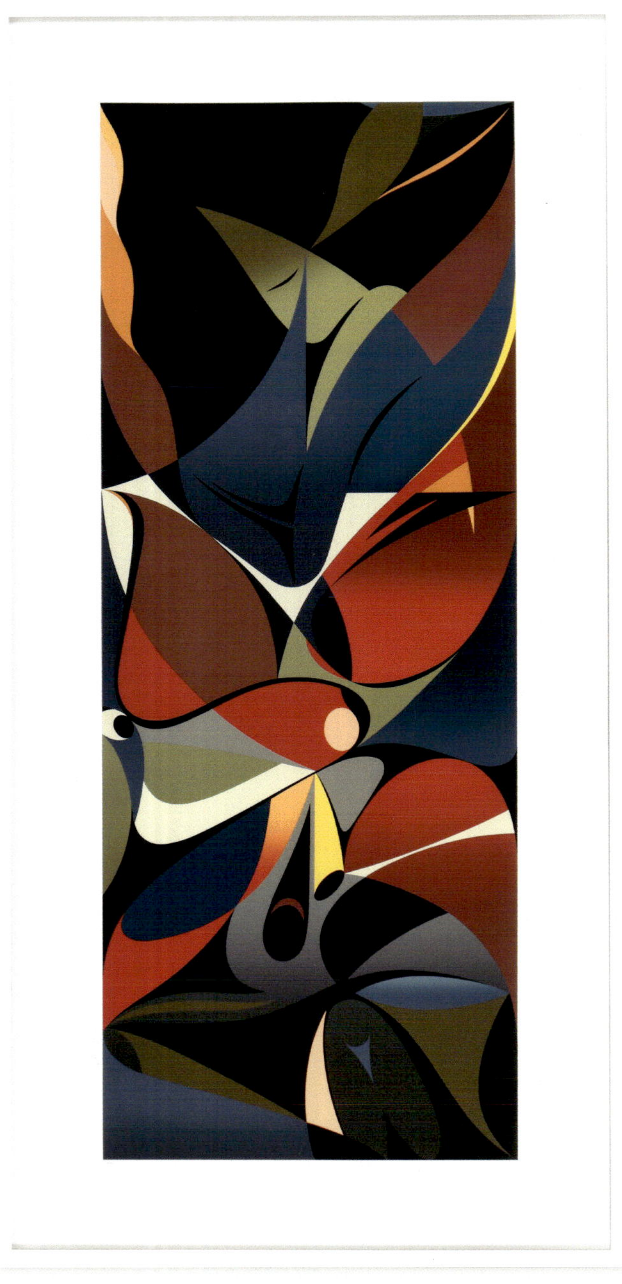

FIGURE 14.1 *Into the Light* by kind permission of Musqueam artist Susan Point, who owns the copyright to the artwork. Image Copyright: Coastal Peoples Fine Art Gallery, Vancouver, BC, Canada.

Commentary: Beneath the Pavement, the Beach?

I mentioned in the Introduction that I took part in student protest movements in the late 1960s. Then, we were against American imperialism and the Vietnam War. But more broadly, some of us were inspired by the French student protests, which weren't just 'anti', but were about dreaming of another, better world. '*Sous les pavés, la plage*'. Underneath the pavement, the beach. 'The beach' was a future that was something other than capitalism. A new world surely lay underneath the old one. What did it look like? Who was going to build it?

These were and still are core questions for me. I have sought answers in Fabian socialism, Euro-communism, and my own invention, 'Cosmopolis' (Chapter 4). I even found Cosmopolis in the Collingwood Neighbourhood House in Vancouver. There was beauty, diversity, joy, and compassion in that local institution, in that neighborhood. It's a model of what is emergent and can and must be built on. Today, my students are influenced by # Me Too, Black Lives Matter, Idle No More, and 'Land Back'. Fifty-five years on, the beach is strewn with plastic and trash, sewage and oil, thanks to the 'gobble gobble economy' (in Kenneth Boulding's felicitous phrase). Ecological destruction is rampant. Our 21st-century world is descending into aggression, fear, and separation in the face of multiple crises: racism, poverty, genocide, rising inequality, pandemic, mental illnesses, and so on.

Is there another world inside this one, waiting to be discovered? What can we, community planners, do to help birth a healthier future? Why and how have I found purpose and hope in this vocation?

★

My very first published essay, 'Capitalism and the Environment: The Failure of Success', appeared in an edited collection, *Essays in the Political Economy of Australian Capitalism*, in 1975. And the last essay I published before embarking on this book was 'Once Upon a Planet: Planning for Transition from Ego-driven to Eco-driven Economies', in another edited collection (Savini, Ferreira and Schonfeld 2022). That essay began as a keynote for the AESOP (Association of European Schools of Planning) conference in Venice in 2019. I was then invited to turn it into that book chapter. As I was working on it, I was groping to explain connections that I felt intuitively, but couldn't articulate: between self and systems, between community planning and societal change. It wasn't until I read Margaret Wheatley's mind-expanding survey of 'the new sciences' that my groping shifted to a grasping of something vitally important, which I'll get to later in this Commentary. I am including a revised version of that Post-growth essay (Savini, Ferreira and Schonfeld 2022) as the conclusion for this book. It is where my learning from Indigenous communities and their worldviews meets my concern for the kind of world my baby-boom generation is bequeathing to Greta Thunberg and her cohort, and, behind them, to 3-year old Ida May Hemphill Campbell and 1-year-old Zebie Graham-Reeves, my dear friends' toddlers, to whom I've dedicated this book.

When I lost my sense of purpose in planning, in the mid-1980s (Introduction), it was disillusion with a critical, purely cerebral approach: the ghostly parade of bloodless categories that characterized my writing from within a critical political economy perspective. My earliest hopes were to be part of top-down change, working with reformist federal or state governments, inspired by my mentor Hugh Stretton. But when I was doing research in actual cities and neighborhoods for my PhD in the early 1970s, my new sources of inspiration were Ruth and Maurie Crow, with their *People's Plan for Melbourne*, and Jack Mundey's charismatic leadership of the construction workers in Sydney, who were placing 'Green Bans' (refusals to build) on projects they deemed socially or environmentally damaging. Here were three contrasting theories of action/change: working within the state, working with 'communities of resistance' (hooks 1990), and direct action. The return of right-wing governments across Australia from 1975 until 1983 meant that I put my goal of working through the state on hold and 'sheltered' in academia, never imagining that would be a life-long choice. In academia during that time, I was both captivated and captured by a radical but masculinist/positivist discourse that was entirely disconnected from actual people in actual communities – hence the crisis I faced by the mid-1980s, when I could not see how to write up the research that I was doing with unemployed steel workers and coal miners in a way that did justice to their lived experience.

My move to Los Angeles, to UCLA, and to film school opened two new windows for me. The first was the language of story (Part 3). The second was the discovery in theory and practice of communities of resistance, of a concept of planning as community building (as opposed to city- and nation-building).[1] The students who came from (what we then called) minority communities inspired me. The colleagues working in those communities inspired me. And this led to a more grounded theorizing and a new intellectual, political, and social project for planning: the Cosmopolis Project. That decade of rethinking in LA also laid the foundation for the following 25 years of work with actual communities, by far the most satisfying period of my life. From the Collingwood Neighbourhood House in Vancouver to the Cheslatta Carrier and Ts'il Kaz Koh Indigenous Nations in north central BC, and finally to the Musqueam and Haida Nations, I experienced the joy of connecting with people and supporting them in their aspirations to repair, heal, transform their communities and tell their stories. Planning in partnership with these communities and Nations became less an intellectual project[2] and more an embodied, involving, emotional, and socio-political project. I have shared tears of pain *and* tears of pure joy with folks in these communities, along with food, laughter, and friendship. The collective practices through which citizens create their own meaningful life spaces in the face of larger social historical forces is the basis of the hope I have for the future.

Throughout my life, I've tried various ways to make a difference through the field of planning: writing books, writing or speaking through mainstream media, making films, doing consulting work, working on government advisory bodies, innovating curriculum, joining street protests, joining political parties. By far the most satisfying has been working with communities. That is the theme of the final chapter, bringing together what I've learned from these communities, about ways of being and knowing, living and loving: finding purpose, joy, beauty, and hope in and through community planning.

*

We live now in a time of grief that's personal, political, and planetary. Millions of people have been personally affected by the covid pandemic. Pre- and post-pandemic, Greta Thunberg represents the grieving of youth over planetary dis-ease. Margaret Wheatley (2018) and Valarie Kaur (2021) both talk of the need for 'warrior-sages', for 'warrior training', to protect and preserve the human spirit in the face of our current descent into war, aggression, and fear. Wheatley (2017) writes about the importance of working locally, in place, even though the issues are global. Those we grieve with are ultimately those we organize with and advocate for. 'When a critical mass of people come together to wonder about one another, grieve with one another, and fight with and for one another, we begin to build the solidarity needed for collective liberation and transformation – a solidarity rooted in love', writes Valarie Kaur. Her

'Revolutionary Love Project' begins with seeing no stranger: a cultivated practice of looking upon the face of anyone who is unfamiliar and choosing to say 'You are a part of me I do not yet know' (Kaur 2021, 311). With such a practice, strangers can become neighbors. bell hooks embraces a spirit of neighborliness: of fellowship, care, and respect for the dignity and wellbeing of the 'beloved community'; a way of thinking and behaving which honors principles of 'inter-being and interconnectedness'.[3] Drawing on all of these wise ones, young and old, this is how I want to reimagine the soul of planning: as a social and spiritual project in support of neighborliness and the beloved community, integrating heart, mind, and spirit; connecting inner knowing with outer practice; integrating masculine and feminine energies; holding in tension future vision and present realities.

In the final chapter, I embrace of a theory of action which can be summarized very simply: making a difference, one community at a time. I've arrived here after decades toiling in the fields of theoretical discourses, each more convoluted than the previous: from Fabian socialism to Euro-communism, neo-Marxism to postmodernism, intersectionality, and decoloniality. Where I've landed is based not on esoteric readings in esoteric journals but on experience, on learning by doing, while also paying attention to and trying to understand the root causes of the multiple global crises noted above. To be clear, community planning cannot solve these global crises, at least not at global scale. The choice now is to do what we can, locally, to make life more sustainable, to create possibility in the midst of global decline – islands of sanity amidst the chaos and breakdown.

In this book I have provided shining examples of what is possible at the local level: from the Collingwood Neighbourhood House (CNH) to the work of Indigenous communities throughout British Columbia, developing their community-driven plans in the service of sustainability and self-determination. Wheatley (2017) provides a similar example to that of the CNH, in the work of the Neighborhood Centers Inc in Houston,[4] and in the stories in an earlier book, tracing the transformation of seven communities, from Mexico to India, South Africa and beyond, as they walked out of limiting beliefs and assumptions and walked on to create healthy and resilient communities, using their own ingenuity to solve seemingly insoluble problems (Wheatley and Frieze 2011). There is a broader theory of change underpinning this deceptively simple approach, best expressed by Margaret Wheatley in her exploration of the relevance of 'the new science' to our understanding of how to make a difference in the world. She argues (based on systems thinking, chaos theory, quantum physics, and more) that changes in small places also affect the global system, not through incrementalism, but because every small system participates in an unbroken wholeness. Activities in one part of the whole create effects that appear in distant places.[5] According to this way of understanding systems and interconnectedness, there is potential value in working anywhere in a system. 'We never know how our small activities will affect others through the invisible fabric of our connectedness' (Wheatley 2007, 45).

From the early Greek Heraclitus to the most recent thinking in science, life is described as a process, a process of becoming (Prigogine 1998, 10). A major challenge for planners now is learning to live in a process world, participating with things as they unfold, expecting to be surprised, unsettled. In a process world, sensitivity to context is everything; falling back on rote procedures or 'best practices' will not work. Process work demands presence, a quality of absolute attention in the moment. This is familiar to surfers and jazz musicians – not so much to planners, wedded as we are to our plans, programs, evaluation measures. These are not obsolescent, but they need to be seen not as completed objects but as processes that enable a group to keep clarifying its intent and strengthening its connections: building relationships and learning how to be together. Process work is relational work. It depends on the *skills* of listening, conversation, participation, creativity, and community-building. It depends on *qualities* such as caring, generosity, openness, curiosity, compassion, and humility.

New ways of understanding often appear simultaneously in widely separated places and in different disciplines. These synchronicities are what strike me now. The same concepts are appearing in many places. These concepts are our *zeitgeist*, a way of thinking that characterizes a generation or a time period. The contemporary *zeitgeist* is an awareness (both new and ancient) that 'we live in a world of exquisite interconnectedness' (Wheatley 2007, 158). Images from quantum physics describe our interconnectedness at a cosmic scale. It can also be seen at microscopic scale, as well as in everyday objects and everyday life. I was filled with astonishment a few years ago at an exhibition of Australian Indigenous women's paintings, 'Marking the Infinite', at the Museum of Anthropology in Vancouver. In the central room, I gazed into the soft pink and rosy hues of the largest painting in the exhibition, a circular image on a rectangular canvas, radiating from a central point and extending out to the edges, in the tiniest, replicated detail, apparently to infinity. It took my breath away. I said to myself, OMG, she's painted the cosmos. I was mesmerized. But when I stepped up close to read the adjacent card naming the artist, Regina Pilawuk Wilson, the title of the painting (*syaw*), and an explanation of its meaning, I was stunned to learn that what she had painted was a fishing net (*syaw*). She had depicted with absolute accuracy the detail of a traditional, cylindrical, hand-woven fishing net. Her bush community had lost the knowledge of how to make such a net after a century of being forced to live on a Christian mission, remote from 'country'. Moving back to her homelands in the 1970s, she started an artists' collective. She then walked 100 miles to a neighboring tribal community where they taught her the lost stitching art.[6] Then she painted it, so it would never be lost again.

Now, systems thinking, or ecological thinking, shows us the webs of interconnections that weave the world together: *from the fishing net to the cosmos*. We are more aware that we live in relationship, connected to everything else: the primacy of relationships, the importance of context, learning to work with the wholeness of life, requires that we bring our whole selves to the work of community planning. No more separation of the inner and the outer self – the divided self (Palmer 2004). I have encountered this new/old understanding in people as diverse as management professor, consultant, and practicing Buddhist Margaret Wheatley; social movement lawyer and Sikh spiritual warrior Valarie Kaur; Chinese American civil rights activist Grace Lee Boggs (2011); African American visionary and feminist theorist bell hooks (1990, 2000); community organizer, educator, and Quaker working on spirituality and social change, Parker Palmer (2018); physicist Fritjof Capra (Capra and Luisi 2014); scientist and environmental activist David Suzuki (Suzuki and McConnell 1999); and of course, Indigenous knowledge holders (some of whom are cited in Chapter 14).

Faced with the myriad contemporary crises already noted above, all of these diverse thinkers from different faith and cultural traditions ask us to act, not to withdraw, not to be defeated by a still-dominant Western worldview based on competition, individualism, and a mechanistic (Newtonian) way of understanding life. All of them believe, as I do, that every one of us can make a difference, in each small act of defiance or new way of acting. That is how societal change happens, not by traditional power politics or high-level leaders – although there are leaders who have led by example, from Vaclav Havel to Nelson Mandela to Jacinda Ardern in Aotearoa/New Zealand. As community planners, we are uniquely situated to help birth this world in waiting, this beach under the pavement. We do this work of community planning because it is rich in meaning and purpose, because it is in tune with the *zeitgeist*, because it is worth doing. And because, as my late husband would have said, we don't know enough to say 'It's too late'.

The future is not yet written.

Notes

1 I had glimpsed this possibility in the Crows' work in Melbourne in the early 1970s, but had left it behind for the top-down world of urban social policy and planning.
2 My biggest divergence from John Friedmann's approach to planning was that John had, from an early age, committed to making planning 'an intellectual project' modelled on European thinkers he admired, such as Karl Mannheim (see Friedmann 2017). For me, planning has always been far more than an intellectual project.
3 'Spiritual life is first and foremost about commitment to a way of thinking and behaving that honors principles of inter-being and interconnectedness' (hooks 2000, 77)
4 www.neighborhood-centers.org
5 Meteorologist Edward Lorenz first drew public attention to this with his now famous 'butterfly effect' (Wheatley 2007, 121)
6 Regina Pilawuk Wilson was born into the Ngan'gikurrungurr language group. She was a founding member of the Peppimenarti Community in 1973, located 320 kilometres south west of Darwin in the Daly River's tropical wetland and floodplain country. Her painting, *golden yellow Syaw (fishnet)* won the General Painting Award at the National Aboriginal and Torres Strait Islander Art Awards in 2003. 'A key message that Wilson portrays in her work is to remember the past to understand the present and hope for the future ... The woven forms take on a cosmic dimension with thousands of rippling lines radiating from a central core or spreading across a surface'. (https://en.wikipedia.org/wiki/Regina_Pilawuk_Wilson)

References

Boggs, G. L. 2011. *The Next American Revolution: Sustainable Action for the 21st Century* (Berkeley: University of California Press).

Capra, F. and Luisi, P. 2014. *A Systems View of Life* (Cambridge: Cambridge University Press).

Friedmann, J. 2017. Planning as a vocation: the journey so far. In: Hasselsberger, B. (Ed.) *Encounters in Planning Thought* (London: Routledge).

hooks, b. 1990. *Yearning* (Boston: South End Press).

hooks, b. 2000. *All About Love* (New York: Harper Collins).

Kaur, V. 2021. *See Stranger, N. A Memoir and Manifesto of Revolutionary Love* (New York: Random House).

Palmer, P. 2004. *A Hidden Wholeness: the Journey toward an Undivided Life* (San Francisco: Jossey-Bass).

Palmer, P. 2018. *On the Brink of Everything: Grace, Gravity, and Growing Old* (Oakland: Berrett-Koehler Publishers).

Prigogine, I. 1998. *The End of Certainty: Time, Chaos, and the New Laws of Nature* (New York: The Free Press).

Savini, F., Ferreira, A. and Schonfeld, K. (Eds.) 2022. *Post-growth Planning: Cities beyond the Market Economy* (London: Routledge).

Suzuki, D. and McConnell, A. 1999. *The Sacred Balance: Rediscovering Our Place in Nature* (Vancouver: Greystone Books).

Wheatley, M. 2007. *Leadership and the New Science* (2nd ed) (Oakland: Berrett-Koehler Publishers).

Wheatley, M. 2017. *Who Do We Choose to Be?* (Oakland: Berrett-Koehler Publishers).

Wheatley, M. 2018. Warriors for the human spirit. Available at: https://margaretwheatley.com/warrior-training-2022/

Wheatley, M. and Frieze, D. 2011. *Walk Out Walk On: A Learning Journey into Communities Daring to Live the Future Now* (Oakland: Berrett-Koehler Publishers).

14

ONCE UPON A PLANET[1]

Reimagining the Soul of Planning (2022)

Introduction: From 'Dare to Know!' to 'How Dare You!'

In a 1930 essay, 'Economic Possibilities for our Grandchildren', John Maynard Keynes speculated that by 2030, capital investment and technology would have raised living standards as much as eightfold, creating a society so rich that people would work less than 15 hours a week, devoting the rest of their time to 'non-economic' purposes. As striving for greater affluence faded, he predicted, 'the love of money as a possession … will be recognized for what it is, a somewhat disgusting morbidity' (Cassidy 2020, 23).

Yet here we are, almost 100 years later, working 40–70 hours a week, while most economic policy makers and politicians are still committed to maximizing the rate of economic growth, and the majority of planning practice is committed to serving that goal, underpinned by a techno-rational worldview that has dominated thinking, valuing, and acting since the Enlightenment. 'Dare to Know!' was the rousing and, for its time, radical call of the German philosopher Emanuel Kant in 1780, expressing unbounded faith in the liberatory potential of scientific and technical knowledge – the dark side of which, as we now know, has been humankind's assertion of the right to dominate nature for our own material ends for the past two centuries.

Once confined to the margins, the ecological critique of economic growth has been accumulating attention for several decades. At a UN climate change summit in September 2019, the teenage environmental activist Greta Thunberg declared: 'We are at the beginning of a mass extinction, and all you can talk about is money and fairy tales of eternal economic growth. How dare you!' (Cassidy 2020, 24)

Provoked by Thunberg, this chapter makes four interconnected arguments. Firstly, planning theory and practice must undergo a paradigm shift from the techno-rational worldview that has dominated our field as it evolved from the Enlightenment frame of mind. A shift to more relational ways of being, knowing, and doing is necessary if we are to be the change agents whose work is relevant to today's most urgent global crises. Secondly, at a more grounded level, drawing on work with Indigenous communities in north-west Canada, the lessons derived from Indigenous worldviews outline an ethic of care and relational accountability

DOI: 10.4324/9781003325451-20

towards fellow humans, alongside an ethic of stewardship of their lands and waters, based on their understanding of the interconnectedness of all life forms. This worldview overlaps with an ecological perspective. Drawing further on this ecological perspective, the third argument is not for an end to growth (an ecological impossibility), but for a shift to *qualitative growth*. And finally, the argument returns to lessons learned at a more personal level, through the work with Indigenous communities, that suggests the need to redefine the soul of planning. I argue that in order for our profession to respond to the interconnected planetary crises of climate change, racism, oppression and injustice, mental health and violence, and ongoing struggles against colonization, the most important quality we can bring to the work is a *way of being* that connects our hearts, minds and spirits, foregrounding relationships, trust, humility, deep listening, patience, kindness, and love.

Colliding Worldviews: From Possessive Individualism to Relationship and Interdependence

It is by now a well-rehearsed critique that the origins of contemporary planning dwell in both the Enlightenment project (Friedmann 1987; Sandercock 1998) and the related European colonization project (Jacobs 1996; Jackson 1998; Porter 2010, 2021). I take these critiques as my starting point, as a given, rather than repeating them here. What I'm more interested in, as we try to imagine a 'post-growth future', is the worldview and attendant ontologies and social values underpinning these twinned or interdependent projects, their toxic and destructive impacts on both human community and Mother Earth, and the challenge of re-ordering our very ways of being and hierarchy of social values to ensure the survival of our one precious planet as well as the deep spiritual needs of human beings.

It was in the writings of Thomas Hobbes (1651) that the notion of what political theorist C.B. McPherson termed 'possessive individualism' as the moral basis of society was first made explicit.[2] Hobbes asked two questions: What is human nature? And how shall society be organized to accommodate this nature? Starting with his materialist model of human nature, Hobbes deduced the necessity of a market model of society, sidestepping the whole question of human bondedness, attachment, need for connection and belonging. This emerging worldview was linked with two equally powerful ideas: John Locke's view of land as property that must not lie fallow but must be improved for human use; and the idea/l of progress or 'development' through the application of reason, the vision of the political philosophers of the 18th century. Simply put, these emerging *beliefs* coalesced into a worldview that totally changed people's image of their place in the Universe. 'From being servants of God, each contributing to the life of His Church on Earth, they became isolated competing agents … self-centered consumers of pleasure, the sole purpose of whose Reason was to further its own existence' (Clark 1989, 269). Individualism, competitiveness, and materialism have been nurtured in us (children of the West), woven implicitly into the fabric of our worldview. And once accepted, this view creates the very kind of human behavior it assumes as inevitable. Our worldview becomes its own self-fulfilling prophecy, and we create systems to perpetuate it.

Over the past three centuries, the attributes of possessive individualism (materialism, independence, competition, property rights, individual conscience) that were first designed to free people from oppressive social constraints, have evolved into a bizarre concept of freedom that dominates 21st-century thinking in the West. The notion of individual freedom which began emerging at the end of the Middle Ages, as freedom from arbitrary control, has gradually been converted into the right to do whatever one pleases, nowhere more clear and more damaging than the recent anti-mask and anti-vax movement in the midst of a global pandemic. This idea of absolute freedom, which has reached its most extreme expression in the materially richest country in the world, the USA, now signifies not just the absence of a few specified evils but also freedom from all culture and tradition, resulting in the isolation of the individual from all connection with others and from what has historically given life meaning: tradition, affection, personal and local ties, natural roots and sympathies.

It is tempting but too easy to blame all this on capitalism, as though it were a system out there over which we have no control. In fact, we are all part of this now-global system that fetishizes economic growth, and we help to reproduce it every day in all of our choices, unless we're one of the very few who've made the choice to live self-sufficiently, off-grid, and outside the market economy.

Could there be a more toxic worldview than the one that has increasingly dominated the planet since the 17th century? Disconnected from others, people must independently invent their own life meaning while remaining seemingly helplessly trapped by social institutions devoid of moral direction. In seeking absolute freedom, we leave ourselves alienated from everything – from 'Nature' and from one another. With its central focus on growth and the consumption of material wealth, Western society, or rather the worldview that has emerged from the West and now pervades the rest (and I include here Russia and China, both equally intent on pursuing material wealth at all costs) is inexorably destroying both the environment and the soul or spirit of humanity. We have lost our capacity for mystery, wonder, awe, reverence. The self is the only sacred object that remains.

The foundational belief of the Enlightenment, that we can master all of Nature for our own ever-expanding 'needs', has turned out to be an illusion which has resulted in the destruction of the very life systems that support us – air, water, soil – without concern for the future, and is leading us to the brink of mass extinction, as Greta Thunberg warns us. Our survival as a species demands that we acknowledge the relational nature of our existence, that we are absolutely dependent on the connections that support our existence and make us whole. We must recognize, through a shift in our worldview, our utter dependence on and embeddedness in the natural world.

To adequately articulate this shift, we would need to redefine the secular, that is all those institutions that organize the material and social foundations of life, as well as redefining the sacred. We could turn to the writings of such (Western) thinkers as Aldo Leopold's or Wendell Berry's work on the meaning of land; E.F. Schumacher's Buddhist economics; Fritjof Capra's systems view of life; ecological economists such as Paul Hawken, Amory Lovins, Herman Daly; James Lovelock's Gaia hypothesis; Mary Clark's exploration of the essentially cooperative nature of human nature; and the growing number of advocates of bio-regionalism. For this essay, I will simply focus on what we in the field of planning scholarship and community planning practice can do to address this crisis, and I will pay particular attention to the teachings of contemporary Indigenous scholar/planners and to current approaches to Comprehensive Community Planning being developed by Indigenous Nations with whom we are working in the School of Community and Regional Planning in what is now known as British Columbia, the western-most province of what is now known as Canada.

Working in Partnership: Being in Community

'That land is a community is a basic concept of ecology, but that land is to be loved and respected is an extension of ethics. That land yields a cultural harvest is a fact long known, but latterly often forgotten' (Leopold 1949, viii).

> Today we live in a globally interconnected world in which biological, psychological, social and environmental phenomena are all interdependent. To describe this world appropriately we need a new paradigm, a new vision of reality – a fundamental change in our thoughts, perceptions, and values. The beginnings of this change, or the shift from the mechanistic to the holistic conception of reality, are already visible ... The gravity and global extent of our crisis indicates that the current changes are likely to result in a transformation of unprecedented dimension, a turning point for the planet as a whole.
>
> *(Capra 1982, 29–30)*

'On every side we can now see people busily engaged in sawing off the branches on which they (along with many others) are sitting, intent only on getting those branches to market before the price of timber falls' (Midgley 1985, 135).

To register that physicist Fritjof Capra wrote these words in 1982 and philosopher Midgley in 1985, while Leopold's words go back to 1949, is to understand the increasing anger and impatience behind Greta Thunberg's protests. Her anger now pervades academia, among students who see nothing but planning's complicity in environmental devastation and systemic racism. Students who want to be change agents are increasingly skeptical about the prospects of planning as a tool of social transformation. Our field is at a crossroads. Unless we address these global challenges with our students, we will lose their respect and trust. We will lose them.

So there has to be a paradigm shift from the techno-rational (Euclidian/Cartesian) worldview that has dominated thinking, valuing, and acting in our field since the Enlightenment to more relational ways of being, thinking, and acting. I want to draw on my own experience in Indigenous Community Planning over the past 15 years to propose that we devote as much effort to *ways of being in the world* as we have in the past to *ways of knowing and ways of doing*. Another way of saying this is that we need to balance the inner and the outer, 'internalizing radical interdependence' in Patricia Wilson's wise words (Wilson 2019), because the conscious integration of these two enables the creativity, and risk-taking, and repair/healing work that we need in order to address the most pressing issues facing this one precious planet: such issues as the refugee/migration crisis, increasing social polarization, systemic racism, and impending socio-environmental catastrophe.

I want to focus on the role of the planner as change agent: a catalyst and facilitator looking beyond technical problem solving and the efficiency of policies, programs and delivery systems to addressing issues of social and environmental justice. There are other roles that planners fulfil (see Willson 2018), but this is the generative edge of our field.

If, as scientists argue, the world is continuing down a path to socio-environmental catastrophe, then we have to ask: how do we become more effective societal change agents? What are the attributes of planners as change agents? What are the underlying skills and sensibilities? What would it take to make our work truly transformative?

The kind of planning for transition that I want to advocate is, as foreshadowed above, an evolutionary shift: from the techno-rational Cartesian worldview to a more relational and holistic view of life. For the past 15 years, I have been working with colleagues and students to support Indigenous communities in the north-west of Canada, and two things stand out from this experience. The first is that, thanks to a new, community-driven approach, we are seeing significant transformation on the ground, one community at a time, each one inspiring another to embark on a holistic/comprehensive community planning process, the goals of which are self-determination and sovereignty. The second is that Indigenous worldviews share certain philosophical elements that have something to offer the global crises we're in the midst of: namely, an ethic of care and of relational accountability towards fellow humans (an emphasis on the 'we' and not the 'I'); and an understanding of the interconnectedness of all life forms on the planet, which leads to an ethic of stewardship of their/our lands and waters.

My work for over four decades has focused on social justice and sustainable community development. But my understanding of how to do that work has changed radically over the past two decades or so. In the 1990s, part of my mission was to de-center the Enlightenment epistemology at the heart of planning and foreground an epistemology of multiplicity (Chapter 2). The evolution I now want to suggest is that, in addition to focusing on different ways of knowing and doing, we need to make an inner or self-reflexive turn and focus more on *ways of being*.

Twelve years ago, I initiated a process of curriculum reform in our School of Community and Regional Planning, a process that various scholars across the settler societies of the New World (Jojola 2013; Matunga 2013, 2017, Porter 2017) refer to as the decolonizing and indigenizing of planning theory and practice. I initiated a community–university partnership in the design and delivery of a new Master's degree program in Indigenous Community Planning (ICP), approaching the Indigenous nation on whose unceded territory (that is, stolen land) my university is located. The Musqueam Nation had, at that time, just completed their first *contemporary* Sustainable Community Development Plan, which subsequently won a UN Habitat award. (I emphasize 'contemporary', because one of the things we learn from Musqueam and other First Nations in this program is that Indigenous communities have been doing their own planning for sustainable development, based on their own worldviews, laws, and traditional knowledges, since time immemorial.)

I now co-teach, with various Musqueam knowledge holders, a core course within this program, *Indigenous Community Planning: Ways of Being, Knowing, and Doing*. We spend half of our class time on Musqueam's reserve, listening to their Elders talking about their worldview, their history and culture, and the devastating, near-genocidal impacts of colonization on their community. Then, for the second year of the two-year program, students (Indigenous and non-Indigenous together) spend about 400 hours working with an Indigenous community on their reserve, outside the metropolitan area, working in support of the goals established by that community through extensive community engagement processes, co-designed and co-facilitated by the students, with their Indigenous host. Through this ongoing curriculum, we have now partnered with 20 Indigenous communities throughout the province, and what follows is the learning about social transformation the students and I have gleaned from this immersive experience.

The single most important learning for me on this journey is that the inner practice of community engagement is arguably more important than (although inseparable from) the outer practice; in other words, the *way of being* that one embodies in community engagement and community building (literally *how* one shows up in community) is every bit as, if not more important than the *ways of doing*. In fact, it is the *being skills* that make the *doing skills* possible. The *being* skills are about relational awareness, humility, patience, presence, intentionality, and courage.

To summarize what we've learned about working in partnership and *being in community* through the ICP program:

- 'Stop talking and listen' would be the most common piece of advice we've received. Implicit in this advice is that we need to set aside our Western ways of knowing and doing and set out with a beginner's mind.
- Planning cannot begin before trust is established, and trust will only be established if people feel heard, valued, and respected. Focus on building relationships, getting to know people.
- Social transformation moves at the pace of trust. Trust the wisdom of the community. They will collectively find the answers, once they have a shared sense of purpose and a sense of safety.
- There's no limit to the creativity of communities. It's our job to nurture that creativity through creative forms of engagement.
- Learn to work at the speed of the community, not at the speed of university or other institutional or funding deadlines.
- Learn to trust the process. Learn to improvise when things go sideways. And that implies an ability to always be in the moment.
- Let go of control. Let go of outcomes. Learn to thrive in uncertainty.
- Learn to feel safe not knowing.
- Learn from your own mistakes. Own your mistakes and ask how to make amends.
- Recognize your own blind spots. Know your own insecurities and defense mechanisms.
- Always be fully present, 'with an open heart and an open mind', as the tag line of Musqueam's community plan reads.
- Learn to discern when and how to contribute your knowledge and expertise to the knowledge residing in the community, and when to step back, to let go.
- Instead of focusing on what's wrong, (the 'damage-centered approach'), focus on the community's gifts and strengths, and that will tell you what's possible.[3]

All of the above speak to the importance, arguably the primacy, of both emotional and relational intelligence in planning work. This connects with an emergent stream of thought in our field (Marris 1975; Baum 2015: Erfan 2017; Sandercock 2018; Willson 2018) arguing for a way of conceptualizing planning as the work of repair or healing: therapeutic planning, as I once labeled it (Sandercock 2000). Peter Marris's beautiful book *Loss and Change* (1974), written almost half a century ago, made the case for acknowledging loss or grief as an inevitable part of the work we do as planners in managing change processes. All change, even change that's widely perceived as desirable, involves some loss, and unless we pay attention to that, instead of sweeping it under the carpet, unless we're prepared to respect it as a legitimate response and work it through, we risk massive pushback.

The 'planner as change agent' embodies self-awareness and relational awareness. This is an evolution that builds on the communicative and collaborative approaches developing since the 1970s. The works of Friedmann, Schön, Forester, Innes, Healey, and myself emphasized the importance of listening and of dialogue. We challenged the dominant Enlightenment epistemology and proposed an epistemology of multiplicity, of multiple ways of knowing. We added communicative and creative skills to the technical and design toolkits previously used by planners. And we also emphasized (thanks to Schön) the importance of reflection: reflection on action and reflection in action. The so-called communicative paradigm shift was an important shift in how some of us teach and practice *ways of knowing and ways of doing* in planning. But it did not turn the gaze inwards, to examine our *ways of being* in the world. That is the next step in the transformation of planners who aspire to be change agents.

So, now, to ground all this in one community: our Indigenous Community Planning program worked in partnership with the Skidegate Haida community for four years on an extensive community engagement process, which led to their Comprehensive Community Plan, *Gud Ga Is*, which means, literally, 'coming together to talk' (Skidegate Indian Band 2017). All of the qualities outlined above were essential in doing this work together, and in the process, we learned the Haida laws that have governed their planning since time immemorial:

> Respect. All acts must be done with respect.
> Ask first. All acts must be done with consent.
> Everything depends on everything: all things are connected
> Sharing of wealth
> Only take what you need
> Putting back into the ocean.
> When the tide is out, there is food on the table.
> The tide waits for no one.
> The food is our medicine.
>
> (Skidegate Indian Band 2017)

Indigenous Elder and scholar Umeek (E. Richard Atleo), a hereditary chief from the Nuu-chah-nuulth Nation on the west coast of Vancouver Island, has expressed a similar worldview in his book *Principles of Tsawalk: An Indigenous Approach to Global Crisis* (2011). 'Tsawalk' translates as 'everything is one', just as the Haida say 'everything is connected'. According to the Nuu-chah-nuulth worldview, all living things (humans, plants, animals) form part of an integrated whole brought into harmony through constant negotiation and mutual respect. In this book, Umeek argues that contemporary environmental and political crises, and the related ongoing plight of Indigenous Peoples, reflect a world out of balance, a world in which Western/techno-rational approaches for sustainable living are not working. The global crisis is one of 'relational disharmony'.

Likewise, Potawatomi scholar Robin Wall Kimmerer, in her beautiful book *Braiding Sweetgrass* (2013), embraces the notion that plants and animals are our oldest teachers. She uses this unique lens to show that an awakening of a wider ecological consciousness requires the acknowledgement of our reciprocal relationship with the rest of the living world.

> Berries are always present at our ceremonies. They join us in a wooden bowl. One big bowl and one big spoon, which are passed around the circle, so that each person can taste the sweetness, remember the gifts, and say thank you. They carry the lesson, passed to us by our ancestors, that the generosity of the land comes to us as one bowl, one spoon. We are all fed from the same bowl that Mother Earth has filled for us. It's not just

> about the berries, but also about the bowl. The gifts of the earth are to be shared, but gifts are not limitless. The generosity of the earth is not an invitation to take it all. Every bowl has a bottom. When it's empty, it's empty. And there is but one spoon, the same size for everyone. ... The moral covenant of reciprocity calls us to honor our responsibilities for all we have been given, for all that we have taken. ... Whatever our gift, we are called to give it and to dance for the renewal of the world. In return for the privilege of breath.
>
> *(Kimmerer 2013, 382–4)*

Maori scholar/planner Hirini Matunga describes Indigenous planning thus: 'as much an ethic and philosophy as it is a planning framework with a set of approaches and methods' (Matunga 2013, 31). He uses the image of a spiral to convey the conceptual basis of Indigenous planning. At the center of the spiral is *wairuatanga*, the state of being spiritual. Moving from this central anchor, around the spiral, he next identifies relationship and kinship, followed by the responsibilities and principles passed down from the ancestors to take care of the land and resources; respect given to visitors through sharing and caring; self-determination; unity; alignment; and transparency (Matunga et al. 2003). Elsewhere Matunga argues that Indigenous worldviews are essential for reconceptualizing human–environment relationships. The Maori worldview sees land as a community to which we belong and from which we are inseparable (Aldo Leopold's land ethic is a close cousin). In this worldview, furthermore, there is an hierarchy that positions humans below the natural environment, while seeing everything as connected. Maori *whakapapa* is a tradition of genealogy outlining not just kinship between humans but also between humans and all living things and places and the spiritual realm (Matunga et al. 2020).

But how to connect all this work on the ground, community by community, to the bigger picture of paradigm or worldview shift, and see the connection between self and systems? Indigenous planner Jessie Hemphill uses a 'nested systems' concept in her teaching and practice to show the connection from individual to family to community to land to the cosmos (Hemphill 2022). The processes of transition to more sustainable ways of living that our ICP program is a small part of in British Columbia is an approach that might be called a fractal theory of change: a strategy for building complex solutions through relatively small but holistic interventions and interactions. Community by community, Indigenous Peoples living on reserves in British Columbia are transitioning to their own visions of self-determination, sovereignty, and sustainability. The role of the planner is to support these goals. Making a difference, one community at a time, may seem like fiddling while Rome is burning. But in the context of what environmentalist Paul Hawken (2007) has called a 'blessed unrest', perhaps it is not such a fanciful theory of change. Hawken is referring to the coalescence and intertwining of hundreds of thousands of organizations worldwide into a 'movement' without central leadership, command, or control, that has three basic roots: environmental activism, social justice initiatives, and Indigenous cultures' resistance to globalization. Although this phenomenon of 'blessed unrest' is primarily a grass-roots movement of people from every culture and continent, it may also represent a natural self-organizing principle, free of any hegemonic ideology. The philosophy suggested by 'blessed unrest' is that living systems are fundamentally cooperative: that life can only exist in cooperative, sharing relationships with other life, a deeply relational, interconnected worldview. As John Muir expressed it over 100 years ago, 'When we try to pick out anything by itself, we find it hitched to everything else in the Universe' (quoted in Wheatley 2017, 211).

Blessed Unrest: Connecting the Dots

My ecological economics literacy is basic, so I will avoid engaging with debates about the economic and institutional contours of a so-called post-growth world. I prefer to question the very concept of 'post-growth' and, drawing on an ecological perspective, talk instead about *qualitative growth* and what the field of community planning has to contribute to this concept. The key challenge is to shift from an economic system based on the notion of unlimited growth to one that is both ecologically sustainable and socially just. 'From the perspective of the systems view of life, "no growth" cannot be the answer. Growth is a central characteristic of all life. A society, or economy, that does not grow will die sooner or later. Growth in nature, however, is not linear and unlimited. While certain parts of organisms, or ecosystems, grow, others decline, releasing and recycling their components, which become resources for new growth' (Capra and Luisi 2014, 368).

So, let's define growth as that which enhances (rather than destroys) life – as generation and regeneration – and declare that what our planet most needs is *more of that kind of growth*. This notion of 'growth which enhances life' – qualitative economic growth – can be sustained if it involves a dynamic balance between growth, decline, and recycling, and if it also includes the inner growth of learning and maturing, a spiritual growth. So now we circle back to the resurgent practice of Indigenous Community Planning, with its emphasis on caring relationships between people and between people and land, and an internalized awareness of the necessary interdependence of these.

Why is this *way of being* a part of the causal chain of socio-ecological transformation? It is in the work of Patricia Wilson that the causal connection between the inner and the outer, between ways of being and social transformation, has perhaps been best articulated among non-Indigenous/White planning scholar/activists. Drawing from her own practice in the global south, she takes us beyond (Donald Schön's) 'double loop learning' (which still lies in the cognitive domain of the reflective practitioner) to triple loop learning, the capacity to reflect on one's own interior state of being, the *being* that one experiences while *doing*. She then connects this triple loop learning with what Gregory Bateson called 'spiritual knowing': a shift in the place of perception to one's deepest source. 'Such transformative experiences do more than change one's approach to a problem: they change the change agent' (Wilson 2019, 6).[4] She also talks about developing an 'ensemble awareness', an awareness of the wider social field and the inherent possibilities of new patterns of relationship within the particular community and between government and community, like tiny fractals that can replicate and grow. 'These fractals of transformative outer change start with our own inner practice' (ibid, 13). That inner practice, in my experience, is the hard work of connecting our own mind with our heart and spirit; unlearning the separation of reason from emotion, intuition and imagination that has dominated the techno-rational worldview promoted since the Enlightenment (Nussbaum 2001). Practitioners who develop these '*being* skills' that undergird the '*doing* skills' become the 'societal change agents so needed now: the ones whose efforts can knit together a social fabric torn asunder' (Wilson 2019, 14).

There's a further insight that Wilson offers as part of her own credo for practice, which resonates for me:

> I choose to see myself not as outside or against the system I want to change, but as an interconnected part of that evolving system. I seek to respect and understand the living elements of that system – individual and collective, human, and natural. I seek to know,

> understand, and connect with whatever 'other' or enemy (internal or external) I may perceive. I seek to care about the other and the whole, at whatever scale … to see our interconnection'.
>
> *(ibid, 220)*

This felt sense of interconnectedness leads to a sense of responsibility for the wellbeing of the community as a whole, emphasizing the social and spiritual nature of being human.

Building and nurturing community lies at the core of ecological sustainability, according to bell hooks and Wendell Berry (Wickenden 2022), and building and nurturing community lies at the heart of community planning, whether that community planning is serving to support resurgent Indigenous communities on their path to self-determination, or working in the multi-ethnic, multi-racial neighborhoods of cities, addressing systemic racism and other forms of discrimination, and inviting strangers to become neighbors. A community planning that has as its heart a practice of kindness (Forester 2021), loving attachment (Porter, Sandercock and Umemoto 2012), caring for each other and the air, lands, and waters (Musqueam Indian Band 2011; Skidegate Indian Band 2017), requires a shift of attention in our field that *connects ways of knowing and acting to ways of being, as part of the broader socio-ecological paradigm shift embracing connectedness, relationship, and interdependence.*

Notes

1 This felicitous phrase was borrowed from Chapter 15 of Paul Hawken, Amory Lovins, & L. Hunter Lovins *Natural Capitalism* (New York: Little Brown & Company, 1999).

2 Hobbes was not alone. Descartes, Spinoza, Adam Smith, Freud, Camus, and many more revered thinkers believed that self-interest was the fundamental human trait and prime mover of human action.

3 I am struck by the congruence between these 'lessons' and those drawn by Patricia Wilson from her work in communities in the global south, as described in Wilson (2019).

4 Margaret Wheatley (2017) makes a similar argument in *Who Do we Choose To Be?* Leaders must make a philosophical leap and become more human human beings. In order to transform the world, we must transform ourselves.

References

Baum, H. 2015. Planning with half a mind. Why planners resist emotion, *Planning Theory and Practice* 16(4), 498–516.

Capra, F. 1982. A new vision of reality, *New Age*, February, 29–32.

Capra, F. and Henderson, H. 2009. Qualitative growth. In: *Outside Insights*. Available at: fritjofcapra.net (London: Institute of Chartered Accountants in England and Wales) October; posted on www.

Capra, F. and Luisi, P. L. 2014. *The Systems View of Life: A Unifying Vision* (Cambridge: Cambridge University Press).

Cassidy, J. 2020. Steady state, *New Yorker*, Feb 10, 24–27.

Clark, M. E. 1989. *Ariadne's Thread: the Search for New Modes of Thinking* (New York: St. Martin's Press).

Erfan, A. 2017. Confronting collective traumas: an exploration of therapeutic planning, *Planning Theory and Practice* 18(1), 34–50.

Forester, J. 2021. Our curious silence about kindness in planning: challenges of addressing vulnerability and suffering, *Planning Theory* 20(1), 63–83.

Friedmann, J. 1987. *Planning in the Public Domain: From Knowledge to Action* (Princeton, NJ: Princeton University Press).

Hawken, P. 2007. *Blessed Unrest. How the Largest Social Movement in History Is Restoring Grace, Justice, and Beauty to the World* (New York: Viking Press).

Hemphill, J. 2022. *Personal Correspondence*, 21 February, 2022.

Hobbes, T. 1651. *Leviathan: Or the Matter, Forme, and Power of a Commonwealth Ecclesiasticall and Civill*, ed. by Ian Shapiro (Yale University Press).

Jackson, S. 1998. *Geographies of Co-Existence: Native Title, Cultural Difference and the Decolonization of Planning in Northern Australia* (Unpublished PhD (Sydney: School of Earth Sciences, Macquarie University)).

Jacobs, J. 1996. *Edge of Empire. Postcolonialism and the City* (London: Routledge).

Jojola, T. 2013. Indigenous planning: Towards a seven generations model. In: Walker, R., Jojola, T. and Natcher, D. (Eds) *Reclaiming Indigenous Planning* (Montreal: McGill-Queens University Press).

Kimmerer, R. W. 2013. *Braiding Sweetgrass. Indigenous Wisdom, Scientific Knowledge, and the Teachings of Plants* (Minneapolis, MN: Milkweed Editions).

Leopold, A. 1949. *A Sand County Almanac* (London: Oxford University Press).

Marris, P. 1974. *Loss and Change* (London: Routledge Kegan Paul).

Matunga, H. 2013. Theorizing indigenous planning. In: Walker, R., Jojola, T. and Natcher, D. (Eds) *Reclaiming Indigenous Planning* (Montreal: McGill-Queens University Press).

Nussbaum, M. 2001. *Upheavals of Thought: the Intelligence of Emotions* (New York: Cambridge University Press).

Matunga, H. 2017. A revolutionary pedagogy of/for indigenous planning, *Planning Theory and Practice* 18(4), 640–644.

Matunga, H., et al. 2003. The values associated with Maori-centred tourism. *Tourism Recreation Research Centre, Report 35* (Lincoln University).

Matunga, H., et al. 2020. From exploitive to regenerative tourism: Tino rangatiratanga and tourism in Aotearoa New Zealand, *Mai Journal: a New Zealand Journal of Indigenous Scholarship* 9(3), 295–308.

Midgley, M. 1985. *Evolution as Religion* (London: Methuen).

Musqueam Indian band. 2011, *Comprehensive Community Plan*. Available at: https://www.musqueam.bc.ca/community-engagement/ccp/

Porter, L. 2010. *Unlearning the Colonial Cultures of Planning* (Burlington: Ashgate).

Porter, L. 2017. What is the work of non-indigenous people in the service of a decolonizing agenda?, *Planning Theory and Practice* 18(4), 650–653.

Porter, L., Roy, A. and Legacy, C. 2021. Against world-killing silence, *Planning Theory and Practice* 22(1), 111–138.

Porter, L., Sandercock, L., Umemoto, K., Umemoto, K., Bates, L. K., Zapata, M. A., Kondo, M. C., Zitcer, A., Lake, R. W., Fonza, A., Sletto, B., Erfan, A. and Sandercock, L. 2012. What's love got to do With it? Illuminations on loving attachment in planning, *Planning Theory and Practice* 13(4), 593–627.

Sandercock, L. 1998. *Towards Cosmopolis: Planning for Multicultural Cities* (London: Wiley).

Sandercock, L. 2000. When strangers become neighbors: managing cities of difference, *Planning Theory and Practice* 1(1), 13–30.

Sandercock, L. 2003. *Cosmopolis 2: Mongrel Cities of the 21st Century* (London: Continuum).

Sandercock, L. 2018. Finding my way: emotions and ethics in community-based action research with indigenous communities'. In: Christensen, J., Cox, C. and Szabo-Jones, L. (Eds) *Activating the Heart: Storytelling, Knowledge Sharing and Relationship* (Waterloo: Wilfred Laurier University Press).

Skidegate Indian Band. 2017. *Gud Ga Is (Skidegate Comprehensive Community Plan 2012–17)*. Available at: http://www.skidegate.ca/wp-content/uploads/2022/02/ccp2012_2017.pdf.

Umeek (E. Richard Atleo) 2011. *Principles of Tsawalk. An Indigenous Approach to Global Crisis* (Vancouver: UBC Press).

Wheatley, M. 2017. *Who Do We Choose to Be?* (Oakland: Berrett-Koehler Publishers).

Wickenden, D. 2022. Wendell Berry's legacy, *New Yorker*, 28 February.

Willson, R. 2018. *A Guide for the Idealist* (New York: Routledge).

Wilson, P. 2019. *The Heart of Community Engagement* (New York: Routledge).

INDEX